Deep Learning of Climate Change and Extreme Events, Hydrological Processes and Land Use Dynamics Relationships

Deep Learning of Climate Change and Extreme Events, Hydrological Processes and Land Use Dynamics Relationships

Guest Editors

Hanoch Lavee
Jinping Liu

MDPI

Basel • Beijing • Wuhan • Barcelona • Belgrade • Novi Sad • Cluj • Manchester

Guest Editors

Hanoch Lavee
Department of Environment, Planning and Sustainability
Bar-Ilan University
Ramat Gan
Israel

Jinping Liu
College of Surveying and Geo-Informatics
North China University of Water Resources and Electric Power
Zhengzhou
China

Editorial Office
MDPI AG
Grosspeteranlage 5
4052 Basel, Switzerland

This is a reprint of the Special Issue, published open access by the journal *Land* (ISSN 2073-445X), freely accessible at: https://www.mdpi.com/journal/land/special_issues/VVW0J9CFY3.

For citation purposes, cite each article independently as indicated on the article page online and as indicated below:

Lastname, A.A.; Lastname, B.B. Article Title. *Journal Name* **Year**, *Volume Number*, Page Range.

ISBN 978-3-7258-4161-5 (Hbk)
ISBN 978-3-7258-4162-2 (PDF)
https://doi.org/10.3390/books978-3-7258-4162-2

Contents

About the Editors

Hanoch Lavee

Hanoch Lavee is a Professor Emeritus of Geomorphology at Bar-Ilan University, Israel. His research is focused on rainfall-runoff-erosion relationships, hillslope processes in arid and semi-arid areas, overland flow discontinuity, source–sink relationships along climatic transects, the resilience of eco-geomorphic systems, and desertification under climate change. He founded the Laboratory of Geomorphology and Soil at his university and has coordinated about 20 research projects funded by competitive research grants on fire effects on runoff and erosion, soil–vegetation–erosion relations under climate change, ecosystem desertification along a climatological transect, etc. He has acted as a member of task teams of several IGU commissions on specific issues and as a member of the Editorial Board of scientific journals such as *CATENA*, *Land Degradation and Development*, and *Land*. He has spent time in sabbaticals or as a visiting scientist at the London School of Economics, the Catholic University of Leuven, the University of Amsterdam, the University of Aix-en-Provence, and the University of Malaga. Prof. Lavee has supervised almost thirty graduate students for master's and doctoral studies and has published several books, about 120 scientific articles in refereed journals, and over 120 abstracts in international conferences. He has served as the president of the Israel Geographical Association and chairman of the Committee for Curriculum in Geography of the Ministry of Education of Israel. He has also served as a Dean, the Rector of his university, and President of the Kinneret College on the Sea of Galilee.

Jinping Liu

Jinping Liu, Ph.D., works in the North China University of Water Resources and Electric Power and is a visiting scholar at Katholieke Universiteit (KU) Leuven, Belgium. His research activities include cryosphere remote sensing, remote sensing of vegetation, climate change, and hydrological modeling, with particular research experience and a knowledge background in the use of remote sensing and geographic information tools for the identification, impacts, and mechanisms of extreme climatic events, as well as risk assessment. As a supervisor of master's students, he has contributed significantly to academia and mentoring students. Also, he has served as a reviewer for several SCI journals and has organized several Special Issues with international scholars as a Guest Editor.

Preface

This Special Issue, "Deep Learning of Climate Change and Extreme Events, Hydrological processes and Land Use Dynamics Relationships," brings together researchers from diverse fields who are applying deep learning methods and new models to investigate the relationships between climate change drivers and impacts, droughts, desertification, ice sheet melting, land use changes, and ecosystem service value.

In the first article of this SI, Shuokai Wang, Bo Zeng, Yong Feng, and Fangping Cao investigated the effect of agricultural land lease policy in China on the transmission mechanisms of agricultural carbon emissions, a dominant greenhouse gas, using an equilibrium model that accounts for the heterogeneous production efficiency among farmers.

The next three articles deal with droughts. The first, by Jieru Wei, Wensheng Tang, Pakorn Ditthakit, Jiandong Shang, Hengliang Guo, Bei Zhao, Gang Wu, and Yang Guo, deals with drought forecasting in northeastern China. It shows that the Informer model outperforms some other models in terms of drought forecasting accuracy across all time scales.

In the third article, Changfu Tong, Hongfei Hou, Hexiang Zheng, Ying Wang, and Jin Liu investigated the spatiotemporal evolution of vegetation drought in the Inner Mongolia Autonomous Region, China, using a deep learning-based forecasting model.

The fourth one, by Jinping Liu, Mingzhe Li, Renzhi Li, Masoud Jafari Shalamzari, Yanqun Ren, and Esmaeil Silakhori, deals with drought susceptibility in northeastern Iran using land use change modeling and climate change projections. DS maps were developed using agrometeorological sample points and environmental indices, which were used as inputs into a machine learning model.

In the fifth article, Weibo Yin, Qingfeng Hu, Jinping Liu, Peipei He, Dantong Zhu, and Abdolhossein Boali investigated the impact of climate and land use changes on desertification in northeast Iran, using machine learning algorithms and remote sensing indices.

In the sixth article, Caixia Hu, Jie Li, Yaxu Pang, Lan Luo, Fang Liu, Wenhao Wu, Yan Xu, Houyu Li, Bingcang Tan, and Guilong Zhang investigated the spatial heterogeneity of nitrate leaching and its drivers in China by developing and using deep learning models and methods.

In the seventh article, Kaifeng Ma, Jing Han, Zhen Li, Junzhen Meng, Qingfeng Hu, Peipei He, and Changxu Yao investigated the Greenland ice sheet mass changes, using GRACE/GRACE-FO data alongside regional climate model outputs. The findings highlighted the complex interplay of climatic and hydrological processes as drivers of mass loss.

The eighth article, by Tanweer Abbas, Muhammad Shoaib, Raffaele Albano, Muhammad Azhar Inam Baig, Irfan Ali, Hafiz Umar Farid, and Muhammad Usman Ali, uses artificial intelligence tools, especially remote sensing and GIS, to examine the negative impact of population growth in South Punjab, Pakistan, on land use and land cover.

In the ninth article, Dahai Yu and Chang You revealed the dynamic impacts of climate change on ecosystem service value in China by analyzing the effects of climate change on ecosystem service value profits and losses.

The tenth and last article in this Special Issue, by Yajie Shang, Yuanyuan Chen, Yalin Zhai, and Lei Wang, deals with the spatial and temporal interaction patterns between carbon metabolism and ecosystem service value in the northeastern black soil region of northeast China.

Those articles promote understanding and managing the future of our planet and provide new tools for analyzing complex environmental data. These offer a promising avenue for advancing our potential response to environmental challenges.

We would like to sincerely thank the *Land* editorial office for their generous assistance, especially Ms. Nikita Dong, the Section Managing Editor, and the Managing and Assistant Editors.

Hanoch Lavee and Jinping Liu
Guest Editors

land

Article

How Does Agricultural Land Lease Policy Affect Agricultural Carbon Emission? Evidence of Carbon Reduction Through Decreasing Transaction Costs in the Context of Heterogeneous Efficiency

Shuokai Wang, Bo Zeng, Yong Feng and Fangping Cao *

School of Economics and Management, Beijing Forestry University, Beijing 100083, China; wangshuokai@bjfu.edu.cn (S.W.); zengbo@bjfu.edu.cn (B.Z.)
* Correspondence: cfp_888@bjfu.edu.cn

Abstract: Given the increasing environmental pressures, it is essential that agriculture achieves the goal of sustainable and low-carbon development. In 2010, China, as the top carbon emitter, introduced a policy on agricultural land lease (ALL), which has been met with considerable approval from farmers and has resulted in a notable surge in the rate of ALL within the country. Nevertheless, the question of how the ALL policy affects agricultural carbon emissions (ACEs) remains unanswered. What are the transmission mechanisms? To answer these questions, this paper presents an equilibrium model that accounts for the heterogeneous production efficiency among farmers. It offers a theoretical analysis of the impact of ALL policy on agricultural carbon emission reduction (ACER) and presents an empirical test of this impact using a difference-in-differences (DID) model. Our research shows that the ALL policy gives impetus to ACER. This conclusion persists even after conducting the robustness and endogeneity tests. The mechanism posits that the policy achieves ACER through reducing the proportion of rural agricultural employees. Heterogeneity analysis indicates that the policy effect is significant in both the northern and southern regions of China. Nonetheless, the effect is only observable in economically developed areas, regions with high chemical fertilizer application rates, and areas with restricted agricultural progress. This study elucidates the connection between land transfer and agricultural carbon emissions, offering empirical evidence to support the advancement of green and low-carbon agricultural development.

Keywords: agricultural land lease policy; agricultural carbon emission reduction; heterogeneous production efficiency; transaction costs; difference-in-differences model

1. Introduction

Greenhouse gas (GHG) emissions from agricultural land account for 10–20% of total global anthropogenic GHG emissions, making agriculture the second largest source of carbon in the world [1–3]. China's agricultural carbon emissions (ACEs) account for 11–17% of the country's total emissions, with an average annual increase of more than 5% in some years [4,5]. This is mainly due to the extensive use of chemical fertilizers and pesticides in agricultural production and the energy consumption of agricultural machinery and equipment [3,6]. In the past decade, with the abandonment of traditional raw material production and management methods in China's agricultural production, the growth rate of ACE has slowed down, but the total amount is still increasing [7]. China is committed to achieving carbon peak by 2030 and carbon neutrality by 2060, and the "dual-carbon" goal has added new content and new requirements to the green development of agriculture [8]. Therefore, it is imperative to promote ACER [9,10].

Studies have shown that one of the factors contributing to the high ACE in China is the status of small-scale agricultural operations [11,12]. First of all, the land management mode with low scale and fragmented land distribution makes it difficult for modern agriculture, characterized by mechanization, to play the role of efficiency improvement and carbon reduction [13]. Second, this state of operation leads to an increase in farmers' part-time activities and an intensification of agricultural sideline farming, which in turn prevents the widespread promotion and application of new agricultural technologies and production methods and also leads to farmers' tendency to invest in intensive chemicals and produce excess carbon emissions [14].

In 2010, China launched the first national standard text of ALL. The release of this text marks the effective and rapid development of China's ALL market. As indicated in the China Rural Management Statistics Annual Report (CRMSAR), the national average ALL rate from 2005 to 2009 was 9.04%. China's ALL market has developed effectively and rapidly, with the national average ALL rate increasing from 17.92% from 2010 to 2012 to 33.06% from 2013 to 2020. These changes have increased the scale of land management [15] and promoted the development of new agricultural management entities such as professional farmers and large planters [16].

The positive role of ALL in promoting agricultural production has been recognized by most scholars, but the academic community has different views on whether ALL can achieve ACER. On the one hand, the ALL may reduce the ACE [17–27]. On the other hand, the ALL may increase the ACE [13,28–31]. The specific perspectives of the relevant studies are shown in Table 1.

Table 1. Studies on the effect of ALL on ACE.

ALL Reduce the ACE	
Wu [17] and Li [18]	ALL reduces the use of chemical fertilizers.
Gao [19], Ntakirutimana [20], Lu [21], Li [22], and Jia [23]	ALL increases the use of organic fertilizers.
Zhou [24]	After ALL, the scale of land increases, and new agricultural technologies have been adopted at a higher level.
Zhang [25]	ALL has improved the efficiency of agricultural mechanization.
Tang [26] and Adamopoulos [27]	ALL enabled farmers to better match their field management level with the actual cultivated area.
ALL Increase the ACE	
Cheng [28] and Guo [29]	For the sake of profit, farmers after ALL apply more fertilizer to increase agricultural yields.
Li [13] and Qi [30]	After ALL, intensive farming accelerates soil depletion, forcing landowners to use more fertilizers.
Tesfaye [31]	After ALL, to prevent pests and diseases, landowners use more pesticides.

With the deepening of the research on ALL and ACER, scholars have gradually realized that the heterogeneity of farmers may be one of the key factors leading to the different results of ALL on ACER. Song [32] and Yu [33]'s study found that older, less-educated farmers transferring land to younger, well-educated farmers may contribute to ACER, but their study does not further explore whether this phenomenon is driven by heterogeneity in farmers' productivity. Yasmeen [34]'s research shows that farmers with higher productivity have significantly lower ACE than farmers with lower productivity. Chari [35]'s research further found that farmers with higher productivity tend to flow from land, while farmers with lower productivity tend to flow out of land. However, there is no direct confirmation of whether ALL based on the heterogeneity of production efficiency among farmers can effectively promote ACER.

In addition, as the first national-level formal contract for ALL, this text clearly and specifically defines the rights and obligations of both parties to the ALL, provides a solid guarantee mechanism for the standardized implementation of ALL, and effectively reduces the transaction costs in the process of ALL. Jiang [36] sees this policy as a quasi-natural experiment and finds that the ALL policy promotes the development of agricultural finance activities. Song [37]'s research further reveals the positive effect of ALL policies in reducing the rate of farmland abandonment. Huo [38]'s research focuses on the effect of ALL policies on sustainable agricultural production in China. The existing literature has shown that in the industrial and urban sectors, the reduction in transaction costs can effectively contribute to carbon emission reduction [39]. However, there is no clear answer to the question of whether the transaction costs reduced by the ALL policy can also promote ACER.

In order to answer the above questions, this study constructs a theoretical model to explore the behavioral motivation of farmers to choose to participate in or exit the ALL market in the context of heterogeneous production efficiency, further clarify the relationship between farmers' production efficiency and ACER, and construct an analytical framework to analyze how the decrease in transaction costs caused by ALL policy affects ACER through farmers' heterogeneous production efficiency. This study verifies the theoretical model through empirical analysis, aiming to provide a theoretical basis and policy implications for improving the ALL market, optimizing the allocation of rural resources, and promoting low-carbon and high-quality development in the agricultural field.

The marginal contribution of this study is mainly reflected in two aspects: first, this study innovatively applies Melitz [40]'s heterogeneous production efficiency model to the field of ACER and theoretically analyzes how farmers with different production efficiency can achieve ACER through ALL. Second, this study combines ALL policy with ACER for the first time, filling the gap in the research on reducing transaction costs in the field of ACER.

2. Theoretical Analysis and Research Hypotheses

This paper elucidates how the decrease in transaction costs resulting from the ALL policy enhances ACER, utilizing Melitz [40]'s heterogeneous production efficiency model. It constructs a theoretical framework linking the ALL to farmers' production efficiency, grounded in the constant elasticity of substitution (CES) utility function, thereby proposing a research hypothesis.

2.1. Consumption Choice

Assumptions:

(1) There is a large and continuous set of goods Ω, given that consumers have the same utility function, and representative consumer preferences are in the form of CES utility functions as follows: $U = [\int_{\omega\in\Omega} q(\omega)^{\rho} d\omega]^{\frac{1}{\rho}}$; where $q(\omega)$ represents the demand for a good ω, there is substitutability between goods such that $0 < \rho < 1$, with the elasticity of substitution between any two goods being $\sigma = 1/(1-\rho) > 1$.

(2) The consumption expenditure function of the representative consumer for all goods is $R = \int_{\omega\in\Omega} p(\omega)q(\omega)d\omega$, where $p(\omega)$ represents the price of the ω.

The solution gives a demand function for ω as follows: $q(\omega) = R\frac{p(\omega)^{-\sigma}}{\int_{\omega\in\Omega} p(\omega)^{1-\sigma} d\omega}$. Let the price summation among commodities be $P = \left[\int_{\omega\in\Omega} p(\omega)^{1-\sigma} d\omega\right]^{\frac{1}{1-\sigma}}$, then the demand function of commodity ω is $q(\omega) = Rp(\omega)^{-\sigma}P^{\sigma-1}$. The total number of goods consumed by the representative consumer is $Q = \frac{R}{P}$. The share of goods ω in the total number of goods consumed and the share of expenditures are, respectively, as follows:

$$\frac{q(\omega)}{Q} = [\frac{p(\omega)}{P}]^{-\sigma} \tag{1}$$

$$\frac{r(\omega)}{R} = [\frac{p(\omega)}{P}]^{1-\sigma} \tag{2}$$

2.2. Agricultural Supply and Carbon Emissions

Assumptions:

(1) Agricultural production uses Θ as an input factor and thus causes carbon emissions. Individual farmers produce the required number of inputs θ and cause carbon emissions φ, with the standardized unit price of inputs being 1.

(2) Agricultural products ω are supplied by a large number of farmers with heterogeneous production efficiencies, where τ represents the production efficiency of each farmer. The heterogeneity of farmers follows the exogenously given initial production efficiency, which is distributed as $g(\tau)$, where $g(\tau) \in [0, \infty]$ and $G(\tau)$ represent the cumulative probability distributions.

(3) A profit-maximizing farmer follows a marginal cost markup strategy, with elasticity of substitution between commodities σ and markup ratio $\sigma/(\sigma-1) = \rho$.

(4) Factor inputs and outputs are linearly related: $q = \theta\tau$, and the more efficient a farmer is, the less factor inputs are required for the same output.

The production cost function of the farmer is $c = f + \theta = f + \frac{q}{\tau}$, where f is the fixed input cost of the farmer's production. From the markup ratio ρ, the price of agricultural products is $p(\varphi) = \frac{1}{\rho\tau}$. In the following sections, $r(\omega(\tau))$ is simplified to $r(\tau)$. Similar treatments include $q(\omega)$ being simplified to q, etc. The agricultural income function and the agricultural profit function of the farmer are $r(\tau) = p(\tau)q$ and $\pi(\tau) = r(\tau) - c$, which can be obtained by combining Equations (1) and (2) with the cost function and the price function as follows:

$$r(\tau) = p(\tau)q = R(P\rho\tau)^{\sigma-1} \tag{3}$$

$$\pi(\tau) = r(\tau) - c = \frac{R(P\rho\tau)^{\sigma-1}}{\sigma} - f \tag{4}$$

Since $\frac{\partial p(\tau)}{\partial\tau} < 0$, $\frac{\partial r(\tau)}{\partial\tau} > 0$, $\frac{\partial \pi(\tau)}{\partial\tau} > 0$, increased production efficiency correlates with reduced prices of agricultural products, elevated agricultural income and profits for farmers, and enhanced motivation to undertake and expand agricultural production.

More importantly, the input factor θ generates carbon emissions φ, $\frac{\partial\varphi}{\partial\theta} > 0$, that is, the more factor inputs, the more agricultural carbon emissions. Since $\frac{\partial c}{\partial\tau} = \frac{\partial\theta}{\partial\tau} < 0$, then:

$$\frac{\partial\varphi}{\partial\tau} = \frac{\partial\varphi}{\partial\theta}\frac{\partial\theta}{\partial\tau} < 0 \tag{5}$$

Consequently, more efficient farmers generate less ACE for equivalent yields. In addition, as China has not yet regulated ACE, farmers' production and management decisions only take into account agricultural profits π.

2.3. Farmer Choice and Equilibrium

For analytical purposes, it is presumed that the farmer's business cycle in relocating to new land is infinite. It is assumed that there are transaction costs f_g and per period land rental costs f_e. The supply of agricultural products takes place in a perfectly competitive market environment, that is, anyone who pays a fee is free to operate.

For those who vacate their lands, f_e returns are obtained in each period. For the transferor, the profit is $\pi'(\tau) = \pi(\tau) - f_e$ per period, while each period faces a certain probability of risk occurrence δ resulting in the farmer's withdrawal from land operations. Then, the expected profit value of the transferred land obtained by the farmer in the whole life cycle is $\sum_{t=0}^{\infty}(1-\delta)^t\pi'(\tau) = \frac{\pi'(\tau)}{\delta}$. Whether the farmer continues to operate on land after transferring to the new land depends on whether it is profitable or not as follows: $v(\tau) = max\left\{0, \frac{\pi'(\tau)}{\delta}\right\}$.

If $\pi'(\tau)$ is negative, the farmer will not relocate to new land; $\pi(\tau)$ will also be negative at this point, and the farmer will choose to leave the business. Since $\frac{\partial\pi'(\tau)}{\partial\tau} > 0$, the more efficient the agricultural production, the higher the farmer's profit. There exists a certain value of production efficiency at which $\pi'(\tau)$ is zero. More efficient farmers will have a greater $\pi'(\tau)$, and the farmer with positive $\pi'(\tau)$ has an incentive to relocate to a new land.

Assuming that the minimum efficiency of exiting farmers is τ^*, since farmers below this efficiency choose to exit the business, the distribution probability of the efficiency of continuing farmers $\mu(\tau)$ is the conditional distribution probability of the initial distribution probability $g(\tau)$: $\mu(\tau) = \begin{cases} \frac{g(\tau)}{1-G(\tau^*)}, & \tau \geq \tau^* \\ 0 , & \tau < \tau^* \end{cases}$. After ALL, the average production efficiency of all farmers is $\tilde{\tau}$: $\tilde{\tau}(\tau^*) = [\frac{1}{1-G(\tau^*)}\int_{\tau^*}^{\infty} \tau^{\sigma-1}g(\tau)d\tau]^{\frac{1}{\sigma-1}}$.

After ALL, the actual operating farmers change from the initial $[0,\infty]$ distribution to $[\tau^*,\infty]$. At the same time, the average productivity $\tilde{\tau}(\tau^*)$ of farming households increases accordingly. At this time, the average income $r(\tilde{\tau})$ and the average profit $\pi'(\tilde{\tau})$ of the farmer transferring to a new land are both determined by τ^*: $\frac{r(\tilde{\tau})}{r(\tau^*)} = (\frac{\tilde{\tau}}{\tau^*})^{\sigma-1}$ and $\pi'(\tilde{\tau}) = (\frac{\tilde{\tau}}{\tau^*})^{\sigma-1}\frac{r(\tau^*)}{\tau} - f'$, where $f' = f + f_e$. Since $\pi'(\tau^*) = \frac{r(\tau^*)}{\sigma} - f' = 0$:

$$\pi'(\tilde{\tau}) = f'[(\frac{\tilde{\tau}}{\tau^*})^{\sigma-1} - 1] \tag{6}$$

It can be shown that $\frac{\partial\pi'(\tilde{\tau})}{\partial\tau^*} < 0$, defining this relationship is represented by the surface I in Figure 1. The surface I represents the profitability of farmers transferring to a new land. As the productivity of the least efficient farmer increases, the relative advantage of the efficient farmer diminishes. Again, since the least efficient farmer is in a zero-profit position, the real profitability resulting from the transfer of the efficient farmer to a new land declines; in this case, $\frac{\partial\varphi}{\partial\pi'} = \frac{\partial\varphi}{\partial\tau}\frac{\partial\tau}{\partial\pi'} > 0$. In order to achieve higher profits, farmers increase planting intensity, which in turn increases ACE [28,29].

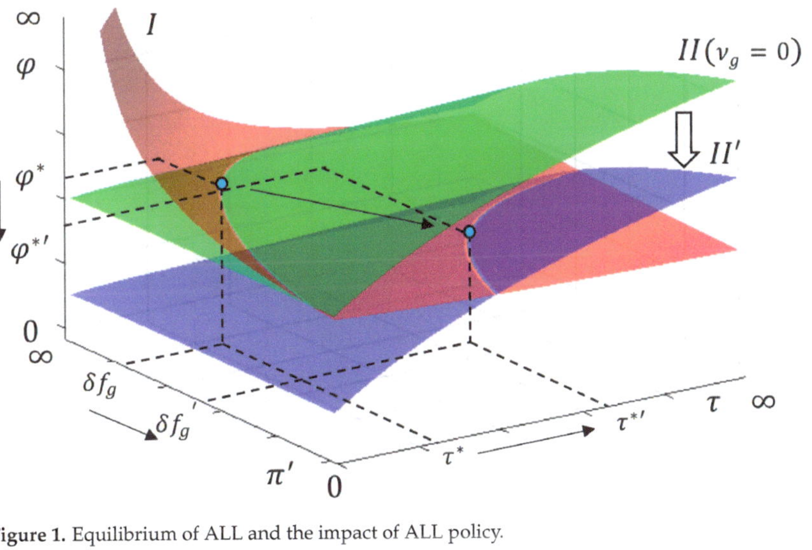

Figure 1. Equilibrium of ALL and the impact of ALL policy.

Since there is a transaction cost f_g, the farmer considers the net value of the land transferred, v_g: $v_g = [1 - G(\tau^*)]\frac{\pi'(\tilde{\tau})}{\delta} - f_g$. This is obtained in a perfectly competitive state with $v_g = 0$:

$$\pi'(\tilde{\tau}) = \frac{\delta f_g}{1 - G(\tau^*)} \tag{7}$$

It can be shown that $\frac{\partial \pi'(\tilde{\tau})}{\partial \tau^*} > 0$. Defining this relationship is represented by surface II. The surface II represents the zero net value situation of the transferred land. Highly productive farmers possess elevated profit expectations and exhibit a greater willingness to pay for land transfers. Since $v_g = 0$ under perfect competition, the zero net value of land corresponds to a higher level of profitability as the least efficient farmers become more productive. In this case, $\frac{\partial \varphi}{\partial \pi'} = \frac{\partial \varphi}{\partial \tau}\frac{\partial \tau}{\partial \pi'} < 0$. A greater zero net value indicates that only proficient farmers possess the capacity to transition to a new land, and efficient farmers have better ACER performance.

This leads to two relationships between φ, $\pi'(\tau^*)$, and τ^*, as shown in Figure 1. The intersection of the two surfaces is the minimum operating efficiency τ^*, below which farmers will opt out of the business.

2.4. Impact of ALL Policy and Research Assumptions

In China, the core principle of agricultural land distribution schemes is equity, which usually means that land is distributed evenly according to the size of the farmer family, taking into account the fertility and size of the land. However, this distribution pattern can lead to a loss of agricultural production efficiency and exacerbate the ACE. The implementation of the ALL policy aims to expand the area of land operated by reducing transaction costs and encouraging some farmers to lease more land. This change will help to optimize the spatial distribution of farmland use, reduce the degree of restriction of land resources and the probability of abandonment, and improve the matching degree between farmland and farmers' production and operation capabilities. In this way, the ALL policy is conducive to the promotion and application of new agricultural technologies and enriches the comprehensive benefits of land. Specifically, the ALL policy promotes the rational circulation of land and makes more efficient use of land resources, which not only helps to improve agricultural production efficiency but also reduces the damage to land, water resources, and biodiversity and maintains ecological balance through the adoption of environmentally friendly agricultural technologies and management measures. In addition, the ALL policy also helps to promote the development of green agriculture to achieve resource conservation and environmental friendliness, which is of great significance for reducing food contamination and improving the health level and quality of life of the people.

By promoting the optimal allocation of land resources and the application of new agricultural technologies, the ALL policy not only improves agricultural production efficiency but also helps to promote the development of green agriculture and achieve the dual goals of sustainable agricultural development and ecological environmental protection. Based on this, the following assumptions are proposed:

H1: *When the ALL policy is proposed, it will have a positive impact on ACER.*

The ALL policy has led to a reduction in δf_g (transaction costs). On the one hand, the policy makes it easier to negotiate ALL. Unlike previous policies, the ALL policy introduced in 2010 for the first time paid attention to the operability of ALL activities and issued specific guidelines on grassroots coordination, information communication, contract execution, and price assessment. The release and acquisition of ALL information has become more convenient, and both parties or even multiple parties have more reliable anchors for ALL negotiations, so information and negotiation costs are relatively lower, leading to a reduction in f_g.

On the other hand, the implementation of the ALL policy, as evidenced by the official publication of the standard text of ALL contracts, has led to a decrease in the incidence of ALL disputes. Compared with the previous non-standard oral contracts, the template makes a standard definition of the responsibilities and obligations of the two parties for the transfer, which makes the land transfer contract more standardized and more adequately protected at the law. In this way, the implementation of the land transfer contract becomes smoother, which lowers the probability of leaving the business in each period δ.

When δf_g declines, surface II flattens downward to surface II′, as shown in Figure 1. The ALL policy reduces the transaction cost and weakens the barriers that occur in ALL. More intense competitions make more non-highly productive farmers be squeezed out of the land operation, and the minimum efficiency of exiting farmers shifts rightward from τ^* to $\tau^{*\prime}$, which makes the land resources to be operated by more efficient farmers through ALL. As the exit threshold increases, $\widetilde{\varphi}(\tau^*)$ rises accordingly.

What is more, since $\frac{\partial \varphi}{\partial \tau} < 0$, as τ^* and $\widetilde{\tau}(\tau^*)$ are increased, the ACE decreases accordingly. The influence of ALL policy on ACER can be encapsulated as follows: "ALL policy—reduces transaction costs—enhances the average production efficiency of farmers—attains ACER," as illustrated in Figure 2. However, at the macro scale, it is difficult to observe the changes in the production efficiency of microfarmers.

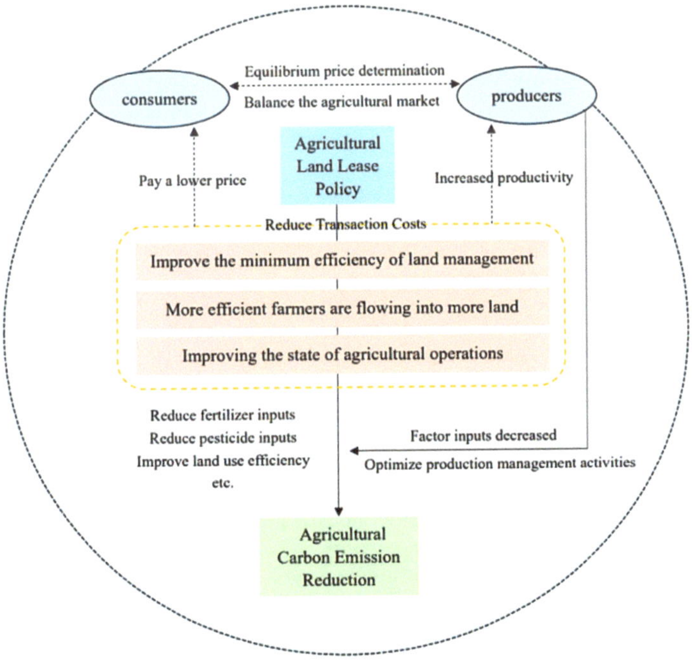

Figure 2. Framework showing how ALL policy implementation promotes ACER.

At the macro level, it is indeed difficult to observe the changes in the production efficiency of microfarmers. However, according to Wang [41] and Yang [42], they used microfarmer data to find that when agricultural labor is transferred to the non-agricultural sector, farmers' agricultural productivity actually increases. This finding suggests that the decline in the number of agricultural workers may be a measure of the improvement in farmers' productivity. In addition, Lewis [43]'s dualistic structure theory proposed in 1954 also states that a decrease in the number of agricultural workers may lead to an increase in the productivity of surplus labor.

In order to cope with the impact of the decline in agricultural workers, farmers are more inclined to adopt modern agricultural techniques and management methods in the production process. As the shift of agricultural labor to the non-agricultural sector has led to a narrowing of the income gap between agriculture and the non-agricultural sector, the surplus agricultural labor force has been able to obtain more skills training and capital allocation, leading to more efficient production methods. This structural change will help to increase the productivity of the surplus agricultural workforce, which in turn will drive the development of agricultural production methods in a more efficient and environmentally friendly direction and ultimately contribute to the realization of ACER. Based on the above analysis, the following hypotheses are proposed:

H2: *ALL policies reduce carbon emissions by reducing transaction costs, increasing the minimum efficiency value of existing operations, and reducing the number of rural agricultural workers.*

3. Research Design and Data

3.1. Model Establishment

In order to explore the impact of ALL policy on ACER, according to Fortson [44], the DID model was used for our assessments in this paper as follows:

$$lnCO2_{it} = \alpha_1 + \beta_1(treat_{it} \times post_{it}) + \gamma \boldsymbol{X_{it}} + \eta_i + \mu_t + \varepsilon_{it} \tag{8}$$

where $CO2_{it}$ is the explanatory variable representing carbon emissions from agriculture in the year t in the province i; $treat_{it} \times post_{it}$ is the core explanatory variable, where $treat_{it}$ represents the land transfer in the year t in the province i and $post_{it}$ represents whether the province i is affected by the land transfer policy in the year t. If the policy is implemented in the year t, it takes the value of 1. Otherwise, it takes the value of 0. η_i and μ_t represent the province fixed effect and time fixed effect, respectively; ε_{it} represents the random disturbance term affecting agricultural carbon emissions; β_1 represents the effect of the land transfer policy on agricultural carbon emissions; and $\boldsymbol{X_{it}}$ represents a set of control variables.

3.2. Description of Variables

Explanatory variable ACE. This paper refers to Du [5], Jin [6], Lal [45], Tian [46], and Wang [3]'s approach and sets the formula for measuring ACE as follows: $ACE = \sum C_j T_j$, where C_j is the carbon emission coefficient of the jth category of carbon sources, and T_j is the absolute amount of consumption of the jth category of carbon sources.

Explanatory variable ALL policy: The ALL policy in this paper is expressed by the interaction term of the proportion of transferred area of family-contracted farmland and the dummy variable of the time of the response to the ALL policy in each province. Given that there are significant time differences in the degree of response to the policy in each region, this paper takes 2010 as the base year for the implementation of the policy and the year in which the provincial ALL rate exceeds the national average transfer rate as the point in time when the region responded to this policy.

Control variables: (1) size of land (SL)—characterized by the per capita sown area of the agricultural labor force [47]; (2) financial support to agriculture (FA)—characterized by the per capita financial expenditure on agriculture of the agricultural labor force [48]; (3) degree of mechanization (DM)—characterized by the total power of agricultural machinery per unit of sown area [49]; (4) structure of the agricultural industry (SA)—characterized by agricultural output as a share of agricultural, forestry, livestock and fisheries output [50]; (5) human capital in agriculture (HA)—characterized by the average number of years of education of the rural population [51]; and (6) rural population structure (RS)—characterized by the share of the rural labor force in the total rural population [52].

Mechanism variables: agricultural employment status (AE)—characterized by the share of agricultural employees in the rural labor force.

Other variables: agricultural carbon intensity (ACI)—characterized by agricultural carbon emissions per unit of cultivated area.

3.3. Data Source and Description

This paper uses panel data from 30 provincial-level administrative regions in China, excluding Tibet, Hong Kong, Macau, and Taiwan, from 2005 to 2020. The ALL data in this paper are acquired from the National Statistics on Rural Economic Situation and the Annual Report of China's Rural Management Statistics, while the rest of the data comes from the China Rural Statistical Yearbook, the China Statistical Yearbook, and the China Population and Employment Statistical Yearbook of previous years. Missing individual data were supplemented using the moving average method, and all data, with the exception of the ALL policy, have been logarithmically transformed. Descriptive statistical analyses of the variables are presented in Table 2.

Table 2. Descriptive statistics of the variables.

Variable Name	Observations	Mean	SD	Min	Max
ACE	480	14.947	1.023	11.899	16.404
ALL policy	480	0.242	0.428	0	1
SL	480	4.111	0.428	3.040	5.625
FA	480	2.296	0.370	0.758	3.015
DM	480	0.600	0.253	0.211	1.416
SA	480	0.523	0.086	0.342	0.746
HA	480	2.019	0.092	1.637	2.276
RS	480	0.431	0.096	0.207	0.748
AE	480	0.582	0.159	0.135	0.988
ACI	480	−0.147	0.265	−0.678	0.564

4. Results

4.1. Benchmark Regression Results

Based on 30 provincial panel data in China from 2005 to 2020, this paper selects the difference-in-differences model with double fixed effects of province and time to test whether the ALL policy effectively promotes ACER. The empirical regression results are shown in Table 3.

Table 3. Benchmark regression results.

Variable Name	ACE	
	(1)	**(2)**
ALL policy	−0.103 **	−0.091 **
	(0.050)	(0.040)
Cons_	14.972 ***	14.14 ***
	(0.012)	(1.089)
Control variable	NO	YES
Province fixed	YES	YES
Time fixed	YES	YES
Observations	480	480
R^2	0.991	0.993

Note: (i) ***, **, and * denote 1%, 5%, and 10% significance levels, respectively. (ii) Robust standard errors in parentheses. The same description holds for the following tables.

From columns (1) and (2) of Table 3, it is evident that the ALL policy is significantly negative at the 5% significant level despite adding or excluding control variables. The results of the benchmark regression preliminarily indicate that the ALL policy reduces ACE, preliminarily verifying H1.

4.2. Parallel Trend and Dynamic Test

In employing the difference-in-differences model to assess policy outcomes, it is crucial for the control and experimental groups to demonstrate parallel trends prior to policy implementation. If the same trend is absent, the policy effects identified by the model may be inaccurately determined. This paper utilizes Nunn's method [53] to develop a novel testing model:

$$InCO2_{it} = \alpha_2 + \sum_{i=-5}^{10} \zeta_t D_{it} + \gamma_2 \mathbf{X}_{it} + \eta_i + \mu_t + \varepsilon_{it} \tag{9}$$

where D_{it} is a dummy variable indicating whether the province i is affected by the ALL policy in the year t. If the policy is implemented in the year t, it is designated as 1 and inversely as 0. The rest of the variables have the same connotations as in Equation (8).

This paper consolidates the data to the period -5, as fewer provinces were impacted by the policy prior to five years ago. Figure 2 shows that the coefficients are insignificant in all periods before the policy. Subsequent to the policy's implementation, ACE in the impacted regions was markedly lower than in other areas, with a discernible trend of gradual escalation in this effect. Figure 3 shows that the ALL policy passes the parallel trend test whether or not control variables are added.

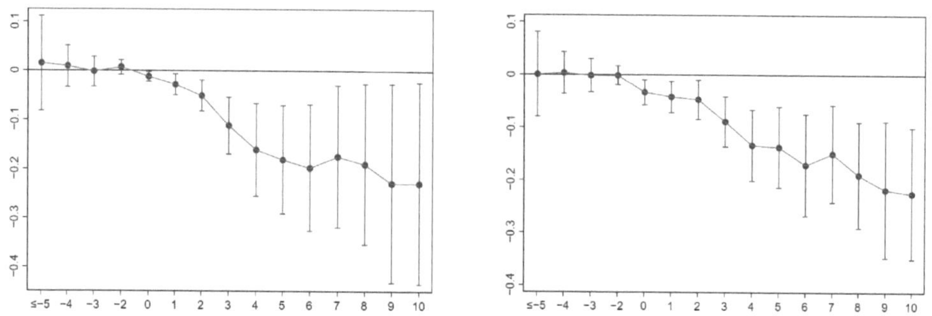

(**a**) No control variables added (**b**) Add control variables

Figure 3. Parallel trend tests and trends in the impact of policy dynamics. Note: (i) Vertical lines indicate 90% confidence intervals for the parameters. (ii) A period before 2010, the year of policy implementation, is used as the base group, so there is no -1 period (i.e., 2009) on the horizontal axis.

4.3. Placebo Test

To prevent the regression results from being influenced by unobservable factors that may induce significance, this paper employs a Monte Carlo simulation, conducted 1000 times, to randomly generate the provinces and years impacted by the ALL policy shocks, following the methodology of Wang [54]. In this paper, 14 provinces and policy response years were randomly selected each time. The p-statistic distribution of the 1000 simulated regression coefficients β is shown in Figure 4. The findings indicate that the p-statistic from the simulation closely aligns with a normal distribution, exhibiting a mean value of 0, and most results are greater than 0.1, that is, most of the simulations do not pass the 10% significance level. This outcome suggests that the significance represents a low-probability event in the placebo test, and the baseline regression results are not attributable to unobservable factors. Accordingly, the ALL policy contributes significantly to the reduction in ACE.

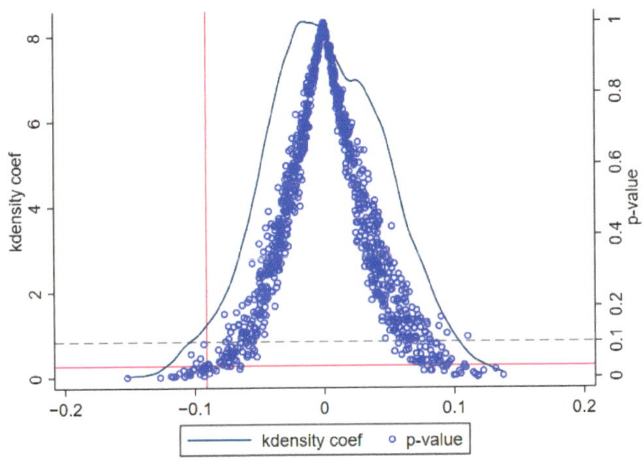

Figure 4. Distribution of p-statistic values for 1000 random simulations. Note: (i) The right vertical axis represents the value of the simulated p-statistic, calculated once per point of the simulation. (ii) The horizontal line represents the *p*-value of the benchmark result at 0.043, and the vertical line represents the benchmark result at −0.091.

4.4. Sensitivity Test

Sensitivity tests: On the basis of the uncontrolled variables in regression (1) in Table 3, the regression (1) to regression (6) in Table 4 includes SL, FA, DM, SA, HA, and RS, respectively [55]. Compared with regressions (1) and (2) in Table 2, there was no significant change in the coefficient and significance of ALL.

Table 4. Sensitivity test results.

Variable Name	ACE					
	(1)	(2)	(3)	(4)	(5)	(6)
ALL policy	−0.095 ** (0.044)	−0.094 ** (0.044)	−0.090 ** (0.041)	−0.089 ** (0.042)	−0.090 ** (0.042)	−0.091 ** (0.040)
SL	0.326 ** (0.136)	0.337 ** (0.134)	0.374 *** (0.135)	0.372 *** (0.133)	0.372 *** (0.133)	0.391 *** (0.130)
FA	--	0.098 (0.092)	0.102 (0.089)	0.103 (0.089)	0.103 (0.085)	0.086 (0.085)
DM	--	--	0.238 * (0.115)	0.236 * (0.116)	0.237 * (0.118)	0.245 ** (0.114)
SA	--	--	--	0.107 (0.355)	0.107 (0.353)	0.036 (0.334)
HA	--	--	--	--	0.022 (0.406)	−0.452 (0.401)
RS	--	--	--	--	--	−0.539 ** (0.243)
Cons_	13.630 *** (0.557)	13.360 *** (0.585)	13.050 *** (0.578)	13.010 *** (0.602)	12.960 *** (1.022)	14.14 *** (1.089)
Control variable	YES	YES	YES	YES	YES	YES
Province fixed	YES	YES	YES	YES	YES	YES
Time fixed	YES	YES	YES	YES	YES	YES
Observations	480	480	480	480	480	480
R^2	0.992	0.992	0.993	0.993	0.993	0.993

In the controlled variables, only SL, DM, and RS are significant, this result is not surprising. First, previous research indicates that expanding SL leads to a shift in agricultural production from labor-intensive to capital-intensive production, accompanied by increased inputs such as fertilizers and pesticides, which in turn lead to an increase in ACE [30,31]. As the controlled variables increase, the impact of SL on ACE gradually rises, and its significance also increases accordingly. Second, an increase in DM often coincides with higher consumption of fossil fuels, which directly contributes to ACE through the consumption of diesel or gasoline [49]. Lastly, with the improvement in RS, the rural population structure is optimized, meaning that the proportion of young and middle-aged laborers in rural areas increases [52]. These individuals typically have higher labor productivity and the ability to adopt new agricultural technologies. As the proportion of young and middle-aged populations increases, agricultural production efficiency may improve, contributing to achieving ACE.

This paper also found that the effects of FA, SA, and HA on ACE are complex, and these effects are not always significant. The following is a detailed analysis of the impact of these control variables:

Impact of FA: FA has boosted ACE to some extent, but its impact has not been significant. This may be because FA has a twofold effect: on the one hand, FA promotes the advancement of agricultural technology and contributes to the achievement of ACER. On the other hand, FA may lead to factor price distortions, increase the input of machinery and equipment and chemicals in agricultural production, and thus increase ACE. These two opposing effects may cancel each other out, resulting in an insignificant overall effect of FA on ACE.

Impact of SA: The increase in SA may mean an increase in the position of agriculture in the agriculture, forestry, animal husbandry, and fishery industries, which may lead to a greater concentration of resources in the agricultural sector, thereby increasing the ACE. However, if this structural adjustment is accompanied by an increase in agricultural production efficiency, it may partially offset the increase in ACE, resulting in insignificant results. This shows that the impact of agricultural restructuring on ACE is multifaceted and needs to be further studied.

Impact of HA: The improvement in HA has increased farmers' acceptance of new technologies and the efficiency of their application, which may help improve agricultural production efficiency and achieve ACER. However, at the same time, the increase in HA may also lead to the shift of farmers' production focus, causing more farmers to engage in non-agricultural activities, reducing agricultural production efficiency that is not conducive to ACER, resulting in insignificant results. However, after the addition of RS, the coefficient of HA turned negative, suggesting that the effect of HA on improving agricultural productivity was more significant after stripping out the impact of young labor, which contributed to the achievement of ACER.

The analysis of control variables reveals the complex impact of different factors on ACE and provides a new perspective for understanding the driving factors of ACE. These findings highlight the multifaceted factors that need to be considered when formulating agricultural policies and the combined impact these factors can have on ACE.

4.5. Robustness Test

The benchmark analysis demonstrates that the ALL policy is instrumental in promoting ACER, but a series of robustness tests are still needed to exclude the interference of confounding factors. This paper employs sample data screening, substitution of explanatory variables, and endogeneity methods to enhance the credibility of the research findings.

Sample data screening: In this paper, the explanatory variables are truncated by 5% to avoid the impact of sample extremes on the benchmark regression; Equation (6) is subjected to regression analysis. After truncation, the regression results in column (1) of Table 5 show that the positive impact of ALL policy on ACER passes the significance test at the 10% level, which is similar to the results of the benchmark regression.

Table 5. Robustness test results.

Variable Name	Truncated 5%	Replacing Explanatory Variable	Add L.ACE	PSM-DID	Exclude Other Policy
	(1)	(2)	(3)	(4)	(5)
ALL policy	−0.0756 * (0.038)	−0.101 *** (0.051)	−0.061 *** (0.005)	−0.188 * (0.921)	−0.065 ** (0.035)
L.ACE	--	--	0.918 *** (0.016)	--	--
D2015	--	--	--	--	−0.059 * (0.033)
Cons_	15.364 *** (0.948)	0.908 * (0.502)	1.093 *** (0.266)	14.389 *** (1.049)	14.192 *** (1.089)
Control variable	YES	YES	YES	YES	YES
Province fixed	YES	YES	YES	YES	YES
Time fixed	YES	YES	YES	YES	YES
Observations	432	480	450	426	480
R^2	0.993	0.971	0.999	0.995	0.994

Substitution of explanatory variables: Given that ACE faces not only the issue of total quantity but also the issue of intensity, this paper substitutes the explanatory variables with carbon emissions per unit of sown area and regresses Equation (8). The results in column (2) of Table 5 show that the negative effect of the ALL policy on carbon intensity passes the significance test at the 1% level, indicating that this policy also reduces carbon emissions per unit of cultivated area. This again demonstrates the robustness of the previous results.

Consideration of endogeneity: Considering the possible endogeneity problem of the model, we utilized a combination of adding the lagged explanatory variables and propensity score matching (PSM). In Table 5, columns (3) lists the regression results for the addition of lagged explanatory variables. The results showed that the ALL policy still significantly promotes ACER, and the significance level is higher (from 5% to 1% of the basic regression). Column (4) shows the regression results of the PSM-DID method, which indicates that the ALL policy still significantly reduces ACE after changing the regression method. The above test shows the robustness of the previous results.

Exclude other policy implications: In 2015, China introduced a plan to keep increasing the input of pesticides and fertilizers, which may have an impact on ACE [3]. In order to exclude the impact of this program, D2015 was added. Column (5) of Table 5 shows that after considering the impact of this program, the ALL policy still significantly boosted ACER. The above test shows the robustness of the previous results.

5. Discussion

5.1. Mechanism Test

From the previous theoretical analyses, it is clear that the ALL policy may cause low-productivity farmers to exit agricultural production through the crowding-out effect. Since this paper uses macro data for analysis, the most intuitive representation is that the policy achieves its impact on ACE by reducing the number of local agricultural employees. To prevent scale interference with the results, this indicator is denoted as AE.

To identify this mechanism, referring to Jiang [56], this paper constructs the following model:

$$M_{it} = \alpha_3 + \beta_2(treat_{it} \times post_{it}) + \gamma_3 \boldsymbol{X_{it}} + \eta_i + \mu_t + \varepsilon_{it} \quad (10)$$

$$lnCO2_{it} = \alpha_4 + \beta_3 M_{it} + \gamma_4 \boldsymbol{X_{it}} + \eta_i + \mu_t + \varepsilon_{it} \quad (11)$$

where M_{it} is the mechanism variable, denoting AE. The rest of the variables have the same connotations as in Equation (8).

As evidenced in column (1) of Table 6, the coefficient estimate of the ALL policy is −0.088, which passes the 10% significance level. This suggests that the ALL policy increases the critical value of productivity for farm households exiting the business by reducing transaction costs, which in turn leads to a reduction in AE. Column (2) of Table 6 demonstrates that an increase in AE is associated with a concomitant rise in ACE. This relationship is statistically significant at the 1% level. According to the prior theoretical analysis, a higher AE signifies a lower exit threshold, recognizing the variation in farm household productivity. This results in diminished average efficiency of local agricultural production, subsequently causing a less effective utilization of agricultural input factors and an escalation in ACE. In conjunction with the findings in column (1), it becomes evident that the ALL policy effectively eliminates low-productivity farmers, thereby enhancing the local average agricultural productivity and ultimately facilitating the realization of ACER.

Table 6. Mechanism test results.

Variable Name	AE (1)	ACE (2)
ALL policy	−0.088 * (0.051)	--
AE	--	1.327 *** (0.159)
Cons_	1.055 ** (0.498)	12.720 *** (0.723)
Control variable	YES	YES
Province fixed	YES	YES
Time fixed	YES	YES
Observations	480	480
R^2	0.922	0.997

5.2. Heterogeneity Test

The impact of ALL policy may be affected by regional heterogeneity. In this paper, (1) the dividing line between the north and south of China divides the sample into the northern region and the southern region; (2) the average per capita GNP of each province during the sample period was used as the dividing line between high-income and low-income regions; (3) the mean value of fertilizer application intensity in each province was used as the dividing line between high-intensity and low-intensity regions; and (4) the mean value of agricultural GDP in each province was used as a dividing line between agriculturally advantaged and agriculturally disadvantaged regions.

Columns (1) and (2) of Table 7 show the estimated impact of ALL policy on ACE in the north and south, respectively. Regression analysis showed that ALL policy had a significant negative impact on ACE in both the north and the south, and there was almost no significant difference in geography. This finding suggests that although rice is mainly cultivated in southern China, while wheat and corn are mainly cultivated in the north, the impact of ALL policy on ACE is not affected by these geographical and planting structure differences, which may be related to the general applicability and effectiveness of the policies.

Columns (3) and (4) of Table 7 reveal the impact of ALL policy in regions with different levels of economic development. In high-income areas, the impact of ALL policy on ACE was −0.277, and this passed the significance test of 5%. In low-income areas, this effect did not pass the 10% significance test. This result indicates that the impact of ALL policy on ACE is significantly affected by the level of economic development. Sun [57] pointed out that the level of economic development is closely related to the spirit of contract, and the non-market rules are more obvious in economically underdeveloped regions. In regions with better economic development and more active markets, ALL policy can more effectively reduce ACE by reducing transaction costs and raising the threshold for farmers' exit. However, in areas with poor economic performance, the mechanism of ALL policy may be hindered due to weak contract spirit and higher market uncertainty, so the impact on ACE is not significant.

Table 7. Results of heterogeneity test.

Variable Name	North Regions (1)	South Regions (2)	Low-Income Regions (3)	High-Income Regions (4)	Low Intensity of Fertilizer (5)	High Intensity of Fertilizer (6)	Agriculturally Advantaged Regions (7)	Agriculturally Disadvantaged Regions (8)
ALL policy	−0.288 **	−0.240 ***	−0.009	−0.277 **	0.062	−0.351 **	0.030	−0.155 **
	(0.130)	(0.076)	(0.089)	(0.104)	(0.072)	(0.130)	(0.029)	(0.067)
Cons_	13.680 **	13.680 ***	15.620 **	14.220 ***	13.850 ***	15.510 ***	16.050 ***	13.210 ***
	(1.636)	(0.664)	(1.076)	(0.732)	(0.745)	(1.545)	(1.063)	(1.274)
Control variable	YES	YES	YES	YES	YES	YES	YES	YES
Province fixed	YES	YES	YES	YES	YES	YES	YES	YES
Time fixed	YES	YES	YES	YES	YES	YES	YES	YES
Observations	240	240	352	144	208	272	208	272
R^2	0.995	0.997	0.994	0.998	0.997	0.994	0.990	0.992

Columns (5) and (6) of Table 7 provide further analysis of the impact of ALL policy in regions with different fertilizer input intensities. In areas with high fertilizer input, the impact of ALL policy on ACE was −0.351, and this passed the significance test of 5%. However, in areas with low intensity of chemical fertilizer input, this effect did not pass the 10% significance test. This result may be related to the distortion of the input structure of agricultural factors and the low level of green production efficiency in high-intensity regions [58]. In areas with low intensity of chemical fertilizer use, the average efficiency of farmers is higher, the structure of agricultural factors is more reasonable, and the role of ALL policy in improving efficiency is limited, so the ability to achieve ACER is weakened.

Columns (7) and (8) of Table 7 explore the impact of ALL policy in areas with weak and advantageous agricultural areas. In the disadvantaged agricultural areas, the impact of ALL policy on agricultural carbon emissions was −0.155, and this passed the significance test of 5%. However, in the agricultural advantageous areas, this effect did not pass the 10% significance test. Agricultural production facilities in the agricultural advantage areas are more complete, the factor utilization degree is higher, the agricultural green production status is better, and the ACE is lower [59,60], so the ALL policy has a limited effect. Conversely, resources are underutilized in agrarianly disadvantaged areas [61,62]; the ALL policy expands the scope of land management of advantageous agricultural operators through the transformation of operators, improves the level of land management, and finally realizes ACER.

The impact of ALL policy on ACE shows obvious heterogeneity in different regions, which is affected by the level of economic development, fertilizer input intensity, and agricultural development. These findings provide important insights into the mechanism of ALL policy in different contexts and provide targeted implementation suggestions for policymakers.

5.3. Discussion with Relevant Research

Scholars have performed a series of analyses on land tenure, ACE, and environmental policy. On this basis, this paper takes the reduction in transaction costs brought about by the ALL policy as the research object and conducts a theoretical and empirical analysis.

ALL policy relationship with ACE. Scholars have extensively studied the impact of ALL on ACE, but they often look at the perspective of farmer behavior and ignore the impact at the policy level. This paper emphasizes that the core of ALL policy is to improve the normative and legal guarantee of ALL activities, which directly promotes the reduction in transaction costs. Although the role of ALL policy in agricultural finance [36] and farmland abandonment [37] has been noted in previous studies, there is a lack of research on the ACER effect of ALL policy in the context of global warming. The existing literature shows that in the industrial and urban sectors, the reduction in transaction costs can effectively promote carbon emission reduction [39]. This study not only expands the research perspective of ALL and ACE but also expands the study of the effect of transaction cost reduction on ACER from the industrial sector to the agricultural sector.

Application of the Melitz [40] model was performed in the field of agriculture. Based on the heterogeneous efficiency model proposed by Melitz [40] in 2001, most scholars have analyzed the entry and exit decisions of enterprises and the improvement in industry efficiency and have achieved fruitful results. In this study, farmers are regarded as the smallest agricultural enterprises, which breaks through the limitation of only analyzing joint-stock enterprises in the past. Based on the research of Song [32], Yu [33], Yasmeen [34], and Chari [35], this study combines Melitz [40]'s heterogeneous efficiency model and the impact of agricultural profits on farmers' entry (exit) decisions with ACE for the first time, extends Melitz [40] model to the field of farmers and agricultural production, and links the impact of heterogeneous efficiency with ACE, providing a strong theoretical analysis framework for related research.

Impact of non-agricultural environmental policies on ACER. Some studies have focused on the impact of agricultural environmental policies on ACER, but some scholars have found that other agricultural policies can also lead to ACER. For example, the policy of establishing major grain-producing areas [3] and the policy of building high-standard fields [13] have contributed to the achievement of ACER. This paper not only broadens the scope of ALL policy research but also enriches the research on the positive impact of non-agricultural environmental policies on ACER.

6. Conclusions

Starting from the perspective of heterogeneous production efficiency, this paper first conducts a theoretical analysis of the impact of ALL policy and then establishes an asymptotic DID model to estimate the effect of ALL policy on the impact of ACE by using the provincial panel data of China from 2005 to 2020.

The main findings of this paper are as follows: First, the benchmark regression results show that the ALL policy significantly contributes to ACER. Following a series of robustness analyses, the fundamental conclusion that the ALL policy attains ACER remains valid. Conversely, in areas where agriculture is impeded, the complete utilization of agricultural resources is unfeasible [61,62]. Second, the mechanism analysis shows that the ALL policy achieves ACER by increasing the critical value of the minimum efficiency of farmers' exit from land operation and reducing the proportion of agricultural employees. Third, ALL contributes significantly to the achievement of ACER in both the North and the South of China. In regions with advanced economic development, high intensity of fertilizer application, and agricultural underdevelopment, the ALL policy has more effectively promoted ACER.

The following policy implications are obtained from this study: with the implementation of policies such as "separation of powers", farmers' land rights and interests are more stable. To further enhance the vitality of the ALL market, it is more necessary to start by improving the degree of standardization of the ALL market and strengthening the legal guarantee for ALL. Establish an ALL information platform, strengthen the supervision of ALL, optimize the resolution mechanism of ALL disputes, and establish a white list system to reduce the risk of ALL and effectively enhance the market's confidence and expectation of long-term and stable ALL.

This study was validated at the provincial level, but there were some limitations. The heterogeneity analysis indicated that the variability of the natural environment minimally affected the ACER effect of ALL, whereas the technical disparities in agricultural production significantly influenced this effect. There were significant differences in agricultural scale, skill level, financial status, digitalization, and mechanization of individual farmers that led to a heterogeneous production efficiency, which was difficult to observe at the macro level. Second, farmers' productivity may change as a result of leased land. For example, the increase in scale may bring economies of scale, increase productivity, or exceed the farmer's capacity and lead to a decrease in efficiency, which is also difficult to observe at the macro level. While aggregating the data can reflect the overall trends, it may mask important changes at the individual farmer level. In terms of mechanism, although the research of Wang [41] and Yang [42] provides a correlation between the reduction in personnel in the agricultural industry and the change in microefficiency, it is still difficult for provincial data to measure the changes in real efficiency. Therefore, future research can use microfarmer surveys to focus on the specific changes in farmers' business behavior after leasing land, so as to verify and deepen the conclusions of this study.

Author Contributions: Conceptualization, design, data collection, and writing—original draft, S.W.; methodology and software, S.W. and Y.F.; writing—review and editing, Y.F., B.Z. and S.W.; supervision, S.W. and F.C. All authors have read and agreed to the published version of the manuscript.

Funding: This research received no external funding.

Data Availability Statement: Dataset available on request from the authors.

Conflicts of Interest: The authors declare no conflicts of interest.

References

1. Chaudhari, U.; Bhadoriya, A.; Jani, M.Y.; Sarkar, B. A generalized payment policy for deteriorating items when demand depends on price, stock, and advertisement under carbon tax regulations. *Math. Comput. Simul.* **2023**, *207*, 556–574. [CrossRef]
2. Bhadoriya, A.; Jani, M.Y.; Chaudhari, U. Combined Effect of Carbon Emission, Exchange Scheme, Trade Credit, and Advertisement Efforts in a Buyer's Inventory Decision. *Process Integr. Optim. Sustain.* **2022**, *6*, 1043–1061. [CrossRef]
3. Wang, S.; Jin, M.; Liu, L.; Cao, F. Impacts of China's Main Grain-Producing Areas on Agricultural Carbon Emissions: A Sustainable Development Perspective. *Sustainability* **2024**, *16*, 4607. [CrossRef]
4. Li, L.; Khan, U.S.; Guo, C.; Huang, Y.; Xia, X. Non-agricultural labor transfer, factor allocation and farmland yield: Evidence from the part-time peasants in Loess Plateau region of Northwest China. *Land Use Policy* **2022**, *120*, 106289. [CrossRef]
5. Du, Y.; Liu, H.; Huang, H.; Li, X. The carbon emission reduction effect of agricultural policy-Evidence from China. *J. Clean. Prod.* **2023**, *406*, 137005. [CrossRef]
6. Jin, M.; Feng, Y.; Wang, S.; Chen, N.; Cao, F. Can the development of the rural digital economy reduce agricultural carbon emissions? A spatiotemporal empirical study based on China's provinces. *Sci. Total Environ.* **2024**, *939*, 173437. [CrossRef]
7. Yang, H.; Wang, X.; Bin, P. Agriculture carbon-emission reduction and changing factors behind agricultural eco-efficiency growth in China. *J. Clean. Prod.* **2022**, *334*, 130193. [CrossRef]
8. Cui, X.; Cai, T.; Deng, W.; Zheng, R.; Jiang, Y.; Bao, H. Indicators for evaluating high-quality agricultural development: Empirical study from Yangtze River Economic Belt, China. *Soc. Indic. Res.* **2022**, *164*, 1101–1127. [CrossRef]
9. Xiong, L.; Shah, F.; Wu, W. Environmental and socio-economic performance of intensive farming systems with varying agricultural resource for maize production. *Sci. Total Environ.* **2022**, *850*, 158030. [CrossRef]
10. Ma, W.; Zhou, X.; Boansi, D.; Seyram, G.; Horlu, A.; Owusu, V. Adoption and intensity of agricultural mechanization and their impact on non-farm employment of rural women. *World Dev.* **2024**, *173*, 106434. [CrossRef]
11. Lu, H.; Chen, Y.; Luo, J. Development of green and low-carbon agriculture through grain production agglomeration and agricultural environmental efficiency improvement in China. *J. Clean. Prod.* **2024**, *442*, 141128. [CrossRef]
12. Hu, Y.; Li, B.; Zhang, Z.; Wang, J. Farm size and agricultural technology progress: Evidence from China. *J. Rural Stud.* **2022**, *93*, 417–429. [CrossRef]
13. Deng, O.; Ran, J.; Huang, S.; Duan, J.; Reis, S.; Zhang, J.; Zhu, Y.; Xu, L.; Gu, B. Managing fragmented croplands for environmental and economic benefits in China. *Nat. Food* **2024**, *5*, 230–240. [CrossRef] [PubMed]
14. Li, X.; Qiao, S.; Jiang, Y. The causal mechanism of farmers' chemical fertilizer reduction: An empirical perspective from farmland transfer-in and digital extension. Front. *Sustain. Food Syst.* **2023**, *7*, 1231574. [CrossRef]
15. Fu, J.; Ding, R.; Zhu, Y.; Du, L.; Shen, S.; Peng, L.; Zou, J.; Hong, Y.; Liang, J.; Wang, K.; et al. Analysis of the spatial-temporal evolution of Green and low carbon utilization efficiency of agricultural land in China and its influencing factors under the goal of carbon neutralization. *Environ. Res.* **2023**, *237*, 116881. [CrossRef]
16. Zou, J.; Shen, L.; Wang, F.; Tang, H.; Zhou, Z. Dual carbon goal and agriculture in China: Exploring key factors influencing farmers' behavior in adopting low carbon technologies. *J. Integr. Agric.* **2024**, *23*, 3215–3233. [CrossRef]
17. Wu, J.; Wen, X.; Qi, X.; Fang, S.; Xu, C. More Land, Less Pollution? How Land Transfer Affects Fertilizer Application. *Int. J. Environ.* **2021**, *18*, 11268. [CrossRef]

18. Li, B.; Zeng, Q. The effect of land right stability on the application of fertilizer reduction technologies-evidence from large-scale farmers in China. *Sustainability* **2022**, *14*, 8059. [CrossRef]
19. Gao, L.; Sun, D.; Huang, J. Impact of land tenure policy on agricultural investments in China: Evidence from a panel data study. *China Econ. Rev.* **2017**, *45*, 244–252. [CrossRef]
20. Ntakirutimana, L.; Li, F.; Huang, X.; Wang, S.; Yin, C. Green Manure Planting Incentive Measures of Local Authorities and Farmers Perceptions of the Utilization of Rotation Fallow for Sustainable Agriculture in Guangxi, China. *Sustainability* **2019**, *11*, 2723. [CrossRef]
21. Lu, H.; Zhang, P.; Hu, H.; Xie, H.; Yu, Z.; Chen, S. Effect of the grain-growing purpose and farm size on the ability of stable land property rights to encourage farmers to apply organic fertilizers. *J. Environ. Manag.* **2019**, *251*, 109621. [CrossRef] [PubMed]
22. Li, F.; Ren, J.; Wimmer, S.; Yin, C.; Li, Z.; Xu, C. Incentive mechanism for promoting farmers to plant green manure in China. *J. Clean. Prod.* **2020**, *267*, 122197. [CrossRef]
23. Jia, R.; Lu, Q. Land transfer, collective action and the adoption of soil and water conservation measures in the Loess Plateau of China. *Nat. Hazards* **2020**, *102*, 1279–1304. [CrossRef]
24. Zhou, L.; Lu, H.; Wu, B. Identifying the effects of the stability of farmland transfer management rights on the use of organic fertilizer with different farmland scales and crop types. *Front. Environ. Sci.* **2020**, *10*, 1006405. [CrossRef]
25. Zhang, Q.; Tian, J.; Zheng, J.; Abdullahi, N.M.; Huo, X. How Does Land Tenure Security Affect Farm Succession? Evidence from Apple Growers in China. *Land* **2022**, *11*, 1036. [CrossRef]
26. Tang, M.; Hu, F. How Does Land Urbanization Promote CO2 Emissions Reduction? Evidence From Chinese Prefectural-Level Cities. *Front. Environ. Sci.* **2021**, *9*, 766839. [CrossRef]
27. Adamopoulos, T.; Brandt, L.; Leight, J.; Restuccia, D. Misallocation, selection, and productivity: A quantitative eanalysis with panel data from China. *Econometrica* **2022**, *90*, 1261–1282. [CrossRef]
28. Cheng, C.; Zhang, F.; Shi, J.; Kung, H. What is the relationship between land use and surface water quality? A review and prospects from remote sensing perspective. *Environ. Sci. Pollut.* **2022**, *29*, 56887–56907. [CrossRef]
29. Guo, L.; Song, Y.; Tang, M.; Tang, J.; Dogbe, B.; Su, M.; Li, H. Assessing the Relationship among Land Transfer, Fertilizer Usage and PM2.5 Pollution: Evidence from Rural China. *Int. J. Environ.* **2022**, *19*, 8387. [CrossRef]
30. Qi, Y.; Chen, X.; Zhang, J.; Li, Y.; Zhu, D. How Do Rising Farmland Costs Affect Fertilizer Use Efficiency? Evidence from Gansu and Jiangsu, China. *Land* **2022**, *11*, 1730. [CrossRef]
31. Tesfaye, K.; Takele, R.; Sapkota, T.B.; Chhetric, A.K.; Solomon, D.; Stirling, C.; Albanito, F. Model comparison and quantification of nitrous oxide emission and mitigation potential from maize and wheat fields at a global scale. *Sci. Total Environ.* **2021**, *782*, 146696. [CrossRef] [PubMed]
32. Song, M.; Wu, Q.; Zhu, H. Could the Aging of the Rural Population Boost Green Agricultural Total Factor Productivity? Evidence from China. *Sustainability* **2024**, *16*, 6117. [CrossRef]
33. Yu, Y.; Hu, Y.; Gu, B.; Reis, S.; Yang, L. Reforming smallholder farms to mitigate agricultural pollution. *Environ. Sci. Pollut. Res.* **2022**, *29*, 13869–13880. [CrossRef] [PubMed]
34. Yasmeen, R.; Tao, R.; Shah, W.U.H. The nexuses between carbon emissions, agriculture production efficiency, research and development, and government effectiveness: Evidence from major agriculture-producing countries. *Environ. Sci. Pollut. Res.* **2022**, *29*, 52133–52146. [CrossRef]
35. Chari, A.; Liu, M.E.; Wang, S.; Wang, Y. Property rights, land misallocation, and agricultural efficiency in China. *Rev. Econ. Stud.* **2021**, *88*, 1831–1862. [CrossRef]
36. Jiang, M.; Paudel, K.P.; Mi, Y.; Li, J. Farmland transfer and rural financial structure: Evidence from China. *Int. Rev. Financ. Anal.* **2023**, *90*, 102897. [CrossRef]
37. Song, H.; Li, X.; Xin, L.; Wang, X. Do farmland transfers mitigate farmland abandonment? —A case study of China's mountainous areas. *Habitat Int.* **2024**, *146*, 103023. [CrossRef]
38. Huo, C.; Chen, L. The impact of land transfer policy on sustainable agricultural development in China. *Sci. Rep.* **2024**, *14*, 7064. [CrossRef]
39. Hu, Y.; Ren, S.; Wang, Y.; Chen, X. Can carbon emission trading scheme achieve energy conservation and emission reduction? Evidence from the industrial sector in China. *Energy Econ.* **2020**, *85*, 104590. [CrossRef]
40. Melitz, M.J. The Impact of Trade on Intra-Industry Reallocations and Aggregate Industry Productivity. *Econometrica* **2003**, *6*, 1695–1725. [CrossRef]
41. Wang, J.; Cramer, G.L.; Wailes, E.J. Production efficiency of Chinese agriculture: Evidence from rural household survey data. *Agric. Econ.* **1996**, *15*, 17–28.
42. Yang, J.; Wang, H.; Jin, S.; Chen, K.; Riedinger, J.; Peng, C. Migration, local off-farm employment, and agricultural production efficiency: Evidence from China. *J. Prod. Anal.* **2016**, *45*, 247–259. [CrossRef]
43. Lewis, W.A. Economic Development with unlimited supply of labor. *Manch. Sch.* **1954**, *22*, 139–191. [CrossRef]
44. Fortson, J.G. HIV/AIDS and fertility. *Am. Econ. J. Appl. Econ.* **2009**, *1*, 170–194. [CrossRef]
45. Lal, R. Carbon emission from farm operations. *Environ. Int.* **2004**, *30*, 981–990. [CrossRef]
46. Tian, Y.; Lautz, S.; Wallis, A.O.; Lambiotte, R. Extracting complements and substitutes from sales data: A network perspective. *EPJ Data Sci.* **2021**, *10*, 45. [CrossRef]
47. Liu, Y.; Zou, L.; Wang, Y. Spatial-temporal characteristics and influencing factors of agricultural eco-efficiency in China in recent 40 years. *Land Use Policy* **2020**, *97*, 10479. [CrossRef]

48. Farooq, U.; Gang, F.; Guan, Z.; Rauf, A.; Chandio, A.A.; Ahsan, F. Exploring the long-run relationship between financial inclusion and agricultural growth: Evidence from Pakistan. *Int. J. Emerg. Mark.* **2023**, *18*, 1677–1696. [CrossRef]
49. Lu, F.; Meng, J.; Cheng, B. How does improving agricultural mechanization affect the green development of agriculture? Evidence from China. *J. Clean. Prod.* **2024**, *472*, 143298. [CrossRef]
50. Liu, Y.; Sun, D.; Wang, H.; Wang, X.; Yu, G.; Zhao, X. An evaluation of China's agricultural green production: 1978–2017. *J. Clean. Prod.* **2020**, *243*, 118483. [CrossRef]
51. Yang, G.; Li, S.; Wang, H.; Wang, L. Study on agricultural cultivation development layout based on the matching characteristic of water and land resources in North China Plain. *Agric. Water Manag.* **2022**, *259*, 107272. [CrossRef]
52. Chao, P.; Ma, B.; Zhang, C. Poverty alleviation through e-commerce: Village involvement and demonstration policies in rural China. *J. Integr. Agric.* **2021**, *20*, 998–1011.
53. Nunn, N.; Qian, N. The potato's contribution to population and urbanization: Evidence from a historical experiment. *Q. J. Econ.* **2011**, *126*, 593–650. [CrossRef]
54. Wang, F.; Ge, X. Can low-carbon transition impact employment empirical evidence from low-carbon city pilot policy. *China Ind. Econ.* **2022**, *5*, 81–99. (In Chinese)
55. Jani, M.Y.; Chaudhari, U.; Sarkar, B. How does an industry control a decision support system for a long time? *RAIRO-Oper. Res.* **2021**, *55*, 3141–3152. [CrossRef]
56. Jiang, T. Mediating effects and moderating effeets in causal inference. *China Ind. Econ.* **2022**, *5*, 100–120. (In Chinese)
57. Sun, J.; Zhai, N.; Miao, J.; Mu, H.; Li, W. How do heterogeneous environmental regulations affect the sustainable development of marine green economy? Empirical evidence from China's coastal areas. *Ocean Coast. Manag.* **2023**, *232*, 106448. [CrossRef]
58. Mohammadi, A.; Rafiee, S.; Jafari, A.; Keyhani, A.; Mousavi-Avval, S.H.; Nonhebel, S. Energy use efficiency and greenhouse gas emissions of farming systems in north Iran. *Renew. Sustain. Energy Rev.* **2014**, *30*, 724–773. [CrossRef]
59. Wang, Y.; Zhu, Y.; Zhang, S.; Wang, Y. What could promote farmers to replace chemical fertilizers with organic fertilizers? *J. Clean. Prod.* **2018**, *199*, 882–890. [CrossRef]
60. He, P.; Zhang, J.; Li, W. The role of agricultural green production technologies in improving low-carbon efficiency in China: Necessary but not effective. *J. Environ. Manag.* **2021**, *293*, 112837. [CrossRef]
61. Xie, H.; Zhang, Y.; Choi, Y. Measuring the Cultivated Land Use Efficiency of the Main Grain-Producing Areas in China under the Constraints of Carbon Emissions and Agricultural Nonpoint Source Pollution. *Sustainability* **2018**, *10*, 1932. [CrossRef]
62. Ertunç, E. The Effect of Land Consolidation Projects on Carbon Footprint. *Land* **2023**, *12*, 507. [CrossRef]

land

MDPI

Article

Enhancing Drought Forecast Accuracy Through Informer Model Optimization

Jieru Wei [1,2], Wensheng Tang [1,2], Pakorn Ditthakit [3], Jiandong Shang [1,2,*], Hengliang Guo [1,2,*], Bei Zhao [1,2], Gang Wu [1,2] and Yang Guo [1,2]

1 The School of Computer and Artificial Intelligence, Zhengzhou University, Zhengzhou 450000, China; jieruwei@zzu.edu.cn (J.W.); tangwensheng@gs.zzu.edu.cn (W.T.); 202012172013072@gs.zzu.edu.cn (B.Z.); gangwu@zzu.edu.cn (G.W.); guoyang@zzu.edu.cn (Y.G.)

2 National Supercomputing Center in Zhengzhou, Zhengzhou University, Zhengzhou 450000, China

3 Center of Excellence in Sustainable Disaster Management, School of Engineering and Technology, Walailak University, Nakhon Si Thammarat 80161, Thailand; dpakorn@mail.wu.ac.th

* Correspondence: sjd@zzu.edu.cn (J.S.); guohengliang@zzu.edu.cn (H.G.)

Abstract: As droughts become more frequent due to climate change and shifts in land use, enhancing the accuracy of drought prediction is becoming crucial for informed land and water resource management. This study employed the Informer model to forecast drought and conducted a comparative analysis with Autoregressive Integrated Moving Average (ARIMA), long short-term memory (LSTM), and Convolutional Neural Network (CNN) models. The findings indicate that the Informer model outperforms the other three models in terms of drought forecasting accuracy across all time scales. Nevertheless, the predictive capacity of the Informer model remains suboptimal when it comes to short-term intervals. Aiming at the problem of drought forecasting accuracy in a short time scale, this study proposed a drought forecasting model named VMD-JAYA-Informer based on Variational Mode Decomposition (VMD) and the JAVA optimization algorithm to improve the Informer model. This study conducted a comparative analysis of VMD-JAYA-ARIMA, VMD-JAYA-LSTM, VMD-JAYA-CNN, and VMD-JAYA-Informer drought prediction models. The performance of these models was evaluated using the root mean square error (RMSE), Nash–Sutcliffe efficiency coefficient (NSE), and Mean Absolute Error (MAE). The VMD-JAYA-Informer model's forecast for the 1-month SPEI significantly surpasses that of alternative models and demonstrates a robust agreement with the actual data. Simultaneously, the model exhibits equally optimal forecasting performance across different time scales. In order to validate the VMD-JAYA-Informer model, four meteorological stations in the Songliao River Basin were chosen at random. The validation results demonstrate that VMD-JAYA-Informer outperforms the Informer model in terms of prediction accuracy on the 1-month time scale (NSE values of 0.8663, 0.8765, 0.8822, and 0.8416, respectively). Additionally, the model outperforms Informer in terms of prediction performance on other time scales, further demonstrating its generalizability and excellence in drought prediction on shorter time scales.

Keywords: drought forecasting; multi-time scales; Informer; VMD; JAYA; VMD-JAYA-Informer

1. Introduction

Drought is a multifaceted meteorological phenomenon that has an impact on agriculture, economy, ecosystems, and various other aspects [1]. Over the past few decades, particularly since the 1990s, drought has become increasingly severe in most parts of the

https://doi.org/10.3390/land14010126

world. Central and northeastern regions have experienced noticeable drought trends, with some areas experiencing a yearly increase in drought coverage of 3.72%, surpassing the area of humid regions [2]. Mishra et al. identified four distinct drought categories: meteorological, agricultural, hydrological, and socio-economic [3,4]. Drought presents significant challenges to ecosystems, agriculture, and economies, necessitating research to concentrate on prediction, assessment, and risk management. Traditional forecasting methods, such as ARIMA and hydrometeorological models, leverage historical data to identify drought patterns; however, they often fail to accurately capture multidimensional non-linear data structures. The advent of artificial intelligence offers transformative advances for drought prediction, with deep learning models such as LSTM, CNN, and Informer adept at extracting intricate spatiotemporal features from diverse data sources, thus markedly improving predictive accuracy. Despite these advancements, AI-based forecasting faces critical challenges, including its reliance on extensive datasets, limitations in predicting short-term and extreme drought events, and the need for more effective multi-source data integration and model optimization. Addressing these limitations in future research is imperative to improve the accuracy of drought prediction and provide scientifically robust information to support disaster prevention and mitigation efforts. This research aims to predict drought using the Informer model, emphasizing reduction in data noise and errors to enhance prediction accuracy and efficiency. Through the refinement of the model architecture and parameter optimization, this study sought to enhance the model's capacity to capture critical drought characteristics, ultimately providing a more precise and effective tool for drought early warning and risk management.

Prior studies typically have employed a drought index for the accurate measurement of regional drought. This index is extensively utilized for assessing and analyzing drought conditions or its spatiotemporal attributes [5–7]. Commonly utilized drought metrics used in studies are the Z index [8,9], Meteorological Drought Composite Index (CI), Palmer Drought Severity Index (PDSI), Standardized Precipitation Index (SPI), and Standardized Precipitation–Evapotranspiration Index (SPEI). The Palmer Drought Severity Index (PDSI) is a meteorological drought measure introduced by Palmer in 1965. It is used to analyze regional water balance by distinguishing between dry and wet periods [10]. Nevertheless, the PDSI calculation heavily relies on meteorological station data at a specific location, which greatly hampers its capacity to evaluate drought conditions across different areas. The index is inadequate for regionalizing drought and is ineffective in monitoring short-term drought across various climate regions [11]. The SPI utilizes long-term monthly precipitation data within a specific range and incorporates various time scales to measure the excess or deficiency in precipitation. This allows for the identification of drought and wet periods [12–15], as well as the determination of drought severity and duration in a particular region [1]. The SPI is determined by analyzing long-term precipitation data for a specific location. The calculation process is straightforward and can be applied to various climate conditions. However, it does not take into account the impact of temperature [16], wind speed, or other meteorological factors on evapotranspiration. Therefore, it is not suitable for analyzing the impact of climate change on evapotranspiration. The SPEI, a novel drought index introduced by Vicente-Serrano et al. in 2010, aims to overcome the constraints of the aforementioned drought indices [1]. SPEI data comprise time series information that displays various temporal scales. Hence, we used the SPEI drought index in this study.

Drought forecasting is crucial in meteorology and plays a pivotal role in preventing, mitigating, and monitoring drought risks [17]. Drought is a crucial concern in climate change and the management of water resources. Accurately predicting when and how droughts will occur is essential for agriculture, water resource planning, and disaster management [18]. Forecasting approaches for time series data play a crucial role in predicting droughts [19]. Drought forecasting methods utilize mathematical models to predict the occurrence and intensity of future droughts using historical data and meteorological indicators. There are two main types of drought forecasting methods: statistical methods and machine learning methods. Mi et al. employed an ARIMA model, LSTM model, improved LSTM model (ILSTM), and ILSTM model with additional convolutional layers (CLSTM) to forecast forthcoming drought conditions in eight specific areas of China. The findings indicated that the CLSTM reduces the root mean square error by 0.09∼0.33, making it more appropriate for short-term regional drought and climate prediction [20]. In their study on drought forecasting in the Talegan watershed, Tehran province, Iran, Nikbakht et al. discovered that the application of Support Vector Machines (SVMs) yielded more precise predictions during spring and fall [21]. Over the past decade, hybrid models have been utilized to forecast various types of droughts [22]. Ding et al. developed a Complementary Ensemble Empirical Mode Decomposition (CEEMD)-ARIMA model and a CEEMD-LSTM model to forecast drought in the Xinjiang region. The findings indicated that the CEEMD-ARIMA model achieved the highest accuracy in forecasting, and the hybrid model outperformed the individual models in accuracy across all time scales. However, there was still a lower accuracy in forecasting at shorter time scales [23]. Xu et al. integrated the advantages of ARIMA and CEEMD to predict regional droughts in China. They discovered that the combined model yielded superior results compared to using a single model. However, the accuracy of short-term predictions remained relatively poor [24]. Xu et al. conducted a comparative analysis of various forecasting models, including ARIMA, support vector regression (SVR), LSTM, ARIMA-SVR, least square SVR (LS-SVR), and ARIMA-LSTM, to predict the Standardized Precipitation–Evapotranspiration Index (SPEI) in China. The results indicated that the hybrid model had a higher accuracy in forecasting long-term SPEI, but a lower accuracy in forecasting short-term SPEI. Additionally, the ARIMA-LSTM model exhibited the highest accuracy in predicting the SPEI on the 6-, 12-, and 24-month scales, suggesting its suitability for long-term drought forecasting in China [25]. Zhang et al. employed various forecasting models, including ARIMA, random forest (RF), recurrent neural network (RNN), LSTM, and convolutional long short-term memory (ConvLSTM), to predict short-term meteorological droughts. The study revealed that ConvLSTM effectively captures spatiotemporal information and outperforms other models in short-term drought forecasting. Specifically, ConvLSTM exhibited high accuracy in forecasting droughts within a time frame of 1–5 days [26]. Shang et al. employed the Informer model to predict droughts in the Yangtze River Basin. The findings indicated that the Informer model outperforms LSTM and ARIMA in terms of forecasting accuracy. Nonetheless, despite some advancements in forecasting short-term droughts, the general precision of the predictions continues to be inadequate [27].

This work addresses the issue of low accuracy in predicting droughts on a short time scale, based on the scientific background provided above. While the Informer model excels at long-term forecasting, its accuracy in short-term forecasting can be influenced by factors such as excessive noise, seasonality, and periodicity. This study aims to enhance short-term forecasting by concentrating on the Songliao River Basin and integrating Variational Mode Decomposition (VMD) with the Jaya algorithm (JAYA) to improve the Informer model. The resulting model is referred to as VMD-JAYA-Informer. When comparing the predictions of VMD-JAYA-Informer with those of VMD-JAYA-LSTM, VMD-JAYA-ARIMA, and VMD-JAYA-CNN, the VMD-JAYA-Informer algorithm consistently exhibits superior accuracy in forecasting SPEI across all time scales. In this study, the SPEI data from four randomly chosen meteorological stations in the Songliao River Basin were verified.

2. Materials and Methods

2.1. Study Area

The Songliao River Basin is located in northeastern China. The basin's climate is defined by its unique and varied topography, leading to a complex and diverse ecosystem. The spring and winter seasons are dry and cold, while the summer and fall are warm and rainy. There are noticeable variations in precipitation and temperature throughout the year, which are influenced by both seasonal changes and geographic factors. The Basin experiences an average annual temperature ranging from 1 to 5 °C. The map of the study area is shown in Figure 1.

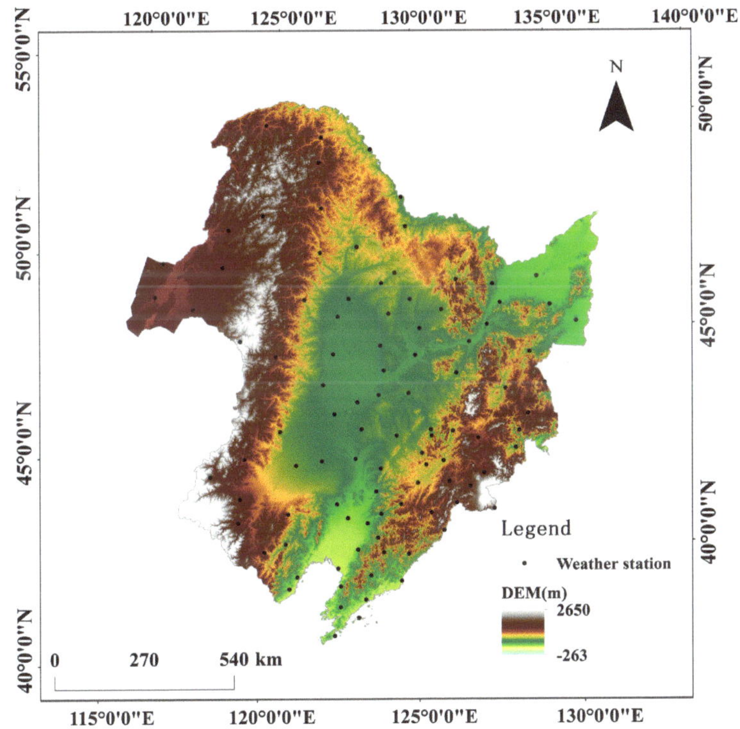

Figure 1. Study area.

2.2. Data Source

The meteorological data for 1980–2019 utilized in this work were obtained from the China Meteorological Science Data Sharing Service (https://www.data.cma.cn/) (accessed on 6 February 2023). In order to maintain the accuracy, consistency, and uniformity of the meteorological data, we omitted years and stations that had more than 5% of missing observations. The meteorological variables consisted of latitude, altitude (m), average wind speed (m/s), average precipitation (mm), sunshine hours (h), maximum temperature (°C), minimum temperature (°C), and average temperature (°C). The drought classification criteria of the SPEI are presented in Table 1.

Table 1. Drought classification based on SPEI.

Level	Type	SPEI
1	No drought	$SPEI \geq -0.5$
2	Mild drought	$-1.0 \leq SPEI < -0.5$
3	Moderate drought	$-1.5 \leq SPEI < -1.0$
4	Severe drought	$-2.0 \leq SPEI < -1.5$
5	Extreme drought	$SPEI \leq -2.0$

2.3. *Methods*

2.3.1. Informer

Vaswani et al. introduced the Transformer model, which utilizes a self-attention mechanism. The model is composed of an encoder–decoder structure that operates in a step-by-step manner. It incorporates a multi-head attention mechanism and a feed-forward neural network [28]. Zhou et al. introduced a model named Informer, which enhances the Transformer model [29]. Figure 2 illustrates the difference between the Informer and the self-attention mechanism in the Transformer network topology. The Informer encoder includes a multi-group multi-head, probabilistic sparse self-attention mechanism and a self-attention distilling operation [30]. The Informer's encoder structure reduces the time dimension of the input data sequence for each layer by minimizing the cascade layers [31]. This method reduces the unnecessary computation and duplication of data storage, theoretically attaining O(LlogL). The idea is illustrated in Figure 3. The decoder architecture of the Informer produces outcomes in a single step, addressing the issue of delayed forecasting and thereby improving forecasting efficacy.

2.3.2. VMD

Dragomiretskiy and Zosso introduced VMD in 2014 as a novel variational technique for signal decomposition [32]. MD effectively decomposes the signal frequency range by taking into account the frequency domain features of the deconstructed signal, resulting in the generation of several Intrinsic Mode Functions (IMFs). The IMF is generally concentrated around its central frequency, with each IMF exhibiting an AM-FM signal [33]. The IMF formula is as follows:

$$u_k(t) = A_k(t)cos(\phi_k(t)), \tag{1}$$

where $A_k(t)$ denotes the instantaneous amplitude of $u_k(t)$, $\phi_k(t)$ denotes the instantaneous phase of $u_k(t)$, and $\phi_k(t)$ is a non-decreasing function.

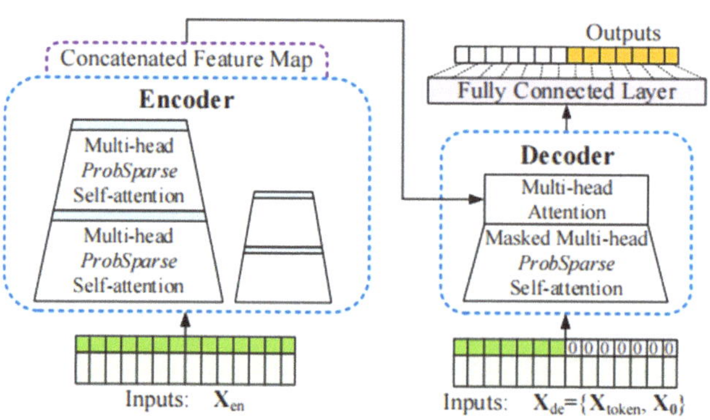

Figure 2. Informer model principle.

VMD requires four parameters to be defined in advance: the modal number K, the modal bandwidth control parameter (or quadratic penalty term) α, the noise tolerance τ, and the convergence criteria tolerance ε. The VMD method can effectively break down the signal into its primary frequency components by selecting an appropriate number of modal K [34]. Despite VMD having a strong theoretical basis and superior decomposition capabilities, parameters such as K and α, which greatly impact VMD, are still determined based on empirical knowledge.

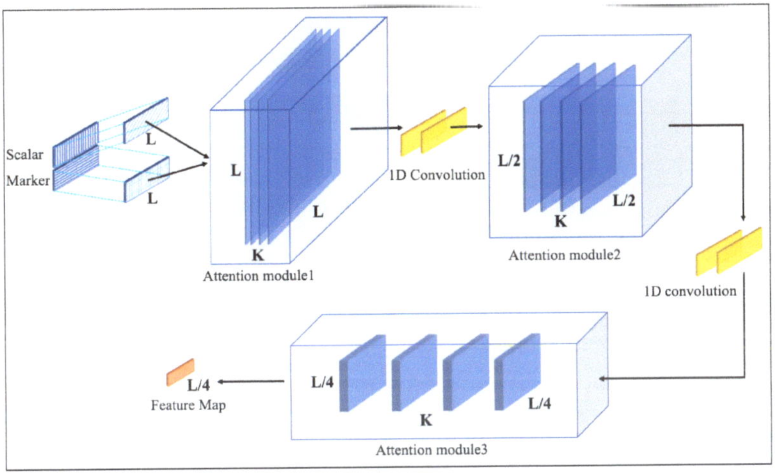

Figure 3. The principle of the encoder.

2.3.3. JAYA

JAYA is a population-based optimization technique, introduced by Rao in 2015 [35]. It is a very efficient algorithm specifically developed to address continuous optimization problems. It is both straightforward and robust, making it a formidable tool for achieving global optimization. The JAYA algorithm employs both optimal and worst solutions to derive a novel solution. A salient characteristic of the JAYA algorithm is its parameter-free nature, indicating that in its standard implementation, there is no requirement to establish conventional parameters (such as crossover rate, mutation rate, etc.). The fundamental idea of the JAYA algorithm is to directly guide the search process toward an optimal solution, adjusting the search direction based on the performance of the individuals [35].

2.3.4. VMD-JAYA-Informer

The VMD-JAYA-Informer model, which combines VMD, JAYA, and Informer, is introduced in this work. The principle framework of VMD-JAYA-Informer is shown in Figure 4 The JAYA method serves as an optimization engine for parameter optimization and the spatial search of the Informer model. The JAYA algorithm utilizes iterative optimization [35] to determine the optimal parameters that improve the forecasting accuracy of the Informer model on SPEI data. VMD-JAYA-Informer enhances the precision and consistency of time series data forecasting by leveraging the decomposition functionality of VMD, the optimization capabilities of JAYA, and the forecasting capabilities of the Informer. VMD-JAYA-Informer is a powerful tool that enhances the accuracy and stability of time series data forecasting. This model accomplishes this by integrating the decomposition abilities of VMD, the optimization proficiency of JAYA, and the forecasting capacity of Informer. This tool is particularly ideal for analyzing and forecasting complex time series.

Initially, the SPEI data are decomposed using VMD, a technique that breaks down the SPEI into several IMFs, where each IMF represents a distinct frequency pattern. Furthermore, the JAYA-Informer model is utilized to train and predict the deconstructed feature data from the VMD, resulting in an improved accuracy and dependability in forecasting SPEI time series. Ultimately, the modal components obtained from the forecasting process are reconfigured to restore the original state of the forecasted results.

The reconstructed data can closely approximate the original signal with a negligible margin of error. This means that the model successfully captures the distinctive features and patterns of the SPEI time series during the decomposition and forecasting procedure, suggesting the VMD-JAYA-Informer model's high accuracy and forecasting power.

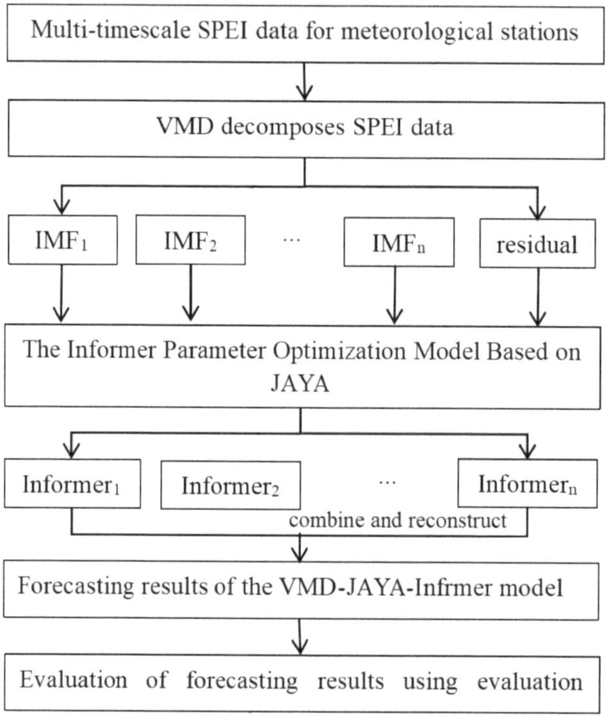

Figure 4. The principle framework of VMD-JAYA-Informer.

2.3.5. LSTM

In 1997, Hochrieter and Schmidhuber introduced a modified neural network model called LSTM, which is built upon the RNN [36]. The LSTM is a network model, depicted in Figure 5. It consists of four interconnected layers arranged in a chain structure. These layers are intended to influence and engage with one another via a logical function. The oblivion gate f_t, in the context of the hidden state h_{t-1}, determines which information should be reset based on the input time sequence x_t at the current time t. It also determines which information should be stored in the cell. Afterward, the two-part input gate determines which new information should be added to the current cell using the sigmoid function. The tanh function is utilized to generate a fresh candidate state, in which the previous cell's state C_{t-1} is modified to the state C_t in comparison to the candidate value $\widetilde{C}_t$. The blue rectangular box labeled "A" indicates an LSTM module or cell. The outcome derived from the past two procedures is utilized as the current state of the unit at the present moment in time. LSTM utilizes bidirectional information transfer and incorporates both forward and backward propagation.

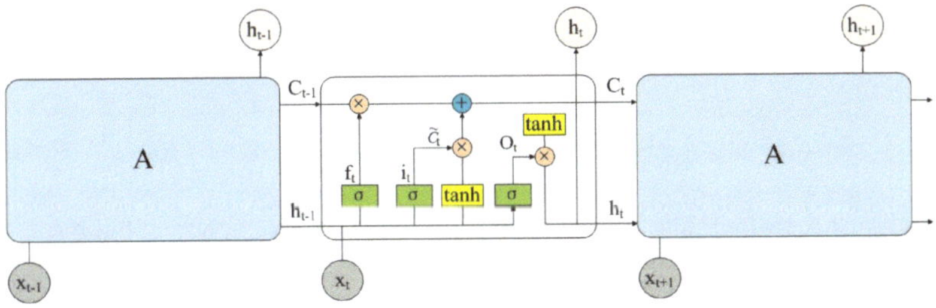

Figure 5. LSTM model structure.

2.3.6. ARIMA

ARIMA is frequently utilized to represent various time series models through parameter simplifications, thereby incorporating numerous exponential smoothing methods [37]. ARIMA is limited by the assumption of linear correlation between time series, making it unsuitable for analyzing non-linear data [38]. The modeling process necessitates numerous intrinsic parameters, resulting in computational complexity and prolonged computation period. This results in an inefficient forecasting process.

2.3.7. CNN

The CNN introduced by LeCun can be classified into 1D-CNN, 2D-CNN, and 3D-CNN based on the dimensions of the input data [39]. Convolution is a distinctive advantage of CNNs compared to traditional neural networks. It involves performing convolution operations on the input data during model training and error transfer. This process enables local connectivity and weight value sharing [40,41]. The schematic representation of its main structure can be seen in Figure 6.

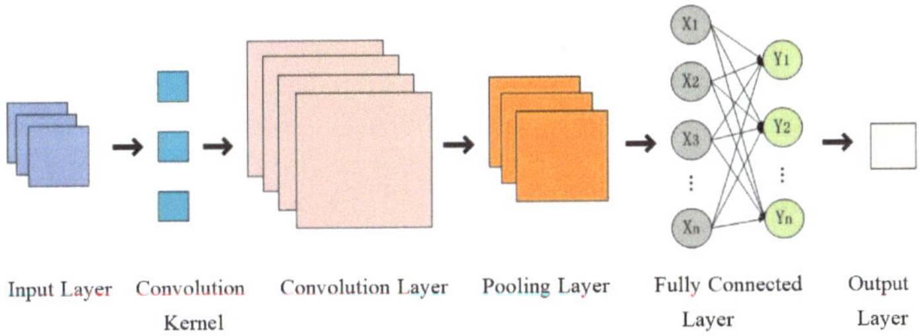

Figure 6. CNN model structure.

2.3.8. Evaluation Metrics

This research utilizes the Nash–Sutcliffe efficiency coefficient (NSE), root mean square error (RMSE), and Mean Absolute Error (MAE) to assess the predictive capability of the model. As the value of the NSE approaches 1, it signifies a stronger alignment between the model and the data. The root mean square error (RMSE) is a numerical measure that falls within the range of $(-\infty, 1)$. A smaller RMSE indicates a higher level of accuracy in the forecasting approach. The MAE is a metric that quantifies the average discrepancy between observed and anticipated values. The formulae utilized to compute the aforementioned metrics are displayed below:

$$\mathrm{RMSE} = \sqrt{\frac{1}{N}\sum_{i=1}^{N}(y_i - \tilde{y}_i)^2} \tag{2}$$

$$\mathrm{NSE} = 1 - \frac{\sum_{i=1}^{N}(y_i - \tilde{y}_i)^2}{\sum_{i=1}^{N}(y_i - \bar{y})^2} \tag{3}$$

$$\mathrm{MAE} = \frac{1}{N}\sum_{i=1}^{N}|y_i - \tilde{y}_i| \tag{4}$$

where y_i indicates the actual value; $\tilde{y}$ indicates the forecasted value; $\bar{y}$ represents the average value of y_i, and N indicates the total value of y_i.

2.3.9. Frequency of Drought

This study examined the Songliao River Basin's drought frequency from 1980 to 2019 on annual and seasonal scales. The following is the formula for the drought frequency:

$$F_i = \frac{w}{W} \times 100\% \tag{5}$$

where F_i is drought frequency at station i (%), w is the number of droughts in the time series, and W is the total number of time series.

3. Results

3.1. Spatial Distribution of Drought Frequency

Figure 7 displays the geographical pattern of drought frequency in the Songliao River Basin, both on the annual basis and during different seasons. Figure 7e shows that the occurrence of drought on an annual scale varies between 30.78% and 35.85%. The regions with a higher drought frequency are mostly located in the western and southeastern parts of the Songliao River Basin. The analysis of drought frequency at the seasonal scale, as shown in Figure 7a–d, reveals that the occurrence of drought in spring ranges from 28.75% to 33.55%. The areas with a high drought frequency are mostly located in the western and northern regions of the Songliao River Basin. The occurrence of drought in summer is more frequent compared to in other seasons, primarily due to elevated temperatures, where the frequency of drought ranges from 31.60% to 36.30%, with areas experiencing high drought frequency in the western and eastern regions of the Songliao River Basin. In autumn, the frequency of drought varies from 27.11% to 33.90%, with a significant concentration of high-frequency droughts in the southwestern, southern, and eastern regions of the basin. The occurrence rate of drought during winter varies from 30.19% to 35.77%, with the region exhibiting the second highest frequency, surpassed only by autumn. This region is predominantly located in the southern and eastern sectors.

3.2. Evaluating Model's Forecasting Performance

3.2.1. Drought Forecasting Based on Deep Learning Models

For the experiment, the main parameters involved in the model and their values are shown in Table 2. The parameter values were predominantly established according to the existing literature [11–13,15,29] and experimental results, whereas other parameters utilize the default values.

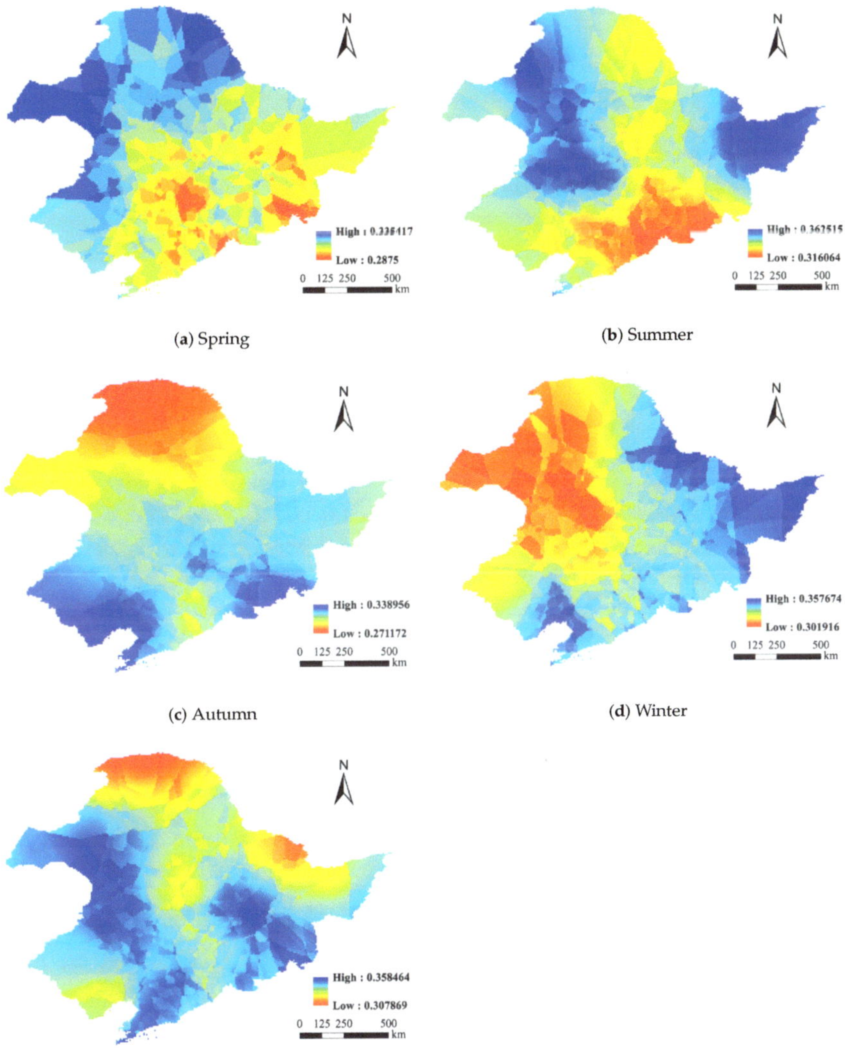

(**a**) Spring (**b**) Summer

(**c**) Autumn (**d**) Winter

(**e**) Year

Figure 7. Spatial distribution of drought frequency on the annual and seasonal scales in the Songliao River Basin.

This study used the SPEI data from 1980 to 2011 at various time scales for the training set. We employed ARIMA, LSTM, CNN, and Informer to predict the SPEI values of the Songliao River Basin from 2012 to 2020. The forecasting results for each model are visualized in Figure 8. The Informer model exhibited enhanced accuracy in aligning predicted SPEI values with actual values and effectively capturing variations in SPEI values, surpassing the performance of the ARIMA, LSTM, and CNN models. The projected Standardized Precipitation–Evapotranspiration Index (SPEI) from the ARIMA, LSTM, and CNN models exhibited significant discrepancies from the actual values over a 1-month interval. Among these three models, the LSTM model demonstrated a notably weak predictive power. When predicting SPEI3, SPEI6, SPEI9, SPEI12, and SPEI24, the discrepancies between the forecasted and the actual SPEI values decreased as the time scales increased. The Informer model generates predicted SPEI values that are nearest to the actual values, succeeded by the ARIMA model. The LSTM and CNN models performed comparatively less accurately. Figure 8 clearly demonstrates that all four models performed poorly in predicting SPEI1 data. This can be attributed to the shorter time scale, frequent oscillations, rapid changes, and the significant influence of meteorological factors on SPEI values. Consequently, a more sophisticated model is required for accurately forecasting SPEI1.This research conducted a comparative analysis of the prediction performance of the ARIMA, LSTM, CNN, and Informer models using three assessment metrics: NSE, RMSE, and MAE. The results of this analysis are presented in Table 3. For the ARIMA, LSTM, and CNN models, the NSE was negative, the RMSE was greater than 0.86, and the MAE was higher than 0.7 at the 1-month time period. In contrast, the Informer model had an NSE value of 0.1542, an RMSE of 0.8583, and an MAE of 0.6964. The RMSE and MAE of the ARIMA, LSTM, CNN, and Informer models decreased as the time scale increased. Conversely, the NSE exhibited the reverse tendency. This fact suggests that the forecasting accuracy of these four models improves as the time scale increases. Through a comparison of the experimental data, the Informer model demonstrates superior forecasting performance compared to ARIMA, LSTM, and CNN across all time scales. This suggests that the ARIMA, LSTM, and CNN models exhibit inferior prediction accuracy due to inherent limitations in their network structures. The Informer model markedly improves the predictive accuracy of the SPEI and adeptly resolves the challenge of insufficiently capturing the relationship between input and output time series data in long-term forecasts, a problem exacerbated by prolonged time scales.

Table 2. LSTM, ARIMA, CNN and Informer Model parameters.

Model	Parameter	Parameter Value
LSTM	Epochs	50
	Batch Size	64
	Activation Function	RELU
	Loss Function	MSE
	Optimizer	Adma
ARIMA	Autoregressive Order (p)	1
	Differencing Order (d)	0
	Moving Average Order (q)	1
CNN	Convolution Kernel Size	1
	Number of Convolutional Layers	4
	Number of Pooling Layers	2
	Epochs	50
	Batch Size	64
	Activation Function	RELU
	Loss Function	MSE
	Optimizer	Adma
Informer	Input Sequence Length of Encoder (seq_len)	3
	Start Token Length of Decoder (Label_len)	1
	Encoder Input Size (enc_in)	16
	Decoder Input Size (dec_in)	8
	Dimension of Model (d_model)	512

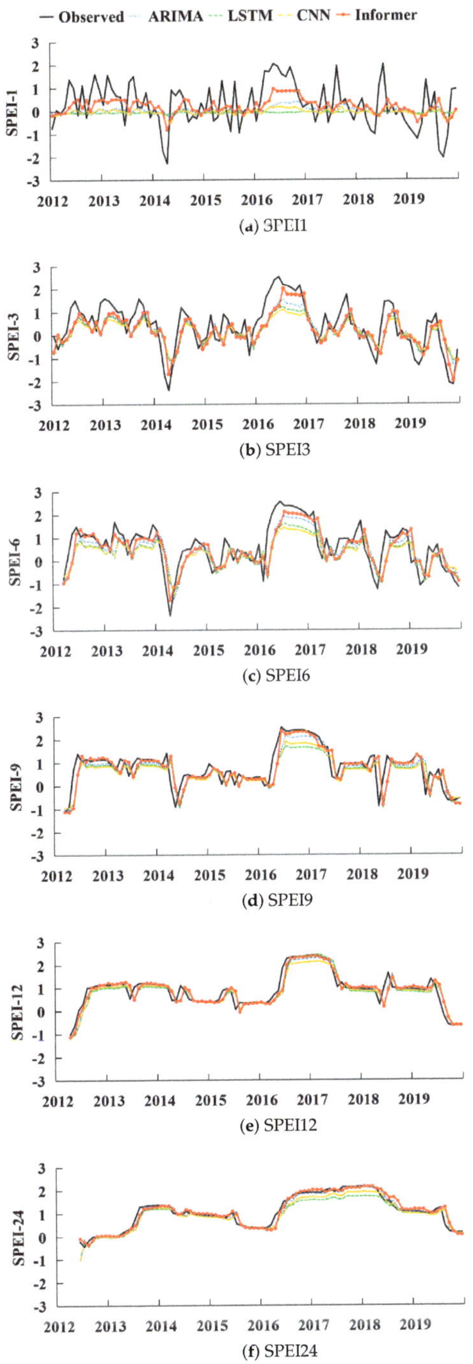

Figure 8. ARIMA, CNN, LSTM, and Informer models accuracy assessment metrics.

Table 3. The statistical criteria of the ARIMA, CNN, LSTM, and Informer models.

SPEI Series	Model	NSE	RMSE	MAE
SPEI1	ARIMA	−0.0181	0.9416	0.7468
	CNN	−0.0388	0.8927	0.7684
	LSTM	−0.1438	0.9846	0.7892
	Informer	0.1542	0.8583	0.6964
SPEI3	ARIMA	0.5086	0.7127	0.6024
	CNN	0.4217	0.6719	0.6494
	LSTM	0.4589	0.7443	0.6316
	Informer	0.5503	0.6818	0.5639
SPEI6	ARIMA	0.5180	0.6787	0.5335
	CNN	0.4351	0.6259	0.5236
	LSTM	0.4450	0.7043	0.5750
	Informer	0.5787	0.6216	0.4751
SPEI9	ARIMA	0.5718	0.5235	0.3503
	CNN	0.5309	0.4342	0.4149
	LSTM	0.4778	0.5339	0.3657
	Informer	0.6050	0.5053	0.3074
SPEI12	ARIMA	0.8325	0.2999	0.1979
	CNN	0.8200	0.2820	0.2257
	LSTM	0.8226	0.3111	0.2173
	Informer	0.8571	0.2785	0.1703
SPEI24	ARIMA	0.9276	0.1845	0.1140
	CNN	0.9009	0.1699	0.1356
	LSTM	0.8537	0.2948	0.2214
	Informer	0.9395	0.1690	0.1080

3.2.2. Decomposition of VMD for SPEI

Constructing the VMD algorithm typically requires determining four parameters, namely, the Number of Modes (K), Penalization Factor (α), Noise Tolerance (τ), and Convergence Criterion Tolerance (ε). Compared to K, α, τ, and ε have less impact on the decomposition effect of SPEI sequences and are usually using their default values [42]. When the K value is smaller, the VMD algorithm is in an under-decomposition state, and the main frequency signal contained in the SPEI series cannot be fully decomposed. Nevertheless, selecting a substantial value for K results in the central frequencies of neighboring modal components being closer together, potentially resulting in modal repetition or increased noise generation. Tang et al. calculated the IMFs' central frequencies obtained for a given range of K values, which resulted in appropriate K values. This paper uses SPEI1 data as an example, and Table 4 shows the IMFs' central frequencies generated with the K values within the established range.

From Table 4, when $K = 7$, the central frequencies of IMF6 and IMF7 are similar, indicating that the decomposition of SPEI data by VMD might be in an over-decomposed state. Therefore, K is set to 6. In this paper, the default values are used for the other parameters of the VMD algorithm.

The results of the SPEI decomposition at 1-, 3-, 6-, 9-, 12-, and 24-month time scales are presented in Figures 9–14. The figures demonstrate that the VMD algorithm decomposes the SPEI sequence across multiple time scales into six components with distinct frequencies. VMD, as an adaptive frequency decomposition method, decomposes the SPEI time series data into six SPEI components. This decomposition enhances comprehension of the frequency characteristics and periodic variations in SPEI data, facilitating the identification of notable frequency components and cycles across various temporal scales. The resulting sequences not only eliminate the noise interference but also more accurately capture the periodic oscillation trends in SPEI.

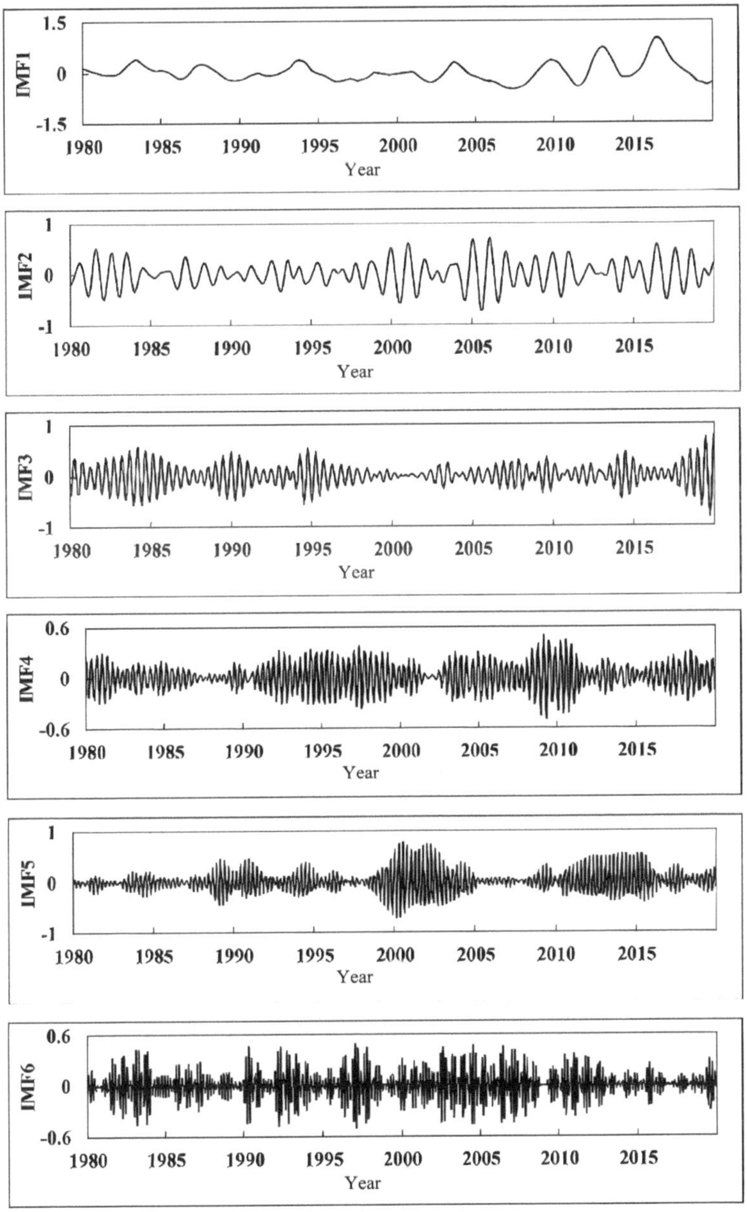

Figure 9. VMD decomposition of SPEI1.

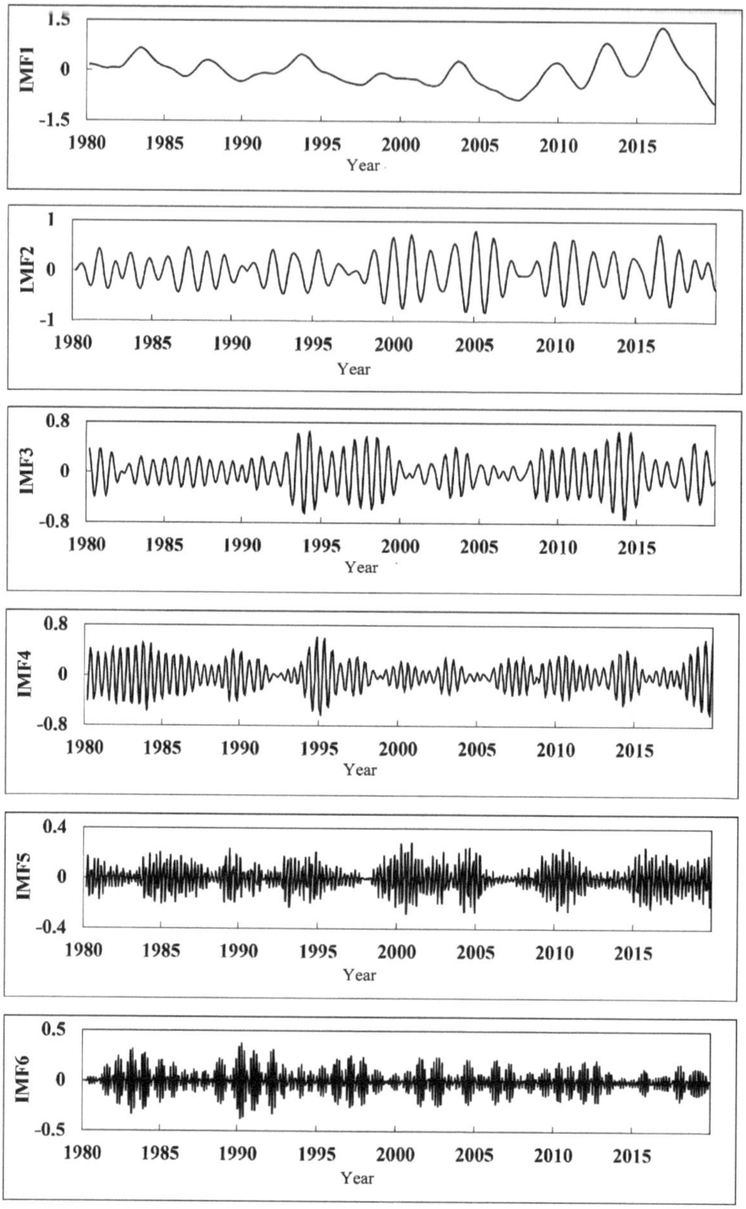

Figure 10. VMD decomposition of SPEI3.

Table 4. The center frequency corresponding to the IMF produced by different values of *K*.

K	Center Frequencies/Hz						
	IMF1	IMF2	IMF3	IMF4	IMF5	IMF6	IMF7
4	69.9	11.4	6	3.3			
5	243.9	47.2	22	12	9		
6	322.6	62.9	29.6	20.7	15.2	11.5	
7	163.9	33.3	16.1	11.4	8.2	6.8	5.8

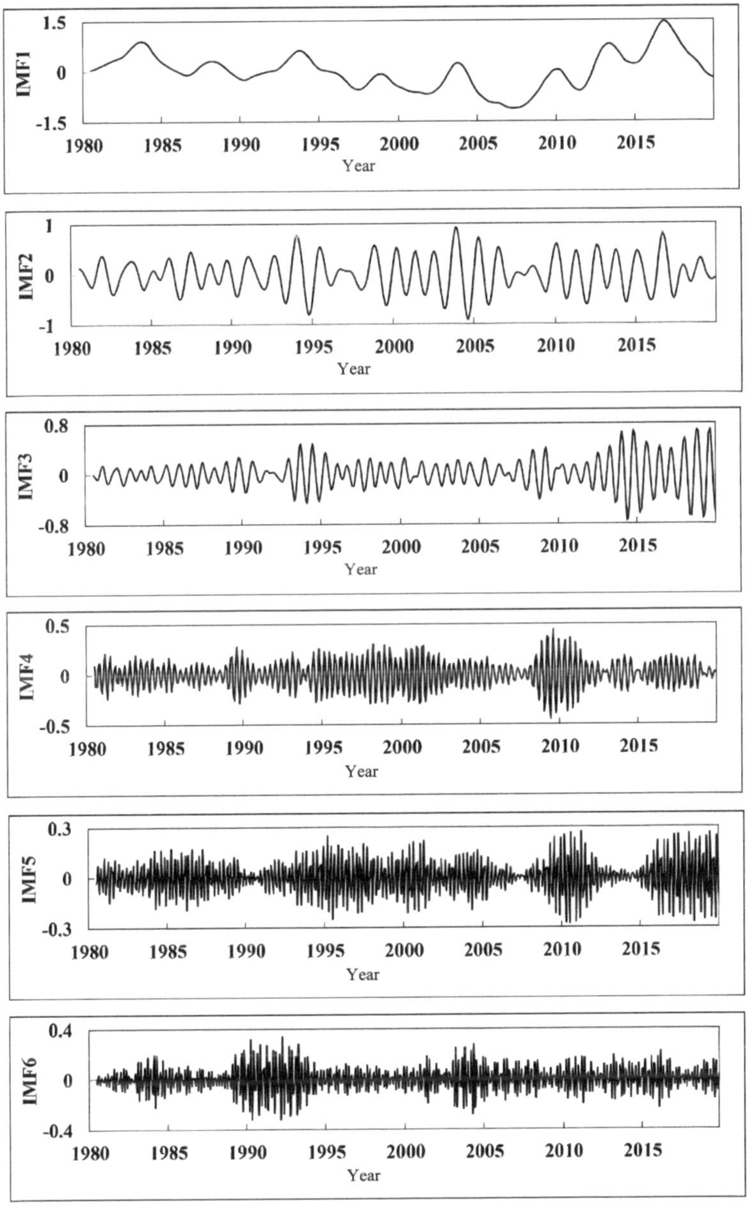

Figure 11. VMD decomposition of SPEI6.

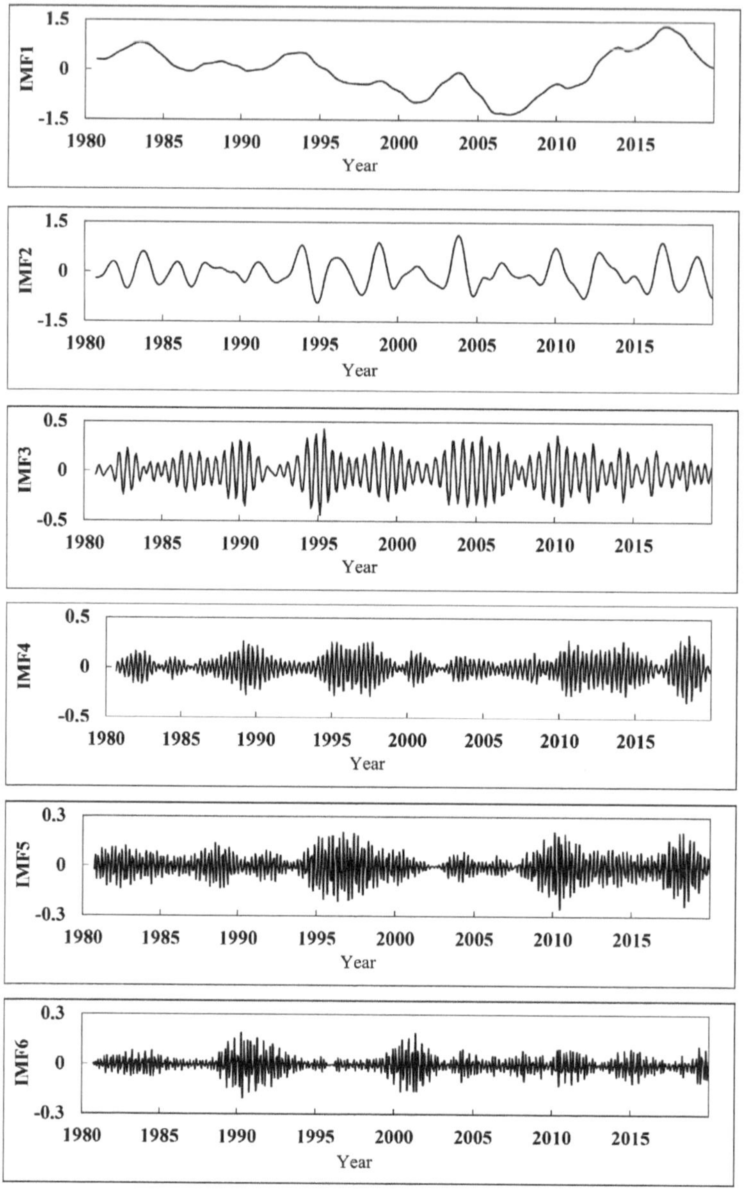

Figure 12. VMD decomposition of SPEI9.

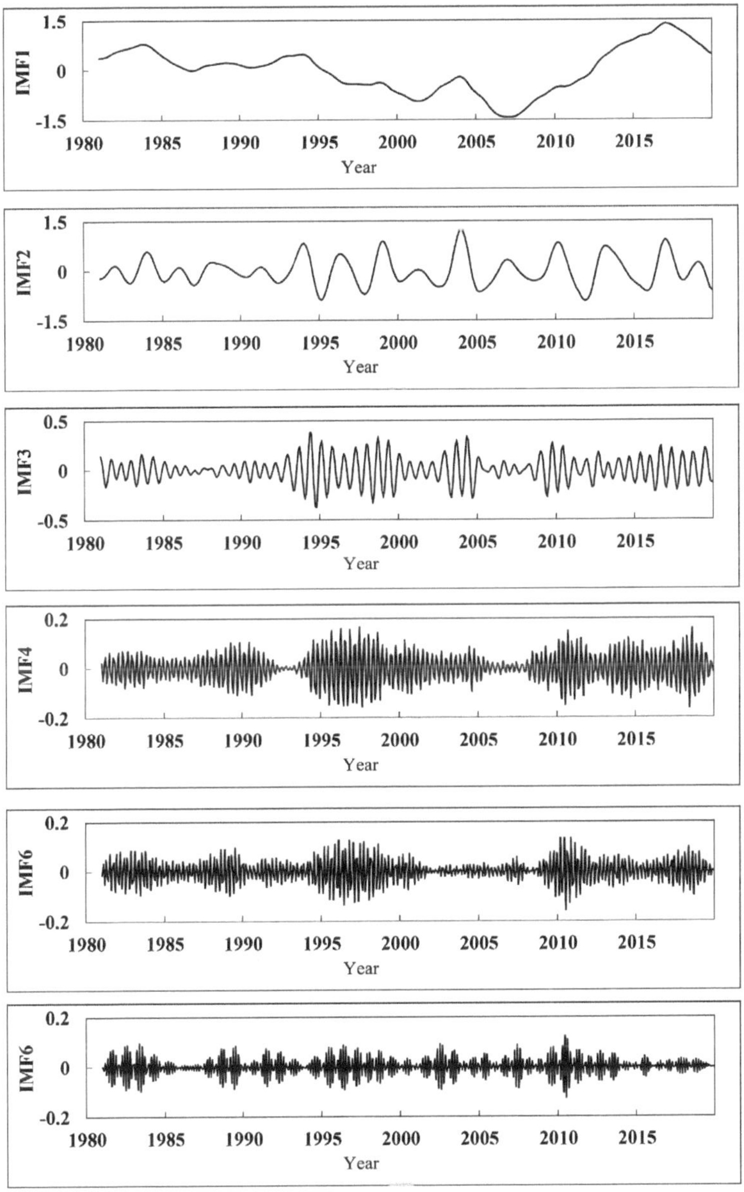

Figure 13. VMD decomposition of SPEI12.

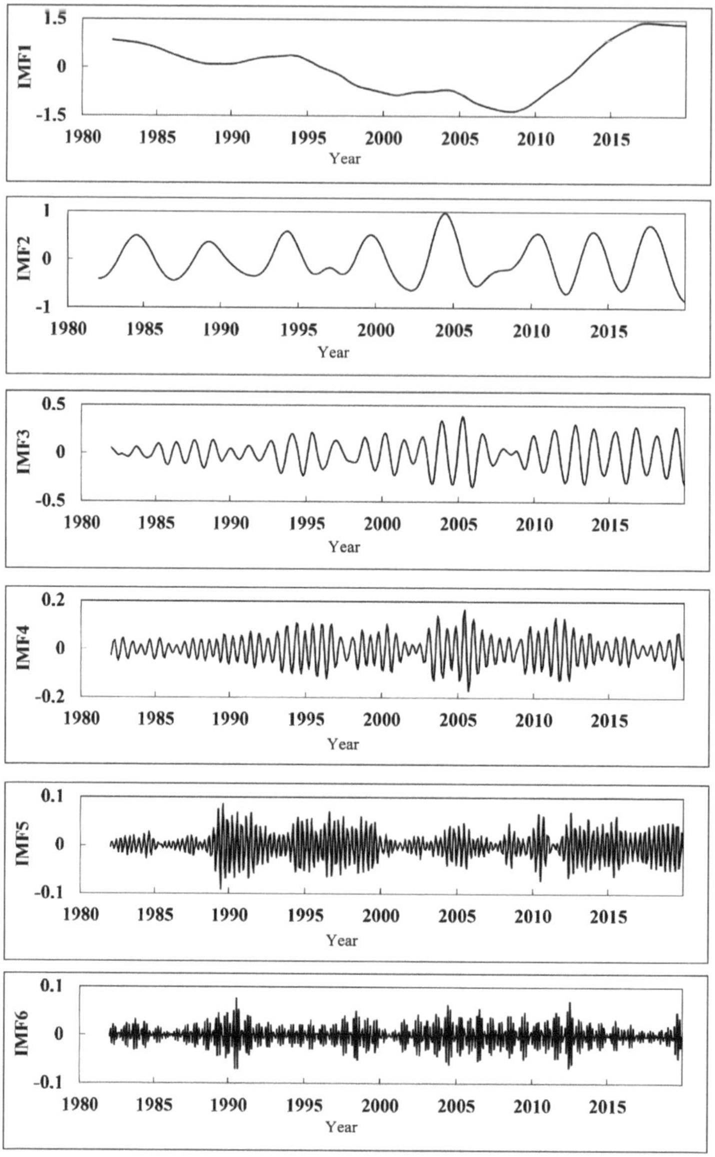

Figure 14. VMD decomposition of SPEI24.

3.2.3. Drought Forecasting Based on VMD-JAYA-Informer

In this study, we employed the JAYA algorithm to optimize key parameters of the LSTM, ARIMA, CNN, and Informer models. The JAYA algorithm is a parameter-free optimization technique that updates parameter values by identifying the differences between the current solution and both the best and worst solutions to achieve global optimality. To ensure the optimal performance of the model, the key parameters optimized, their search spaces, and the optimal parameter values determined using the JAYA algorithm are presented in Table 5.

Table 5. Parameter Optimization List.

Model	Parameter	Range of Parameter Optimization	Optimized Value
VMD-JAYA-LSTM	Epochs	10–200	100
	Batch Size	16–128	64
	Activation Function	Sigmoid, Tanh, ReLU, Leaky ReLU	RELU
	Loss Function	MSE, MAE, Huber Loss, Log–Cosh Loss	MSE
	Optimizer	Adma, SGD, RMSprop, Adagrad, Adadelta	Adma
VMD-JAYA-ARIMA	Autoregressive Order (p)	0–5	1
	Differencing Order (d)	0–2	0
	Moving Average Order (q)	0–2	1
VMD-JAYA-CNN	Convolution Kernel Size	1–7	3
	Number of Convolutional Layers	1–16	8
	Number of Pooling Layers	0–8	4
	Epochs	10–200	100
	Batch Size	16–128	64
	Activation Function	Sigmoid, Tanh, ReLU, Leaky ReLU	RELU
	Loss Function	MSE, MAE, Huber Loss, Log–Cosh Loss	64
	Optimizer	Adma, SGD, RMSprop, Adagrad, Adadelta	Adma
VMD-JAYA-Informer	start token length of decoder (Label_len)	1–64	4
	Start Token Length of Decoder (Label_len)	2–8	3
	Encoder Input Size (enc_in)	4–128	32
	Decoder Input Size (dec_in)	2–64	16
	Dimension of Model (d_model)	64–1028	512

VMD-JAYA-Informer was employed to predict the SPEI values of the Songliao River Basin from 2012 to 2020. The predictions were then compared with the outcomes of the VMD-JAYA-LSTM, VMD-JAYA-ARIMA, and VMD-JAYA-CNN models. The comparative findings are depicted in Figure 15. The SPEI values predicted using VMD-JAYA-ARIMA, VMD-JAYA-LSTM, and VMD-JAYA-CNN deviate substantially from the actual values when considering a 1-month time scale. However, the VMD-JAYA-Informer predictions closely approximate the actual values. The disparity between the projected and the observed values for the VMD-JAYA-ARIMA, VMD-JAYA-LSTM, and VMD-JAYA-CNN models diminishes quickly as the time scale increases. However, the projected values of the SPEI for the VMD-JAYA-Informer model remain in close proximity to the actual values. The VMD-JAYA-Informer model provides a more effective solution to the issue of erroneous forecasting of peaks and valleys in SPEI data within shorter time frames. Furthermore, its forecasting capability surpasses that of the Informer model to a large extent. VMD-JAYA-Informer addressed the issue of the Informer model's limited accuracy in anticipating the SPEI in short time scales.

The performance of the VMD-JAYA-Informer model was assessed using three evaluation metrics: NSE, RMSE, and MAE. The model's performance was compared to that of the VMD-JAYA-ARIMA, VMD-JAYA-LSTM, and VMD-JAYA-CNN models. The results, presented in Table 6, indicate that the evaluation indicators of VMD-JAYA-ARIMA, VMD-JAYA-CNN, VMD-JAYA-LSTM, and VMD-JAYA-Informer are significantly better than those of the ARIMA, CNN, LSTM, and Informer models. This suggests that the improvement has significantly enhanced the predictive capabilities of these four models. At the 1-month time scale, the VMD-JAYA-ARIMA, VMD-JAYA-CNN, and VMD-JAYA-LSTM models exhibit an NSE marginally exceeding 0, an RMSE surpassing 0.55, and an MAE exceeding 0.51. On the other hand, for the VMD-JAYA-Informer model, the NSE, RMSE, and MAE values were 0.9385, 0.1186, and 0.0676, respectively. The results demonstrate that VMD-JAYA-Informer enhanced and streamlined the predictive accuracy of the SPEI at shorter time intervals. The RMSE and MAE of the VMD-JAYA-ARIMA, VMD-JAYA-CNN, VMD-JAYA-LSTM, and VMD-JAYA-Informer models decline as the time scales increase. However, the NSE exhibits the reverse tendency. These trends suggest that the forecasting accuracy of the four models improves as the time scales increase. At the 24-month time scale, the NSE values for the four models are 0.9716, 0.9464, 0.9489, and 0.9903, respectively. All of the models perform optimally for all time scales.

Table 6. VMD-JAYA-ARIMA, VMD-JAYA-CNN, VMD-JAYA-LSTM, and VMD-JAYA-Informer models' accuracy assessment metrics.

SPEI Series	Model	NSE	RMSE	MAE
SPEI1	VMD-JAYA-ARIMA	0.1094	0.6384	0.5145
	VMD-JAYA-CNN	0.0068	0.5554	0.5391
	VMD-JAYA-LSTM	0.0194	0.6699	0.5370
	VMD-JAYA-Informer	0.9385	0.1186	0.0676
SPEI3	VMD-JAYA-ARIMA	0.6469	0.4969	0.4153
	VMD-JAYA-CNN	0.5866	0.5376	0.4549
	VMD-JAYA-LSTM	0.6291	0.5092	0.4290
	VMD-JAYA-Informer	0.9374	0.1478	0.0848
SPEI6	VMD-JAYA-ARIMA	0.5998	0.4844	0.3885
	VMD-JAYA-CNN	0.4249	0.4347	0.4905
	VMD-JAYA-LSTM	0.5686	0.5030	0.4186
	VMD-JAYA-Informer	0.9349	0.1381	0.0746
SPEI9	VMD-JAYA-ARIMA	0.6788	0.3625	0.2731
	VMD-JAYA-CNN	0.6341	0.3869	0.3018
	VMD-JAYA-LSTM	0.6449	0.3811	0.2902
	VMD-JAYA-Informer	0.9751	0.0712	0.0425
SPEI12	VMD-JAYA-ARIMA	0.8869	0.2120	0.1608
	VMD-JAYA-CNN	0.8247	0.2639	0.2145
	VMD-JAYA-LSTM	0.8654	0.2313	0.1844
	VMD-JAYA-Informer	0.9781	0.0659	0.0330
SPEI24	VMD-JAYA-ARIMA	0.9716	0.1175	0.0834
	VMD-JAYA-CNN	0.9463	0.1615	0.1238
	VMD-JAYA-LSTM	0.5686	0.5030	0.4186
	VMD-JAYA-Informer	0.9903	0.0485	0.0264

Overall, the VMD-JAYA-Informer has superior forecasting accuracy compared to the other three models across all time scales. This suggests that VMD-JAYA-Informer greatly improves the accuracy of SPEI forecasting. The predictive capability of VMD-JAYA-Informer is markedly enhanced compared to the Informer.

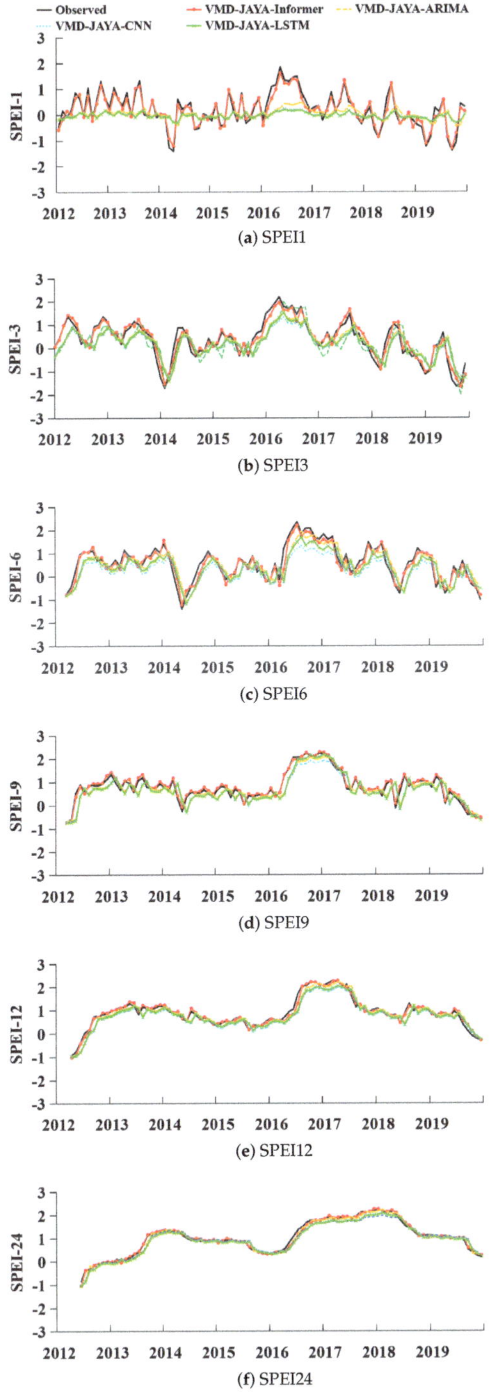

Figure 15. Results of forecasting multi-timescale SPEI values of VMD-JAYA-ARIMA, VMD-JAYA-CNN, VMD-JAYA-LSTM, and VMD-JAYA-Informer.

3.3. Model Validation and Analysis

For this investigation, we chose four meteorological stations, Nenjiang, Yilan, Kailu, and Pankou, randomly located along the Songliao River Basin. Our goal was to test the overall effectiveness of the VMD-JAYA-Informer model and to compare it with the SPEI prediction results of the Informer. Based on Figures 16–19, there is a substantial discrepancy between the projected and the actual SPEI values from the Informer model at the 1-month time scale. The VMD-JAYA-Informer produced SPEI forecasting curves that exhibited a trend comparable to the actual value curves, yet demonstrated a higher degree of accuracy. Furthermore, the model's fitting was found to be superior to that of the Informer model. As the time scale increases, Informer- and VMD-JAYA-Informer-forecasted value curves exhibit an increasingly accurate fit to the true value curves. Regardless of the time scale, VMD-JAYA-Informer consistently outperforms Informer in terms of prediction accuracy.

Table 7 demonstrates that the prediction performance of the Informer and VMD-JAYA-Informer improves progressively as the time scale changes from 1 month to 24 months. Accordingly, the discrepancy between the SPEI prediction and the actual data diminishes gradually, and the change trend becomes increasingly aligned with the actual values. The NSE also increases gradually, approaching its optimal value of 1. Across all time scales, the VMD-JAYA-Informer prediction performance consistently surpasses that of the Informer model, with the difference being most evident at the 1-month time scale.

This research has confirmed the overall effectiveness of VMD-JAYA-Informer for predicting SPEI at multiple scales. The prediction was carried out for the Nenjiang, Yilan, Kailu, and Panshi stations. Compared to the Informer model, the optimized model has superior fitting and accuracy across all time scales. This difference is most obvious for the 1-month time scale, where the VMD-JAYA-Informer exhibits improved forecasting accuracy for both the peak and valley SPEI values.

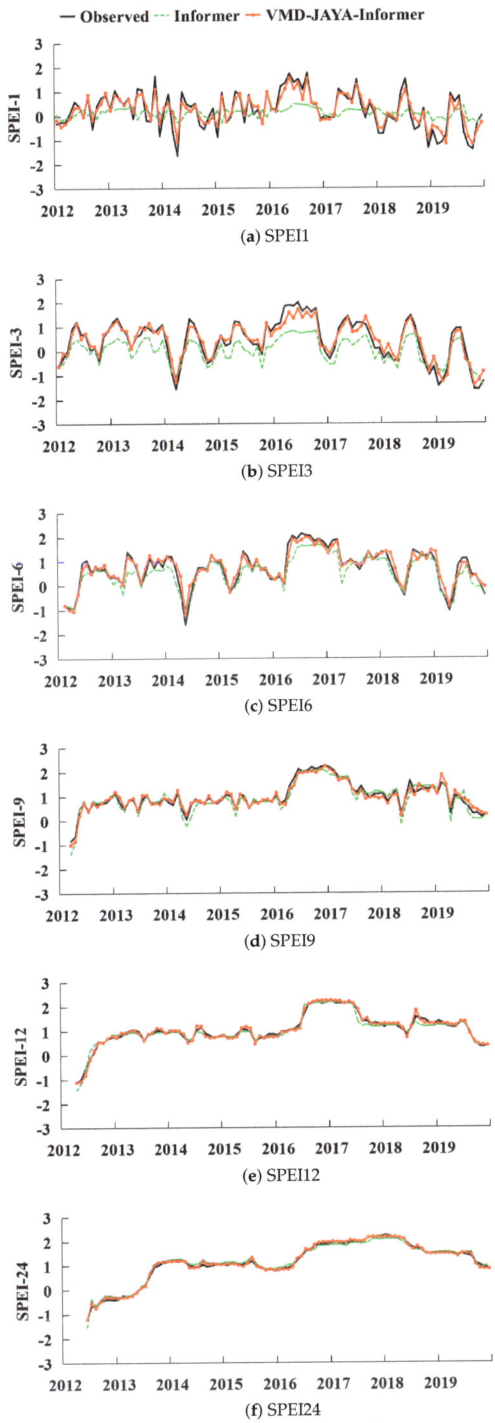

Figure 16. Comparing the results of VMD-JAYA-Informer and Informer in NenJiang.

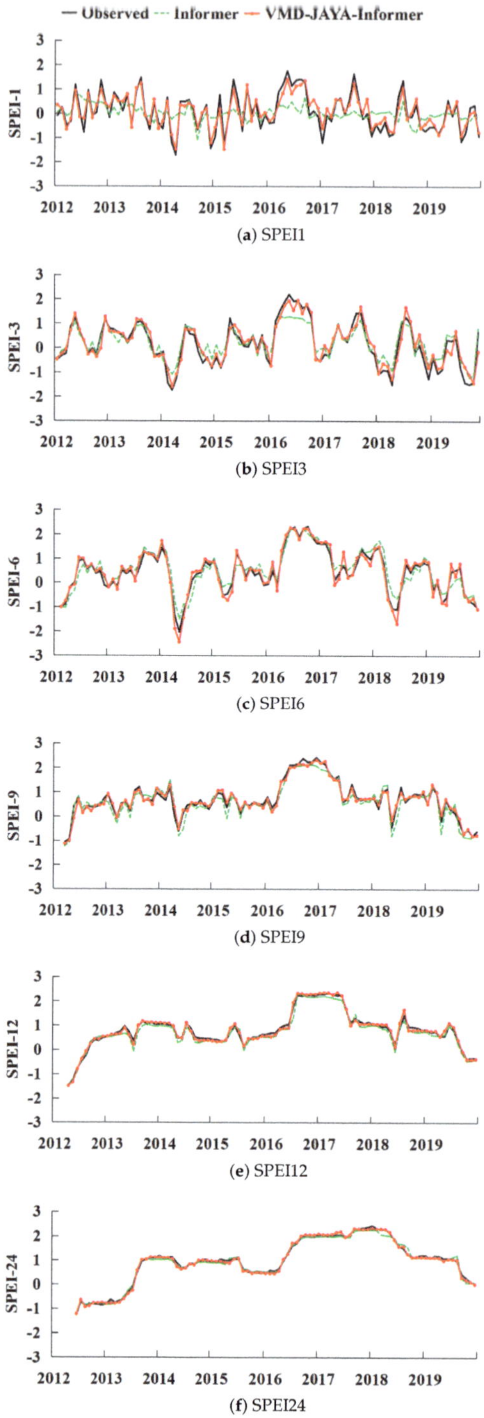

Figure 17. Comparing the results of VMD-JAYA-Informer and Informer in Yilan.

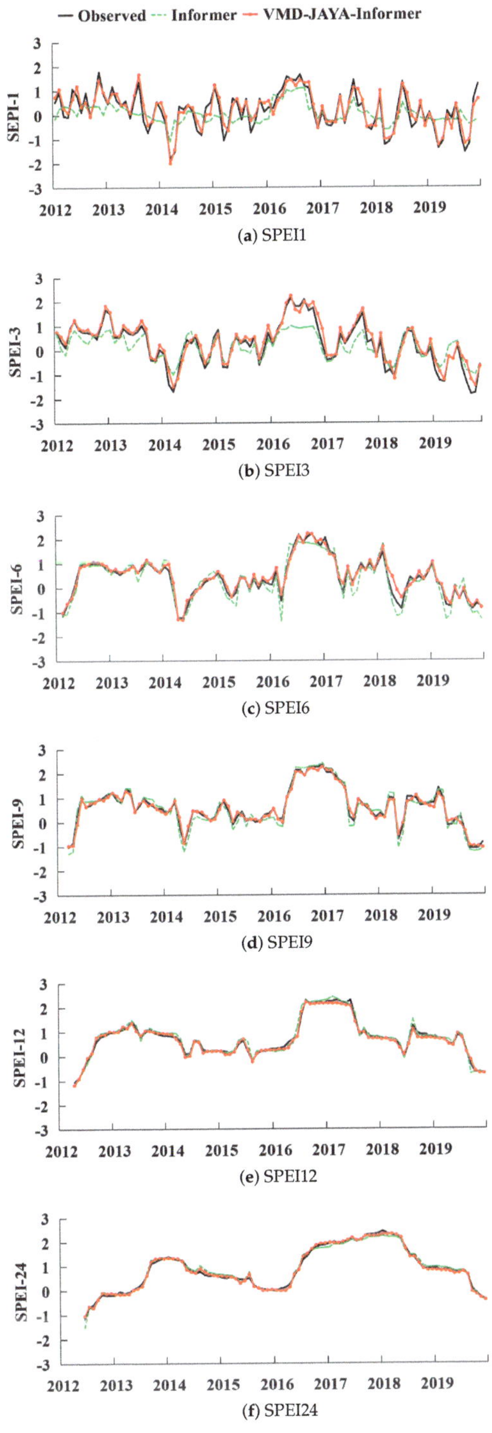

Figure 18. Comparing the results of VMD-JAYA-Informer and Informer in Kailu.

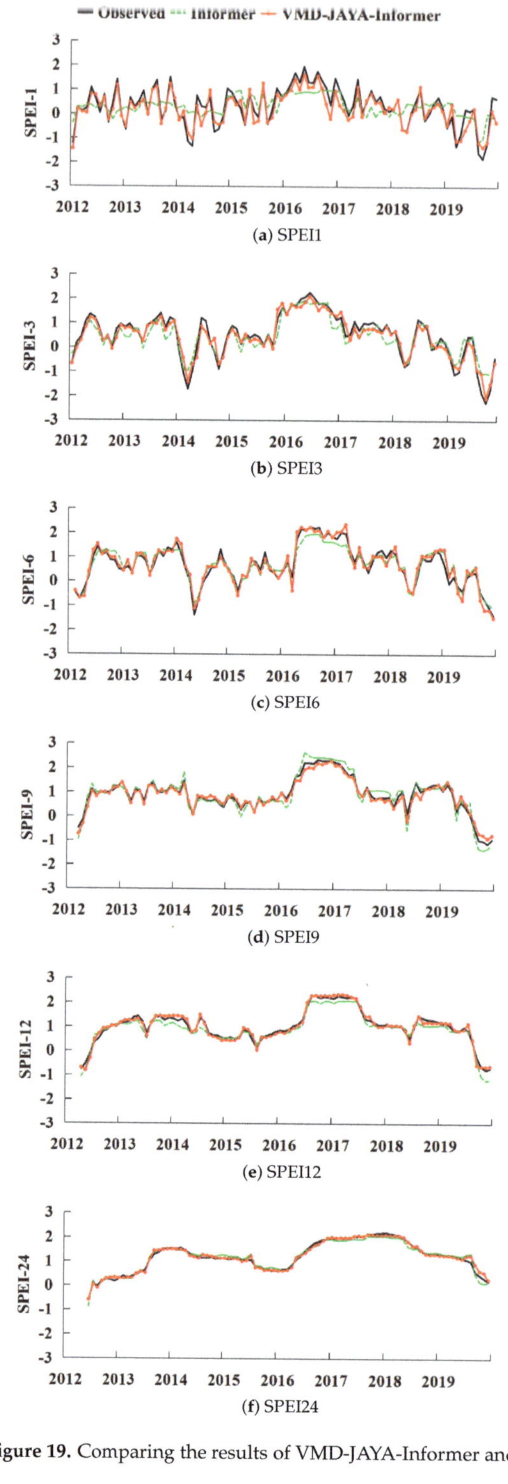

Figure 19. Comparing the results of VMD-JAYA-Informer and Informer in Panshi.

Table 7. The Informer and VMD-JAYA-Informer models' accuracy assessment metrics.

Example Stations	SPEI Series	Model	NSE	RMSE	MAE
Nenjiang	SPEI1	Informer	0.0706	0.6047	0.6206
		VMD-JAYA-Informer	0.5445	0.0809	0.4840
	SPEI3	Informer	0.2519	0.4198	0.6063
		VMD-JAYA-Informer	0.6229	0.0668	0.3760
	SPEI6	Informer	0.2591	0.1264	0.3964
		VMD-JAYA-Informer	0.8150	0.0481	0.2488
	SPEI9	Informer	0.3717	0.1061	0.3346
		VMD-JAYA-Informer	0.8435	0.0470	0.1873
	SPEI12	Informer	0.8594	0.0693	0.1752
		VMD-JAYA-Informer	0.8925	0.0301	0.1579
	SPEI24	Informer	0.9622	0.0322	0.1150
		VMD-JAYA-Informer	0.9785	0.0264	0.1019
Yilan	SPEI1	Informer	0.1312	0.3059	0.5991
		VMD-JAYA-Informer	0.6194	0.0733	0.3845
	SPEI3	Informer	0.3963	0.1518	0.5665
		VMD-JAYA-Informer	0.8640	0.0483	0.2711
	SPEI6	Informer	0.7966	0.0964	0.3245
		VMD-JAYA-Informer	0.8650	0.0271	0.2364
	SPEI9	Informer	0.8330	0.0732	0.2488
		VMD-JAYA-Informer	0.8778	0.0121	0.1919
	SPEI12	Informer	0.8878	0.0453	0.2041
		VMD-JAYA-Informer	0.8942	0.0059	0.1790
	SPEI24	Informer	0.9548	0.0263	0.1259
		VMD-JAYA-Informer	0.9718	0.0018	0.1349
Kailu	SPEI1	Informer	0.2665	0.3707	0.5626
		VMD-JAYA-Informer	0.7261	0.0620	0.3354
	SPEI3	Informer	0.4440	0.1614	0.5437
		VMD-JAYA-Informer	0.7660	0.0390	0.3201
	SPEI6	Informer	0.5661	0.0934	0.3914
		VMD-JAYA-Informer	0.7955	0.0321	0.3159
	SPEI9	Informer	0.6638	0.0243	0.3602
		VMD-JAYA-Informer	0.9464	0.0321	0.1508
	SPEI12	Informer	0.9427	0.0131	0.1456
		VMD-JAYA-Informer	0.9709	0.0119	0.1004
	SPEI24	Informer	0.9473	0.0704	0.1367
		VMD-JAYA-Informer	0.9768	0.0095	0.0924
Panshi	SPEI1	Informer	0.3706	0.2817	0.5307
		VMD-JAYA-Informer	0.6975	0.1028	0.3352
	SPEI3	Informer	0.5110	0.1522	0.5152
		VMD-JAYA-Informer	0.8486	0.0837	0.2792
	SPEI6	Informer	0.7131	0.1121	0.3065
		VMD-JAYA-Informer	0.8612	0.0699	0.2452
	SPEI9	Informer	0.8749	0.0201	0.2216
		VMD-JAYA-Informer	0.8878	0.0187	0.1941
	SPEI12	Informer	0.9224	0.0195	0.1504
		VMD-JAYA-Informer	0.9626	0.0125	0.0997
	SPEI24	Informer	0.9698	0.0064	0.0896
		VMD-JAYA-Informer	0.9880	0.0027	0.0874

4. Discussion

This study investigates the performance of SPEI prediction methods including the ARIMA, LSTM, CNN, and Informer models. To address the challenge of low prediction accuracy at shorter time scales, we proposed a VMD-JAYA-Informer model that integrates Variational Mode Decomposition (VMD) and JAYA optimization with the Informer model, effectively enhancing the Informer's performance in drought prediction.

In this approach, VMD decomposes the SPEI time series into multiple temporal components. This approach reduces noise and captures SPEI's oscillatory patterns with greater accuracy. The decomposed data are then predicted using the Informer model, further enhanced by the JAYA algorithm's optimized parameter selection. This combination enables the model to attain enhanced predictive accuracy, particularly in short-term intervals and for extreme values, where it significantly outperforms the independent Informer model, especially in 1-month forecasts.

The enhanced precision of VMD-JAYA-Informer for short-term SPEI can be attributed to three main factors. First, the adaptive decomposition from VMD enables the breakdown of SPEI data into modal functions across varying frequencies, which reveals both local and global features across multiple time scales. This decomposition better represents SPEI's periodic characteristics and provides a robust foundation for Informer to handle SPEI's complex time series structure. Secondly, the JAYA optimization algorithm refines the Informer's hyperparameters and weights, facilitating a more effective exploration of the parameter space and producing optimal parameter combinations that markedly enhance forecasting performance on SPEI data. Lastly, the Informer model's multi-head self-attention mechanism effectively captures both short-term fluctuations and long-term dependencies in SPEI data. When optimized with VMD and JAYA, Informer utilizes the SPEI's multi-scale information more effectively, resulting in enhanced prediction accuracy, especially for short-term and extreme values.

Our findings align with those of Ding et al. and Xu et al., who found that hybrid models generally outperform single models in drought prediction, particularly for shorter time intervals. Although the CEEMD-ARIMA model has demonstrated efficacy in drought forecasting, it continues to encounter challenges regarding short-term SPEI precision [24,25]. Similarly, Xu et al.'s ARIMA-LSTM hybrid model performs well for achieving long-term accuracy but faces limitations with short-term SPEI predictions [26]. Shang et al. noted enhancements in forecasting precision with the Informer; however, deficiencies in short-term SPEI accuracy persisted [28].

This study's contributions are twofold. First, it applies the Informer model for drought prediction in the Songliao River Basin, and compares its performance with that of the ARIMA, LSTM, and CNN models. While all models exhibit reduced accuracy over shorter time scales, Informer consistently outperforms the others, demonstrating strong adaptability across different time frames. Second, by introducing the VMD-JAYA-Informer approach, this study addresses Informer's short-term accuracy limitations and demonstrates its effectiveness in predicting extreme SPEI values, especially for 1-month forecasts.

This study indicates that VMD-JAYA-Informer improves SPEI forecasting accuracy for short time scales and implies that additional parameter optimization or data decomposition methods may further enhance its efficacy. Future research could explore the model's applicability across various drought indices, potentially broadening its utility under diverse drought prediction scenarios.

5. Conclusions

This study estimated the Standardized Precipitation–Evapotranspiration Index (SPEI) using meteorological data from the Songliao River Basin spanning the years 1980 to 2020 at multiple time scales. Additionally, we made predictions based on these calculations. To improve the Informer's accuracy for short-time-scale SPEI predictions, a drought forecasting approach called VMD-JAYA-Informer was developed. This was achieved by comparing and analyzing the performance of ARIMA, LSTM, CNN, and Informer in drought forecasting. The primary focuses of this study are outlined as follows:

(1) The experimental findings demonstrate that when the time scale diminished, the forecasting ability of ARIMA, LSTM, CNN, and Informer also declined. However, Informer consistently outperformed ARIMA, LSTM, and CNN. As the time scale increased, the forecasting performance of ARIMA, LSTM, CNN, and Informer improved steadily. The NSE also gradually increased and reached its optimal value of 1. The RMSE and MAE exhibited the inverse trends. All four models achieved an optimal predicting performance over the 24-month time period. The NSE values were obtained: 0.9276, 0.8537, 0.9009, and 0.9395.

(2) However, due to the influence of weather changes, the SPEI at short time scales experiences greater fluctuations and is more challenging to predict. Consequently, the Informer model's forecasting accuracy at short time scales has not yet reached the desired level. This study introduces VMD-JAYA-Informer, a method for drought prediction that aims to fix the Informer's poor performance in SPEI prediction at short time scales. The results show that when it comes to predicting the SPEI at the 1-month time scale, VMD-JAYA-Informer outperforms the Informer and aligns better with the real data. It demonstrates that VMD-JAYA-Informer is more effective in accurately predicting the SPEI in short time periods. It also outperforms the Informer model in other time periods.

(3) For this study, we chose four meteorological stations in the Songliao River Basin at random to test the accuracy of the VMD-JAYA-Informer. The results indicate that the VMD-JAYA-Informer outperforms the Informer significantly when it comes to forecasting at the 1-month time scale. The NSE values for the four weather stations were obtained: 0.8663, 0.8765, 0.8822, and 0.8416, respectively. The VMD-JAYA-Informer demonstrates a higher prediction performance compared with Informer across several time scales, highlighting its broad applicability and high accuracy in drought prediction.

Author Contributions: Conceptualization, J.W. and W.T.; methodology, J.W. and J.S.; software, J.W., H.G. and B.Z.; validation, J.W., W.T. and J.S.; formal analysis, J.W.; investigation, J.W.; resources, J.W.; data curation, J.W.; writing—original draft preparation, J.W. and B.Z.; writing—review and editing, J.W. and P.D.; visualization, J.W.; supervision, J.W., W.T., H.G., G.W. and Y.G.; project administration, J.W., W.T., J.S. and H.G.; funding acquisition, J.S. All authors have read and agreed to the published version of the manuscript.

Funding: This research was funded by the Science and Technology Innovation 2030 Major Project Issues, China, grant number "2023ZD0120600"; major science and technology project of Henan Province, China, grant numbers "221100210600"; major science and technology project of Henan Province, China, grant numbers and "201400210100"; and Zhengzhou City Collaborative Innovation Program of Henan Province, China, grant number "22XTZX01235".

Data Availability Statement: The datasets from 1960 to 2019 were obtained from the National Meteorological Data Center (http://data.cma.cn/, accessed on 2 April 2021).

Acknowledgments: We would like to express our heartfelt gratitude to the National Supercomputing Zhengzhou Center. All model code programs in this paper were run on the computational platform of the National Supercomputing Zhengzhou Center.

Conflicts of Interest: The authors declare no conflicts of interest.

References

1. Vicente-Serrano, S.M.; Beguería, S.; López-Moreno, J.I. A Multiscalar Drought Index Sensitive to Global Warming: The Standardized Precipitation Evapotranspiration Index. *J. Clim.* **2010**, *23*, 1696–1718. [CrossRef]
2. Mishra, A.; Singh, V. A review of drought concepts. *J. Hydrol.* **2010**, *391*, 202–216. [CrossRef]
3. Kamruzzaman, M.; Almazroui, M.; Salam, M.; Mondol, M.; Rahman, M.; Deb, L.; Kundu, P.; Zaman, M.; Islam, A. Spatiotemporal drought analysis in Bangladesh using the standardized precipitation index (SPI) and standardized precipitation evapotranspiration index (SPEI). *Sci. Rep.* **2022**, *12*, 20694. [CrossRef]

4. Wang, Q.; Zeng, J.; Qi, J.; Zhang, X.; Zeng, Y.; Shui, W.; Xu, Z.; Zhang, R.; Wu, X.; Cong, J. A multi-scale daily SPEI dataset fo drought characterization at observation stations over mainland China from 1961 to 2018. *Earth Syst. Sci. Data* **2021**, *13*, 331–341 [CrossRef]
5. Li, Y.; Chang, J.; Fan, J.; Yu, B. Agricultural drought evolution characteristics and driving mechanisms in the Yellow River Basin under climate and land use changes. *Trans. Chin. Soc. Agric. Eng.* **2021**, *37*, 84–93.
6. Yang, R.; Geng, G.; Zhou, H.; Wang, T. Spatial-temporal Evolution of Meteorological Drought in the Wei River Basin Based on SPEI_PM. *Chin. J. Agrometeorol.* **2021**, *42*, 962–974.
7. Abbas, A.; Waseem, M.; Ahmad, R.; Khan, K.A.; Zhao, C.; Zhu, J. Sensitivity analysis of greenhouse gas emissions at farm level Case study of grain and cash crops. *Environ. Sci. Pollut. Res.* **2022**, *29*, 82559–82573. [CrossRef] [PubMed]
8. Elahi, E.; Khalid, Z.; Zhang, Z. Understanding farmers'intention and willingness to install renewable energy technology: A solution to reduce the environmental emissions of agriculture. *Appl. Energy* **2022**, *309*, 118459. [CrossRef]
9. Chen, Q.; Gao, L.; Zhong, F.; Zuo, X.; Ma, M. Spatiotemporal Variations of Drought in the Yunnan-Guizhou Plateau, Southwest China, during 1960–2013 and Their Association with Large-Scale Circulations and Historical Records. *Ecol. Indic.* **2020**, *112*, 106041 [CrossRef]
10. Dehghan, S.; Salehnia, N.; Sayari, N.; Bakhtiari, B. Prediction of meteorological drought in arid and semi-arid regions using PDSI and SDSM: A case study in Fars Province, Iran. *J. Arid. Land* **2020**, *12*, 318–330. [CrossRef]
11. McKee, T.; Doesken, N.; Kleist, J. The relationship of drought frequency and duration to time scales. In Proceedings of the 8th Conference on Applied Climatology, Anaheim, CA, USA, 17–22 January 1993.
12. Poornima, S.; Pushpalatha, M. Drought Prediction Based on SPI and SPEI with Varying Timescales Using LSTM Recurrent Neural Network. *Soft Comput.* **2019**, *23*, 8399–8412. [CrossRef]
13. Ren, Y.; Liu, J.; Shalamzari, M.J.; Arshad, A.; Liu, S.; Liu, T.; Tao, H. Monitoring Recent Changes in Drought and Wetness in the Source Region of the Yellow River Basin, China. *Water* **2022**, *14*, 861. [CrossRef]
14. Alahacoon, N.; Edirisinghe, M. A comprehensive assessment of remote sensing and traditional based drought monitoring indices at global and regional scale. *Geomat. Nat. Hazards Risk* **2022**, *13*, 762–799. [CrossRef]
15. Saeed, S.; Mohammadi, G.M.; Saviz, S. Spatial and temporal analysis of drought in various climates across Iran using the Standardized Precipitation Index (SPI). *Arab. J. Geosci.* **2022**, *15*, 1279.
16. Palmer, W.C. *Meteorological Drought*; U.S. Department of Commerce Weather Bureau Research Paper; U.S. Department of Commerce: San Diego, CA, USA, 1965.
17. Narasimhan, B.; Srinivasan, R. Development and evaluation of Soil Moisture Deficit Index (SMDI) and Evapotranspiration Deficit Index (ETDI) for agricultural drought monitoring. *Agric. For. Meteorol.* **2005**, *133*, 69–88. [CrossRef]
18. Liu, J.; Ren, Y.; Tao, H.; Shalamzari, M. Spatial and Temporal Variation Characteristics of Heatwaves in Recent Decades over China. *Remote Sens.* **2021**, *13*, 3824. [CrossRef]
19. Stagge, J.; Tallaksen, L.; Xu, C.; Lanen, H. Standardized precipitation-evapotranspiration index (SPEI): Sensitivity to potential evapotranspiration model and parameters. In Proceedings of the 7th World Flow Regimes from International and Experimental Network Data-Water Conference, Montpellier, France, 7–10 October 2014.
20. Mi, Q.; Gao, X.; Li, Y.; Li, X.; Tang, Y.; Ren, C. Application of Deep Learning Method to Drought Prediction. *J. Appl. Meteorol. Sci.* **2022**, *33*, 104–114.
21. Nikbakht, S.; Zahraie, B.; Nasseri, M. Seasonal meteorological drought prediction using support vector machine. *J. Water Wastewater* **2012**, *23*, 73–85.
22. Fung, K.; Huang, Y.; Koo, C.; Soh, Y. Drought forecasting: A review of modelling approaches 2007–2017. *J. Water Clim. Change* **2019**, *11*, 771–799. [CrossRef]
23. Ding, Y.; Xu, D.; Cao, L.; Guan, X. Applicability of the LSTM and ARIMA model in drought prediction based on CEEMD: A case study of Xinjiang. *Arid Zone Res.* **2022**, *39*, 734–744.
24. Xu, D.; Ding, Y.; Liu, H.; Zhang, Q.; Zhang, D. Applicability of a CEEMD–ARIMA Combined Model for Drought Forecasting: A Case Study in the Ningxia Hui Autonomous Region. *Atmosphere* **2022**, *13*, 1109. [CrossRef]
25. Xu, D.; Zhang, Q.; Ding, Y.; Zhang, D. Application of a hybrid ARIMA-LSTM model based on the SPEI for drought forecasting. *Environ. Sci. Pollut. Res.* **2022**, *29*, 4128–4144. [CrossRef]
26. Zhang, Q.; Miao, C.; Gou, J.; Zheng, H. Spatiotemporal characteristics and forecasting of short-term meteorological drought in China. *J. Hydrol.* **2023**, *624*, 129924. [CrossRef]
27. Shang, J.; Zhao, B.; Hua, H.; Wei, J.; Oin, G.; Chen, G. Application of Informer Model Based on SPEI for Drought Forecasting. *Atmosphere* **2023**, *14*, 951. [CrossRef]
28. Vaswani, A.; Shazeer, N.; Parmar, N.; Uszkoreit, J.; Jones, L.; Gomez, A.; Kaiser, L.; Polosukhin, I. Attention Is All You Need. *arXiv* **2017**, arXiv:1706.03762.
29. Zhou, H.; Zhang, S.; Peng, J.; Zhang, S.; Li, J.; Xiong, H.; Zhang, W. Informer: Beyond efficient transformer for long sequence time-series forecasting. *Proc. AAAI Conf. Artif. Intell.* **2021**, *35*, 11106–11115. [CrossRef]

30. Dong, H.; Sun, L.; Ouyang, F. Prediction of PM2.5 Concentration Based on Informer. *Environ. Eng.* **2022**, *40*, 48–54.
31. Yu, F.; Koltun, V.; Funkhouser, T. Dilated residual networks. In Proceedings of the IEEE Conference on Computer Vision and Pattern Recognition, Honolulu, HI, USA, 21–26 July 2017.
32. Dragomiretskiy, K.; Zosso, D. Variational Mode DecompositionV. *IEEE Trans. Signal Process.* **2014**, *62*, 531–544. [CrossRef]
33. Ding, J.; Xiao, D.; Li, X. Gear fault diagnosis based on genetic mutation particle swarm optimization VMD and probabilistic neural network algorithm. *IEEE Access* **2020**, *8*, 18456–18474. [CrossRef]
34. Zhang, X.; Miao, Q.; Zhang, H.; Wang, L. A parameter-adaptive VMD method based on grasshopper optimization algorithm to analyze vibration signals from rotating machinery. *Mech. Syst. Signal Process.* **2018**, *108*, 58–72. [CrossRef]
35. Rao, R.; Waghmare, G. A new optimization algorithm for solving complex constrained design optimization problems. *Eng. Optim.* **2017**, *49*, 60–83. [CrossRef]
36. Hochreiter, S.; Schmidhuber, J. Long Short-Term Memory. *Neural Comput.* **1997**, *9*, 1735–1780. [CrossRef] [PubMed]
37. Benvenuto, D.; Giovanetti, M.; Vassallo, L.; Angeletti, S.; Ciccozzi, M. Application of the ARIMA model on the COVID-2019 epidemic dataset. *Data Brief* **2020**, *29*, 105340. [CrossRef] [PubMed]
38. Zhang, G. Time series forecasting using a hybrid ARIMA and neural network model. *Neurocomputing* **2003**, *50*, 159–175. [CrossRef]
39. LeCun, Y.; Bottou, L.; Bengio, Y.; Haffner, P. Gradient-based learning applied to document recognition. *Proc. IEEE* **1998**, *86*, 2278–2324. [CrossRef]
40. Wu, J.M.T.; Li, Z.; Herencsar, N.; Vo, B.; Lin, C. A graph-based CNN-LSTM stock price prediction algorithm with leading indicators. *Multimed. Syst.* **2023**, *29*, 1751–1770. [CrossRef]
41. Abdulnabi, A.H.; Wang, G.; Lu, J.; Jia, L. Multi-task CNN model for attribute prediction. *IEEE Trans. Multimed.* **2015**, *17*, 1949–1959. [CrossRef]
42. Tang, G.J.; Wang, X.L. Variational mode decomposition method and its application on incipient fault diagnosis of rolling bearing. *J. Vib. Eng.* **2016**, 29, 638–648.

land

MDPI

Article

A Coupled Model for Forecasting Spatiotemporal Variability of Regional Drought in the Mu Us Sandy Land Using a Meta-Heuristic Algorithm

Changfu Tong [1], Hongfei Hou [1,*], Hexiang Zheng [1], Ying Wang [1,2] and Jin Liu [1,2]

1 Institute of Water Resources for Pastoral Area Ministry of Water Resources, Hohhot 010020, China
2 College of Water Conservancy and Civil Engineering, Inner Mongolia Agricultural University, Huhhot 010018, China
* Correspondence: 12024140142@stu.nxu.edu.cn

Abstract: Vegetation plays a vital role in terrestrial ecosystems, and droughts driven by rising temperatures pose significant threats to vegetation health. This study investigates the evolution of vegetation drought from 2010 to 2024 and introduces a deep-learning-based forecasting model for analyzing regional spatial and temporal variations in drought. Extensive time-series remote-sensing data were utilized, and we integrated the Temperature–Vegetation Dryness Index (TVDI), Drought Severity Index (DSI), Evaporation Stress Index (ESI), and the Temperature–Vegetation–Precipitation Dryness Index (TVPDI) to develop a comprehensive methodology for extracting regional vegetation drought characteristics. To mitigate the effects of regional drought non-stationarity on predictive accuracy, we propose a coupling-enhancement strategy that combines the Whale Optimization Algorithm (WOA) with the Informer model, enabling more precise forecasting of long-term regional drought variations. Unlike conventional deep-learning models, this approach introduces rapid convergence and global search capabilities, utilizing a sparse self-attention mechanism that improves performance while reducing model complexity. The results demonstrate that: (1) compared to the traditional Transformer model, test accuracy is improved by 43%; (2) the WOA–Informer model efficiently handles multi-objective forecasting for extended time series, achieving MAE (Mean Absolute Error) ≤ 0.05, MSE (Mean Squared Error) ≤ 0.001, MSPE (Mean Squared Percentage Error) ≤ 0.01, and MAPE (Mean Absolute Percentage Error) $\leq 5\%$. This research provides advanced predictive tools and precise model support for long-term vegetation restoration efforts.

Keywords: vegetation drought; sparse self-attention mechanism; Whale Optimization Algorithm (WOA); predictive accuracy

1. Introduction

Vegetation is crucial for the functioning of terrestrial ecosystems. Due to the effects of global warming and human activity, there is an increased occurrence of severe weather events, such as droughts, leading to a decrease in vegetation coverage [1]. The ongoing degradation of the ecology has extensive worldwide consequences. Time-series forecasting plays a crucial role in several sectors and is especially significant for extensive, long-term studies of time series [2]. This may aid in comprehending future patterns of change in the dry zone, therefore enhancing the sustainability of the environment [3]. Long-term time-series prediction is a crucial method for analyzing the spatial and temporal dryness of the vegetation in the area and assessing its influence on the biological environment.

Currently, the majority of vegetation-related indicators are specifically developed for addressing short-term issues [4]. Nevertheless, the model's predictive capacity is severely constrained as the series length grows, which has hindered progress in long-term series prediction research [5]. Bishal Roy found that four supervised machine-learning algorithms, specifically support-vector regression, random forest, linear regression, and polynomial regression, successfully predicted vegetation indices over 1 year [6]. However, these algorithms were unable to accurately predict abrupt changes in vegetation indices that occurred during the middle of the year. Furthermore, machine learning has a significant disparity in accurately predicting lengthy time series, leading to suboptimal model performance and a substantial decrease in inference speed [7]. Deep learning is more effective in achieving superior outcomes for Natural Language Processing (NLP) tasks [8]. Nonetheless, the high cost of training makes it challenging to widely adopt deep-learning models. Presently, the self-attention mechanism is a limitation for predicting long time series, and the efficacy of both the self-attention mechanism and the Transformer design is a constraint for applying them to the problem of Long-Sequence Time-Series Forecasting (LSTF) [9]. The computational complexity of the self-attention mechanism is quadratic, meaning that as the length of the sequence rises, both the calculation time and memory needs increase at a square-level rate [10]. Additionally, it is possible to overlook the local information inside the sequence. Zhang et al. introduced an Informer model that addresses the memory bottleneck issue in deep learning and the challenge of long-time series prediction [11]. This model incorporates the ProbSparse self-attention mechanism and the self-attention distillation operation due to an issue of reduced performance. Hence, this work aims to use the enhanced Transformer–Informer model to address the issue of predicting lengthy time sequences.

Deep-learning models frequently encounter difficulties in parameter optimization with insufficient data, particularly when addressing non-stationary data, resulting in suboptimal performance on test sets [12]. Attaining consistency in hyperparameter optimization is challenging. Meta-heuristic optimization strategies are essential for maximizing deep-learning parameters and tackling non-stationarity in long-term time-series predictions [13]. These methods provide three primary advantages: (1) ease of implementation, (2) capacity to circumvent local optima and identify global solutions, and (3) extensive application across many circumstances [14,15]. Numerous modern meta-heuristic algorithms emulate biological processes, including Genetic Algorithm (GA), Evolutionary Strategies (ES), Artificial Bee Colony (ABC), Particle Swarm Optimization (PSO), and Ant Colony Optimization (ACO) [16–19]. PSO, formulated by Kennedy et al., simulates the collective behavior of avian or aquatic species, employing collaboration and information dissemination to identify optimal solutions [20]. It encompasses both individual and collective optimal solutions, rendering it a frequently employed method in deep-learning optimization [21]. Likewise, ACO, created by Dorigo et al., emulates the pheromone trail-following behavior of ants to identify optimal pathways, leveraging positive feedback mechanisms [22]. Although both PSO and ACO provide rapid convergence and efficient global search abilities, they encounter difficulties in multi-objective optimization: PSO is susceptible to local optima, while ACO is significantly dependent on delicate parameter configurations [23–25]. Achieving equilibrium between exploration and exploitation continues to be a significant difficulty for the majority of meta-heuristic algorithms [26]. This study presents the WOA (Whale Optimization Algorithm), which emulates the feeding behavior of humpback whales, as an innovative solution to these issues. The system incorporates the hunting behavior of following prey using either a random or optimum search agent, as well as the use of spirals to emulate the mechanism of humpback whales' bubble net assault [27]. The WOA algorithm is used to optimize the multi-objective prediction of the Informer model, which is proposed to address the issue of forecasting the spatial and temporal variability of a lengthy time series of regional vegetation drought indicators.

This work addresses the present issues of global warming and escalating drought by combining several approaches for analyzing trends in vegetation drought indices over extended periods. The analysis is conducted using the GEE (Google Earth Engine) cloud platform. Additionally, a multi-objective prediction model is proposed to optimize the parameters involved. The system's objective is to comprehensively uncover the geographical and temporal variations in long-term droughts in vegetation at a regional level. This will enable the development of a reliable scientific prediction tool for the establishment of ecological environmental protection measures in arid and semi-arid areas worldwide. The principal objectives of this study are: (1) to examine the advantages of optimizing the parameters of the Informer model; (2) to evaluate the benefits of utilizing the WOA–Informer model for forecasting long-term, multi-objective drought-related vegetation indicators.

2. Materials and Methods

2.1. Study Area

The Mu Us Sandy Land is situated in Ordos City, Inner Mongolia Autonomous Region, within the semi-arid and desert regions of Northwest China, and it is one of the principal sandy locations in the country. The area encompasses over 29,800 square kilometers, with a mean elevation between 1065 m and 1623 m (see Figure 1) [28]. The study area is located inside the administrative confines of Ordos City, encompassing the Uxin Banner, Yijinhuoluo Banner, Otog Banner, and Otogqian Banner. The landscape features include elevated regions in the north and south, with lower areas in the east and west, resulting in a gradual decline from northwest to southeast. The region has a temperate arid and semi-arid climate, categorized as a temperate continental climate, characterized by an average annual temperature of 6 to 8.5 °C and average annual precipitation of 250 to 420 mm, predominantly occurring from July to September. The Mu Us Sandy Land encompasses a variety of natural environments, including mobile dunes, semi-permanent dunes, fixed dunes, low ridges, marshes, alluvial terraces, lakes, rivers, and streams [29]. Soil types are categorized as zonal and non-zonal soils, with chestnut–calcium soil predominating in the northeast and southwest, and brown–calcium soil being prominent in the west. The soils of Mu Us Sandy Land exhibit low organic matter, poor fertility, weak cohesiveness, and low water retention, contributing to desertification and the prevalence of mobile, semi-permanent, and fixed dunes [30]. In conclusion, the biological system of the Mu Us Sandy Land is delicate and dependent on scarce water resources. Consequently, forecasting geographical and temporal variations in drought is highly significant.

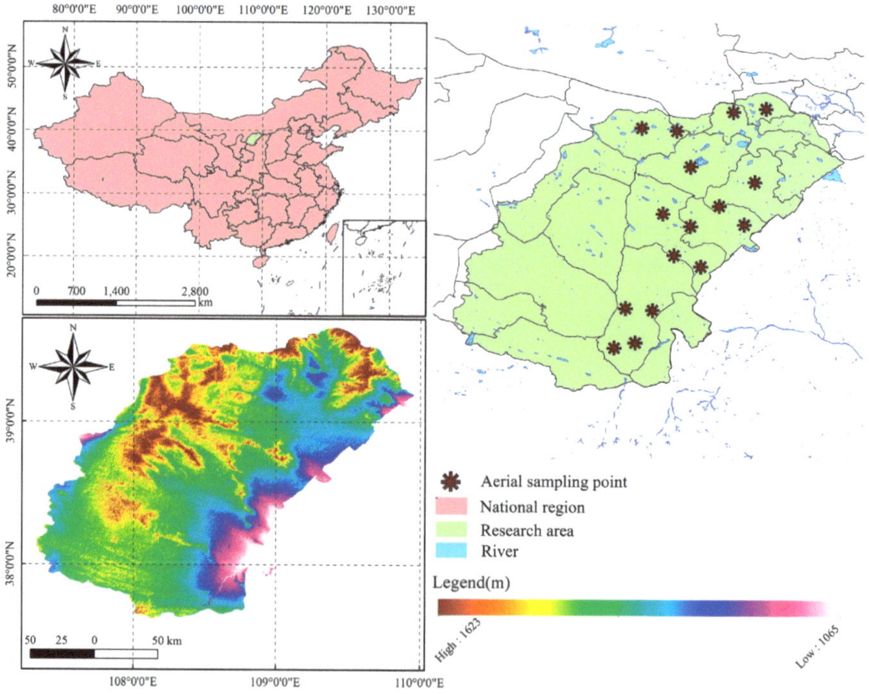

Figure 1. The geographic location of the Mu Us Sandy Land.

2.2. Data and Processing

Data were gathered in the Mu Us Sandy Land region (109°16′–109°26′ E, 38°46′–39°55′ N) in Erdos, Inner Mongolia, from 12 March 2023, to 18 July 2024. The data were acquired with a DJI Wizard 4 Pro multispectral drone, including a lens focal length of 20 mm, an image angle range of −90° to 30°, and a pixel size of 20 μm (Figure 2). The pixel dimension is 20 μm, and the photographic resolution is 400 × 3000 pixels. The UAV navigated an S-shaped trajectory at a height of 95–100 m. The high-resolution drone photographs were used to corroborate the model's predictions by contrasting them with the terrestrial circumstances in the research region.

Figure 2. Aerial detail view.

We derived the NDVI utilizing the NDVI image element dichotomous model from the LANDSAT/LT05/C02/T1_L2, LANDSAT/LE07/C02/T1_L2, and LANDSAT/LC08/C02/T1_L2 datasets, in conjunction with the MOD13A1, MOD13A2, and MOD13Q1 products of MODIS, as well as the LST products MOD11A2 and MOD13A2 as surface temperature variables. Subsequently, we computed NDVI, LST, TVDI, TVPDI, DSI, and ESI indices via the GEE platform (17 June 2024). The indices were used to measure changes in plant cover and drought conditions in the Mu Us Sandy Land from 2010 to 2024, using a spatial resolution of 500 m and a monthly temporal resolution. We picked the maximum pixel values from the monthly picture data for compositing to create time-series images over 15 years. Furthermore, to alleviate the uncertainty arising from the restricted number of picture acquisitions every year, we conducted pixel-by-pixel 95th-quantile compositing analysis on Landsat images obtained annually.

This research utilizes the GEE platform to produce vegetation cover photographs inside the study region. The primary computational procedure comprises [31]:

(1) Specify the scope of the water source: indicate the limits of the region being studied.
(2) Create time-series datasets by combining the LANDSAT/LT05/C02/T1_L2, LANDSAT/LE07/C02/T1_L2, and LANDSAT/LC08/C02/T1_L2 datasets. Apply a filter

to choose data within a certain period and then remove any cloud cover present in the data.

(3) Generate annual median image dataset: aggregate the produced picture datasets by year and compute the yearly median image for each period.

(4) Water body mask processing: Utilize the mask function to exclude water bodies from the picture to minimize their impact on the estimate of plant cover.

Thus, in this research, the NDVI-based image element dichotomy model was used to assess the plant cover in the study area:

$$NDVI = \frac{NIR - R}{NIR + NIR} \tag{1}$$

where NIR and R are the reflectance recorded by the sensor in the near-infrared and red-light regions, respectively.

The $TVDI$ was computed using the following equation (Figure 3):

$$TVDI = \frac{LST - LST_{min}}{LST_{max} - LST_{min}} \tag{2}$$

where LST_{min} refers to the wet boundary with the lowest land-surface temperature (LST) value, while LST_{max} refers to the dry boundary with the highest LST value. LST reflects the monthly land-surface temperature for the same month between 2000 and 2020.

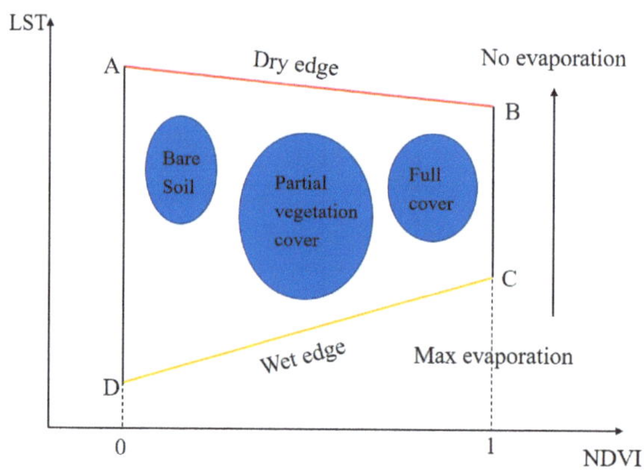

Figure 3. A trapezoid representing the relationship between LST and NDVI at a conceptual level.

The $TVPDI$ is a composite indicator that is calculated using measurements of T_s, $NDVI$, and P. The amalgamation of these three variables encompasses the influence of temperature, the extent of plant cover, and the quantity of precipitation necessary for vegetation development, hence enabling the evaluation of drought conditions in the designated research region [32]. The equation for this is:

$$NT_s = \frac{T_s - T_{smin}}{T_{smax} - T_{smin}} \tag{3}$$

$$NNDVI = \frac{NDVI - NDVI_{min}}{NDVI_{max} - NDVI_{min}} \tag{4}$$

$$NP = \frac{P - P_{min}}{P_{max} - P_{min}} \tag{5}$$

The formula for $TVPDI$ is as follows:

$$TVPDI = \sqrt{(NT_{smax} - NT_S)^2 + (NNDVI - NNDVI_{min})^2 + (NP - NP_{min})^2} \quad (6)$$

where NT_{smax} is the maximum of NT_s, and $NNDVI_{min}$ and NP_{min} are the minimum of the $NNDVI$ and NP.

The DSI used in this research was primarily computed using ET, PET, and $NDVI$. Thus, DSI was determined in this work using ET, PET from the MOD16A2 product, and $NDVI$ from the MOD13A1 product [33]. The calculation may be determined using the following equation:

$$RT = ET/PET \quad (7)$$

$$Z_{RT} = (RT - \overline{RT})/\sigma_{RT} \quad (8)$$

$$Z_{NDVI} = (NDVI - \overline{NDVI})/\sigma_{NDVI} \quad (9)$$

$$Z = Z_{RT} + Z_{NDVI} \quad (10)$$

$$DSI = (Z - \overline{Z})/\sigma_Z \quad (11)$$

where ET and PET are actual evapotranspiration and potential evapotranspiration, respectively, and RT is the ratio of ET to PET, which can be used as an indicator to evaluate the availability of terrestrial water resources and reflect the humid or arid state. RT and $NDVI$ are the average values of RT and $NDVI$, respectively. σ_{RT} and σ_{NDVI} are the standard deviations of RT and $NDVI$, respectively. Z represents the sum of the normalized ratios of RT and $NDVI$; Z and σ_Z are the mean and standard deviation of Z.

In this work, the ESI was computed using ET from the MOD16A2 product. ET is primarily defined as 1 minus the ratio of actual to prospective ET [34]. The formula is simplified as follows:

$$ESI = 1 - \frac{ET}{PET} \quad (12)$$

where ET represents actual evapotranspiration and PET represents potential evapotranspiration.

We used ASTERGDEM 30 M resolution digital-elevation data and, specifically, chose Digital Elevation Model (DEM) data from a geospatial data cloud. Subsequently, we analyzed the DEM data with ArcGIS 10.8 to obtain precise elevation data.

2.3. Research Methodology

2.3.1. Whale Optimization Algorithm

Whales, being marine creatures that rely on obtaining oxygen from the ocean's surface, possess double the number of spindle cells compared to mature individuals of other species. This abundance of spindle cells contributes to their significantly elevated levels of intelligence and empathy [35]. Humpback whales, as an exemplary species of cetaceans, use a distinctive feeding technique known as bubble nets, which involves the creation of "upward spirals" and "double loops" to ensnare krill or tiny fish near the water's surface [36]. The Informer model was enhanced in this work by including the mathematical models of the WOA algorithm for surrounding prey, executing spiral bubble net feeding maneuvers, and hunting for prey.

(1) Encircling prey

The WOA algorithm primarily emulates the hunting behavior of humpback whales to enhance the process of issue resolution by progressively approaching the present ideal answer. The provided equation represents this behavior:

$$\vec{D} = \left|\vec{C}.\vec{X^*}(t) - \vec{X}(t)\right| \tag{13}$$

$$\vec{X}(t+1) = \vec{X^*}(t) - \vec{A}.\vec{D} \tag{14}$$

where t represents the current iteration, $\vec{A}$ and $\vec{C}$ are coefficient vectors, X^* is the position vector of the best solution achieved so far, $\vec{X}$ is the position vector, | | denotes the absolute value, and $\cdot$ is an element-wise multiplication. It is important to note that X^* should be modified in every iteration if a superior solution is found.

$\vec{A}$ and $\vec{C}$ are calculated as follows:

$$\vec{A} = 2\vec{a}\cdot\vec{r} - \vec{a} \tag{15}$$

$$\vec{C} = 2\cdot\vec{r} \tag{16}$$

where $\vec{a}$ is linearly lowered from 2 to 0, while the vector $\vec{r}$ is a random vector within the range of 0 to 1.

The WOA algorithm achieves the optimum solution by iteratively examining the positions of the agents and using Equations (15) and (16) to identify the route that is most similar to the ideal solution [37]. The WOA can investigate the nearest position to the best solution by using the random vector $\vec{r}$. Hence, the WOA achieves optimum exploration by iteratively altering the locations (X, Y) of the search agents to gradually approach the best solution, mimicking the behavior of a whale encircling its prey [38].

(2) Bubble-net attacking method (exploitation phase)

The WOA can replicate two specific behaviors shown by whales while using a bubble net to catch their prey. This simulation allows WOA to effectively find the ideal solution. The shrink-wrap method is accomplished by reducing the value in Equation (15). In other words, the location of the exploring agent may be adjusted to any place that is a certain distance away from the ideal agent position [39]. The location update of the prey is achieved by imitating the spiral motion of humpback whales in the following manner.

$$\vec{X}(t+1) = \vec{D'}\cdot e^{bl}\cdot\cos(2\pi l) + \vec{X^*}(t) \tag{17}$$

where $\vec{D'} = \left|\vec{X^*}(t) - \vec{X}(t)\right|$ and indicates the distance of the i th whale to the prey (best solution obtained so far), b is a constant for defining the shape of the logarithmic spiral, l is a random number in $[-1, 1]$, and . is an element-by-element multiplication.

To more precisely ascertain the position of the ideal solution, we use a simulation of a whale moving in a circular pattern around the prey inside a restricted region. The equation for mathematical modeling is as follows:

$$\vec{X}(t+1) = \begin{cases} \vec{X^*}(t) - \vec{A}\cdot\vec{D} \; if \; p < 0.5 \\ \vec{D'}\cdot e^{bl}\cdot\cos(2\pi l) + \vec{X^*}(t) \; if \; p \geq 0.5 \end{cases} \tag{18}$$

where p is a random number in $[0, 1]$.

2.3.2. Search for Prey (Exploration Phase)

We obtain the best possible answer by representing the process of whale feeding as a stochastic search method that relies on the locations of other whales. The same technique relying on the fluctuation of $\vec{A}$ may be used to hunt for prey [40]. The search agent's location is determined by randomly generated values of A that are higher than 1 or less than −1. When the absolute value of A exceeds 1, the WOA algorithm prioritizes exploration and conducts a comprehensive search throughout the solution space. The following equation represents the mathematical model:

$$\vec{D} = \left| \vec{C} \cdot \vec{X}_{rand} - \vec{X} \right| \tag{19}$$

$$\vec{D}(t+1) = \left| \vec{X}_{rand} - \vec{A} \cdot \vec{D} \right| \tag{20}$$

where $\vec{X}_{rand}$ is a random-position vector (a random whale) chosen from the current population.

The WOA algorithm commences with randomly assigned parameters and iteratively determines the location of the best solution [41]. The search agent is not affected by the continual lowering of the parameter, which controls the range of the best solution (Figure 4). The WOA algorithm can alternate between spiral or circular motion based on the parameter p [42]. The WOA algorithm halts when a certain termination condition is met. The termination criteria may be met by either achieving the maximum number of iterations, discovering a good solution, or satisfying other stated requirements.

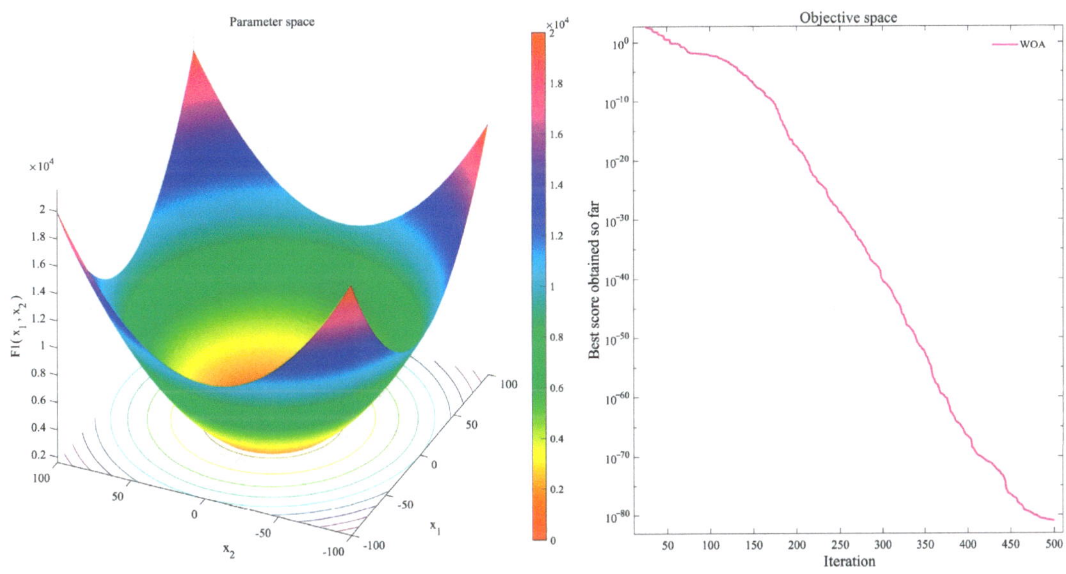

Figure 4. The Whale Optimization Algorithm systematically explores the parameter space throughout the optimization process and progressively approaches the ideal solution.

The WOA method, being a global optimizer, effectively explores and utilizes the search space by combining stochastic and optimum solution searches. It obtains the globally optimal solution by utilizing the fluctuation of parameters and switching between motion modes.

2.3.3. Informer

Currently, short-term time-series forecasting has a restricted range and fails to identify trends over extended periods, making it less efficient in managing intricate system dynamics (see Figure 5). Conversely, long-term forecasting may discern trends, cyclical variations, seasonal oscillations, and enduring nonlinear patterns, making it more appropriate for multi-objective regional precision forecasting. This work employs a deep-learning model to construct an encoder–decoder architecture for addressing the LSTF issue. The flowchart illustrating this design can be seen in Figure 6. Below is the primary data of the model.

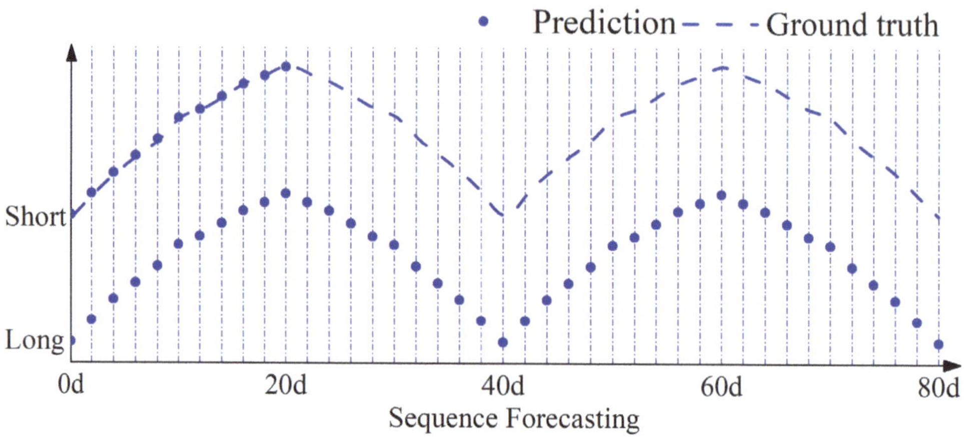

Figure 5. Informer can cover longer periods than short series predictions.

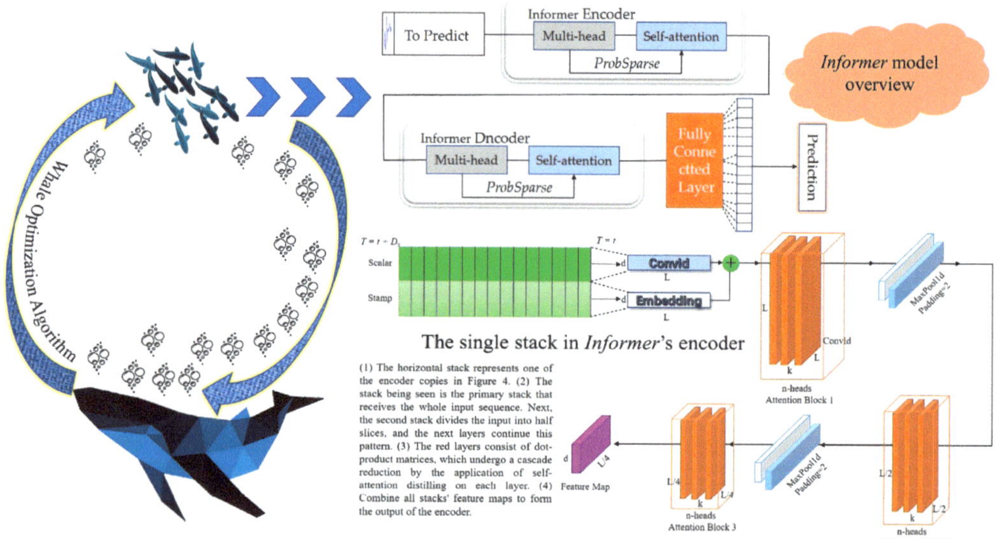

Figure 6. WOA–Informer model architecture (left: WOA model, top right: Informer workflow, bottom right: single stack of Informer's encoder).

The efficient self-attention technique is described using tuple inputs, namely inputs consisting of queries, keys, and values [43]. The attention of the interrogated individual is defined as a kernel smoother in probabilistic form, as shown in Equation:

$$A(Q,K,V) = \sum_j \frac{k(q_i,k_j)}{\sum_l k(q_i,k_l)} v_j = E_p(k_j|q_j)[v_j] \quad (21)$$

where $p(k_j|q_i) = k(q_i,k_j)/\sum_l k(q_i,k_l)$ and $k(q_i, k_j)$ selects the asymmetric exponential kernel $exp\left(q_i k_j{}^T/\sqrt{d}\right)$.

To handle the partial sparsity of the probability distribution in the self-attention mechanism, a sparse approach with high correlation is used to sparsify all $p(k_j \mid q_i)$. Additionally, each cell is designed to attend to the preceding cell in an exponentially rising way. In addition to the prior information, sparse alignment is included to address the issue of individual attention mechanisms [44].

ProbSparse self-focus detects the most significant crucial aspects:

$$A(Q,K,V) = Softmax\left(\frac{\overline{Q}K^T}{\sqrt{d}}\right)v \quad (22)$$

where $\overline{Q}$ is a sparse matrix that has the same dimension as q, and it only includes the Top-u queries based on the sparsity measurement $M(q,K)$.

Measurements of the highest average value:

$$\overline{M}(q_i,K) = max_j\left\{\frac{q_i k_i{}^T}{\sqrt{d}}\right\} - \frac{1}{L_K}\sum_{j=1}^{L_K} \frac{q_i k_i{}^T}{\sqrt{d}} \quad (23)$$

where q_i is the query length, k_i is the bond length, L is query and key length, $\overline{M}(q_i,K)$ is the i th query the sparsity measurement of q_i and key set K.

The encoder of the Informer model consists of two identical polytopic attention layers. These layers are specifically designed to capture highly correlated relationships between different elements of a long sequence [45]. They also prevent the prediction process from decaying too rapidly by dynamically decoding markers and using generations to expand in NLP:

$$X_{j+1}^t = MaxPool\left(ELU\left(Convld\left(\left[X_j^t\right]_{AB}\right)\right)\right) \quad (24)$$

where $[\cdot]_{AB}$ is Attention Block, $Conv1d(\cdot)$ is a 1D convolution filter applied in the time dimension, $ELU(\cdot)$ uses the ELU activation function, X_j^t is the list entries, and *Max-pool* is the maximum pooling layer.

Figure 6 provides a comprehensive explanation. Due to its inherent quality of prioritizing dominating features, the *ProbSparse* self-attention mechanism can effectively identify and focus on the most important elements, minimizing the input sequence as much as feasible [46]. The *ProbSparse* algorithm utilizes CNN techniques while undergoing the process of refinement:

$$X_{de}^t = Concat\left(X_{token}^t, X_0^t\right) \in \mathbb{R}^{(L_{token}+L_y)\times d_{model}} \quad (25)$$

where $X_{token}^t \in \mathbb{R}^{L_{token}\times d_{model}}$ is the start token, $X_o^t \in \mathbb{R}^{L_y\times d_{model}}$ is a placeholder for the target sequence (set scalar as 0), and *Concat* is the output size.

This research evaluates the effectiveness of this strategy by using four different loss functions: *MAE*, *MSE*, *RMSE*, and *MAPE*:

$$e_{MAE} = \frac{1}{n}\sum_{i=1}^{n}|y_i - \hat{y}_i| \quad (26)$$

$$e_{MSE} = \frac{1}{n}\sum_{i=1}^{n}|y_i - \hat{y}_i|^2 \tag{27}$$

$$e_{RMSE} = \sqrt{\frac{1}{n\sum_{i=1}^{n}(|y_i - \hat{y}_i|)^2}} \tag{28}$$

$$e_{MAPE} = \frac{1}{n}\sum_{i=1}^{n}|y_i - \hat{y}_i|/y_i \tag{29}$$

where y_i is the true value of the original data; $\hat{y}_i$ is the predicted value of the original data; n is the test sample size; and i is the test-sample point-sequence number.

The WOA–Informer model combines the WOA with the Informer model to provide automated hyperparameter tuning for time-series forecasting. WOA is used to identify and optimize the critical hyperparameters of the Informer model. Utilizing WOA's global search capacity, the WOA–Informer model can ascertain the ideal hyperparameter combinations for various time-series jobs, thereby enhancing prediction accuracy. WOA modifies the placements of the solutions according to these assessment outcomes, thereby converging towards the best hyperparameter configuration. The WOA–Informer model integrates the robust efficacy of the Informer for long-sequence time-series forecasting with the global optimization proficiency of WOA. The primary benefits are as follows: The model effectively manages long-sequence time-series forecasting by using Informer's sparse self-attention mechanism and hierarchical sampling techniques. WOA allows the WOA–Informer model to adjust to various time-series data formats, hence removing the intricacies of human hyperparameter tweaking and efficiently circumventing local optima. The model may autonomously optimize hyperparameters using WOA's exploration and exploitation processes, identifying the optimal configuration for certain tasks. This amalgamation greatly improves the model's predictive powers and flexibility.

2.4. *Experiment*

2.4.1. Can the WOA Enhance the Functionality and Performance of Informer?

We contend that the use of WOA enhances Informer's capacity to accurately capture the fluctuations in regional non-stationary vegetation drought indices, provided that there is an adequate amount of training data available. To substantiate this, we constructed Transformer, Informer, and WOA–Informer models for the nine sites in this study, utilizing TVDI, DSI, ESI, and TVPDI data from 1 January 2010, to 12 March 2023, (a total of 636 samples) as training input, and forecasting TVDI, DSI, ESI, and TVPDI data for the period from April 2023 to April 2024 for accuracy assessment.

2.4.2. The Need for WOA to Enhance the Informer Model

The linked WOA–Informer strategy is believed to be a viable solution for mitigating the issue of parameter adjustment in the Informer model. WOA employs a random selection process to update the location of the search agent and the best solution, enabling it to conduct a global search and prevent being trapped in a local optimum solution. Furthermore, the adaptive parameter-tuning method included in the WOA algorithm aids the Informer model in identifying the most suitable parameters for various datasets and tasks. The WOA algorithm replicates the feeding behaviors of humpback whales, including the confined encircling and spiral movement. By combining these two behaviors, WOA can effectively obtain the most ideal solution. Implementing this varied search technique may enhance the Informer model's ability to adjust to intricate data patterns. Additionally, it allows for the acquisition of a pre-trained model that is very near to the ideal answer while simultaneously lowering the time required for modeling. About the notion of zones for forecasting the changes in vegetation drought indices in dry and semi-dry regions, we developed three ways for dividing these areas into zones. The purpose was to confirm the importance of WOA in enhancing the Informer model. Therefore, we used distinct zoning strategies for each site to validate performance versus time cost. The research area's data volume was established as 172 months (688 samples from January 2010 to April 2024) to optimize the Informer model. This was done to identify the optimal parameters for various datasets and jobs to enhance the performance of the Informer model.

3. Results

3.1. Advantages of WOA for Optimizing Informer Model Parameters

Our research demonstrates that the Informer model significantly enhances the dynamic prediction of non-stationary data when enough long-time series data are available. The *ProbSparse* self-attention mechanism identifies Q-values with a more informational contribution based on the probabilistic sparsity assumption, hence enhancing the efficiency of long-series data processing and mitigating the internal bottleneck issue in the Transformer model. Simultaneously, the WOA method may dynamically modify the hyperparameters, allowing the Informer model to be adaptively calibrated in response to the fluctuating vegetation drought data, therefore nearing the global ideal solution. Nevertheless, in the context of regional multi-objective prediction, conventional optimization algorithms often do not attain the performance of the Informer model. Integrating WOA with the Informer model enhances prediction accuracy, diminishes time complexity, optimizes resource use, and markedly elevates overall efficiency and performance. In contrast to conventional long-time series models like Transformer and LSTM, Informer is specifically optimized for managing dynamic long-time series multi-objective data, thereby minimizing computational demands while enhancing the extraction of critical information through multi-scale information selection. The WOA method effectively decreases the computational load while preserving prediction accuracy by optimizing the network architecture in conjunction with the *ProbSparse* attention mechanism for handling extensive time-series data. The Informer model employs *ProbSparse* attention for the dynamic prediction of vegetation drought, enabling it to manage extensive time-series data. Additionally, WOA aids the Informer model in identifying intricate non-stationary data characteristics, hence improving the regional model's generalizability. Furthermore, the WOA–Informer model exhibits enhanced robustness while managing non-smooth and variable data. The primary reason is that although the Informer model has resilience in handling non-stationary data, it remains susceptible to local optimum solutions in long-term time-series forecasting, while WOA offers superior global optimization capabilities. Despite Informer's superior predictive capability with intricate dynamic time-series data, a degree of uncertainty persists. WOA may enhance the performance of Informer by delivering more consistent prediction intervals and improved uncertainty estimates. We use WOA with the Informer model, applicable for migration models in multi-task learning, to enhance the model's generalizability. The WOA-optimized Informer architecture enhances the model's capacity for transfer learning and mitigates the generalization deficiencies seen in previous models across other domains. The WOA–Informer model demonstrates a substantial increase in prediction accuracy, with an average MSE of just 0.011 ± 0.012 over many predictions (refer to Figure 7). The integration of WOA and Informer effectively optimizes model parameters, enhances prediction accuracy, manages intricate multi-objective time-series data, and bolsters model robustness while also augmenting the model's generalization capacity for large-scale time-series challenges in complex settings, including climate forecasting, water-quality assessment, and vegetation-health evaluation.

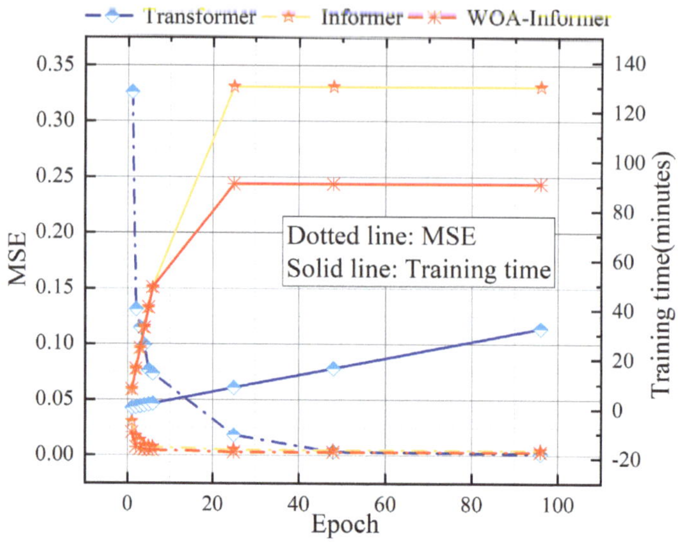

Figure 7. Precision and total running time of the training phase (the dotted line and solid line of the same hue represent the same model).

3.2. Effect of WOA on Informer Model Efficacy

The WOA algorithm markedly enhanced the dynamic prediction ability of the vegetative-drought TVDI index over an extensive time series, exhibiting substantial changes (the trend changes of NDVI, LST, TVDI, TVPDI, DSI, and ESI index from 2010 to 2023 are shown in Figure 8). Nevertheless, in the training of regional multi-objective data, attaining good long-term time-series predictions with modest parameter tweaks to the Informer deep-learning model becomes challenging. The mean MSE value for the 16 locations was 0.174 ± 0.031. Nonetheless, the Informer model optimized using the WOA method consistently surpassed the conventional Transformer model ($p < 0.001$). In contrast to Transformer, which decodes sequences incrementally, Informer uses generative decoding to forecast the full sequence in a single forward propagation. This significantly enhances the inference speed of lengthy sequence predictions and mitigates the prevalent cumulative error issue in the stepwise decoding process. In the vegetation-drought-prediction research, the WOA–Informer model showed superior predictive performance in the study region, with an average MSE of 0.0012 over 16 monitoring stations, whereas the Transformer exhibited the worst performance, with an average MAE of 0.16 (Figure 9). The WOA–Informer model has shown many significant benefits for multivariate predictions: (1) In comparison to the conventional Transformer, the Informer markedly diminishes time complexity and memory consumption, thereby enhancing its efficiency in managing long series prediction tasks; (2) the model is adept at processing ultra-long input sequences, which is essential for the long-term prediction of vegetation drought; (3) Informer substantially enhances predictive capability by proficiently capturing long-range dependencies, demonstrating exceptional performance; (4) the WOA algorithm exhibits robust global search capabilities, preventing the model from becoming trapped in local optima and facilitating the discovery of superior solutions within an expansive search space; (5) the WOA algorithm also possesses rapid convergence during the optimization process, enabling the model to swiftly identify optimal or near-optimal hyperparameters, thereby reducing training duration.

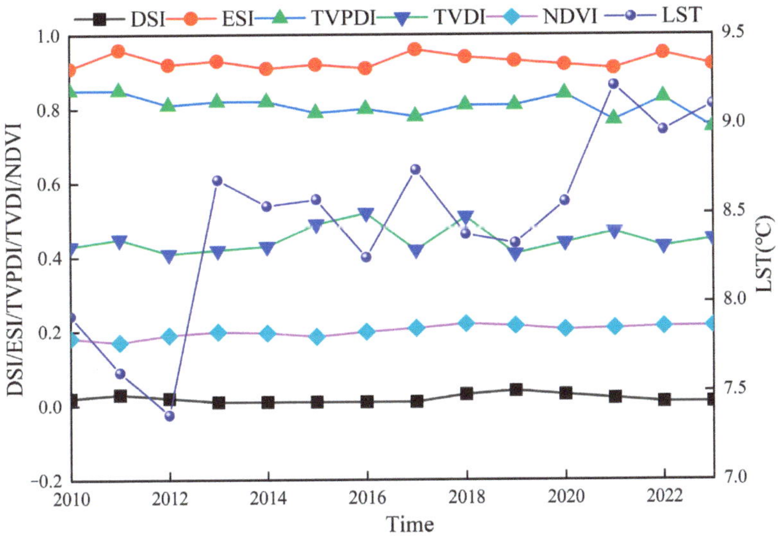

Figure 8. Comparative trends of NDVI, LST, TVDI, TVPDI, DSI, and ESI indices (2010–2023).

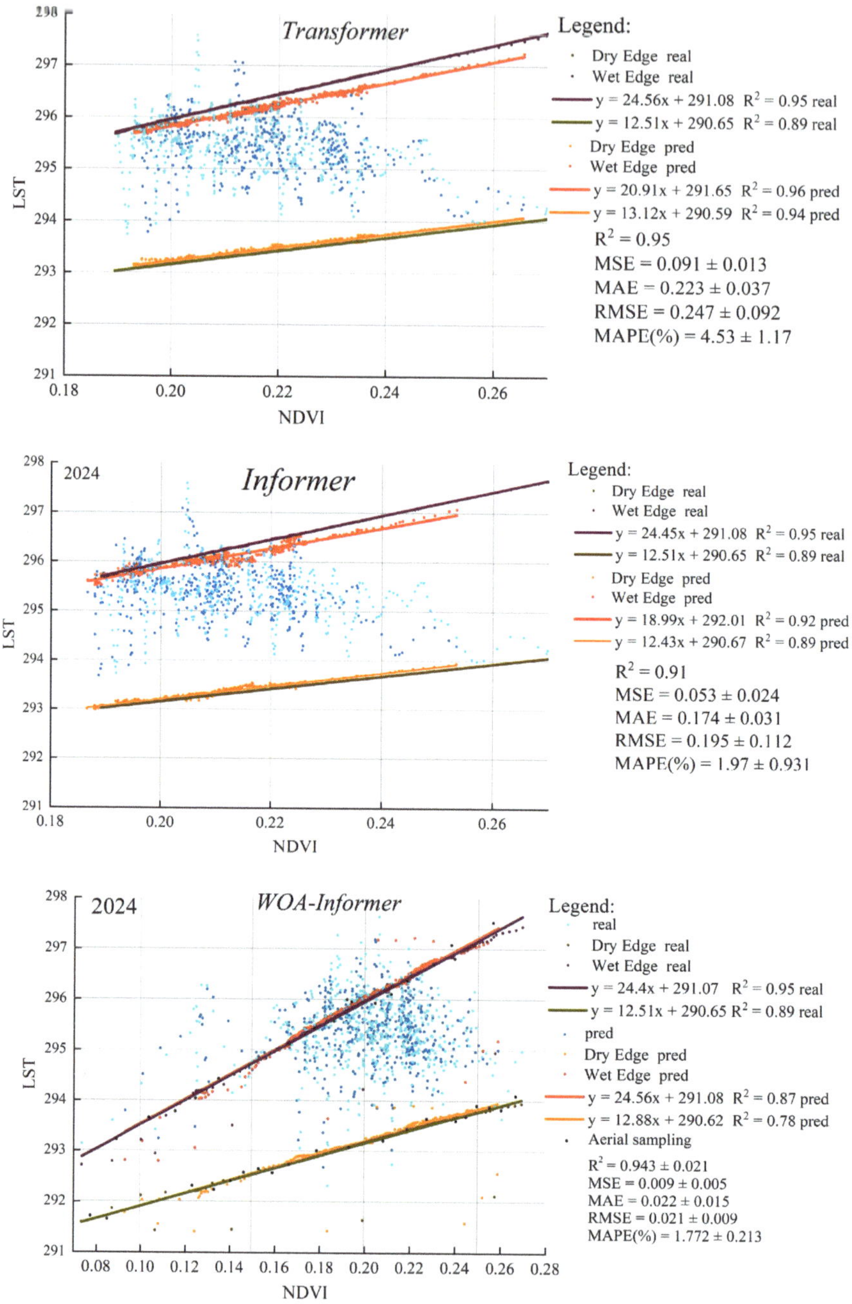

Figure 9. Comparison of prediction accuracy for Transformer, Informer, and WOA–Informer models at Site 1 (April 2023–April 2024).

The research revealed that the Transformer model struggles to accurately capture the intricate relationships among multivariate variables, exhibiting an initial loss of 0.35, but the Informer and WOA–Informer models demonstrate a more rapid reduction in loss throughout training. The WOA algorithm and the sparse-attention mechanism effectively manage non-stationary time-series data, proving particularly useful for multi-scale dynamic information processing in extended time series. The WOA–Informer model requires just 13 epochs to achieve an acceptable loss level, demonstrating rapid convergence during training and proficiency in handling the intricacies of non-smooth time-series data.

3.3. Benefits of the WOA–Informer Model in Forecasting Extended Time Series of Dry Vegetation Indicators

We examine the efficacy of the WOA–Informer model in forecasting the vegetation drought index within the framework of multi-objective optimization for extensive time-series data. Various variables may directly or indirectly influence the performance, training efficiency, and final generalization capability of a neural network model throughout the training process. The criteria include learning rate, network weights, loss function, over-fitting and underfitting, batch size, and activation function (Table 1). The learning rate dictates the magnitude of weight updates in each gradient-descent iteration, the weight-initialization method influences convergence speed and model performance, the batch size impacts training efficiency and generalization capability, and the activation function establishes the level of nonlinearity in the neuron output. Consequently, effective regulation of these elements is essential to enhance the convergence rate, performance, and generalization capacity of the model. To avert the emergence of overfitting or underfitting, we include the Early Stopping module in the WOA–Informer model to oversee the validation set's performance. Nevertheless, in the context of multi-objective data, the WOA–Informer model demonstrates an enhanced capacity for information acquisition, hence augmenting its predictive efficacy. The fundamental mathematical concepts include the following: (1) Constricting Enclosure and Spiral Method: By reducing the proximity to the present ideal solution and meticulously refining the location using the spiral method, it ultimately conducts a global search by randomly deviating from the current optimal solution. Sparse-Attention Mechanism: Enhance model efficiency by reducing computational requirements. Sequence-distillation operation: Incrementally decrease the sequence length to minimize memory and computational burden. Generative decoder: Enhance the inference velocity and mitigate error accumulation. As the volume of source-domain data used to train the WOA–Informer model escalates, the model's predictive accuracy is markedly enhanced. Nonetheless, when the quantity of regional targets escalates to 16, the model's predictive accuracy seems to peak, encountering its growth limitation. This indicates that the efficacy of the WOA–Informer model may be constrained by the computing capacity of the sparse-attention mechanism. Consequently, considering computational efficiency and model performance, we find it fair and suitable to establish the number of targets in the research area at 16.

Table 1. WOA–Informer model hyperparameter.

	Learning Rate	Overfitting	Batch Size	Activation Function
Transformer	0.0001	may	128	Relu
Informer	0.001	no	128	Gelu
WOA–Informer	0.001	no	128	Gelu

3.4. Prolonged Observation and Examination of Vegetation Dynamics and Their Response to Drought

This work analyzes the weight correlations of multi-objective vegetation-drought indicators (TVDI, TVPDI, DSI, and ESI) over an extended time period from 2010 to 2024 to improve the precision of regional forecasts prior to regional forecasting. The findings indicate that the WOA–Informer model, using a regionalized methodology, has considerable superiority compared to locally trained deep-learning models. In the WOA–Informer model, we delineate the weight connection among the goals by formulating a loss function for multi-objective optimization and assessing their interactions (Figure 10). In comparison to the locally trained combinatorial model, the WOA–Informer model excels at capturing multi-objective synergistic interactions. Moreover, the WOA–Informer model is better suited for predicting analogous areas, owing to the simplicity of its parameters, which significantly enhances the model's generalizability. This methodology examines the interconnections among several goals from a broader viewpoint, offers a more appropriate resolution for source domains, and aids in addressing the difficulties of regional forecasting and real-time assessment of vegetation drought.

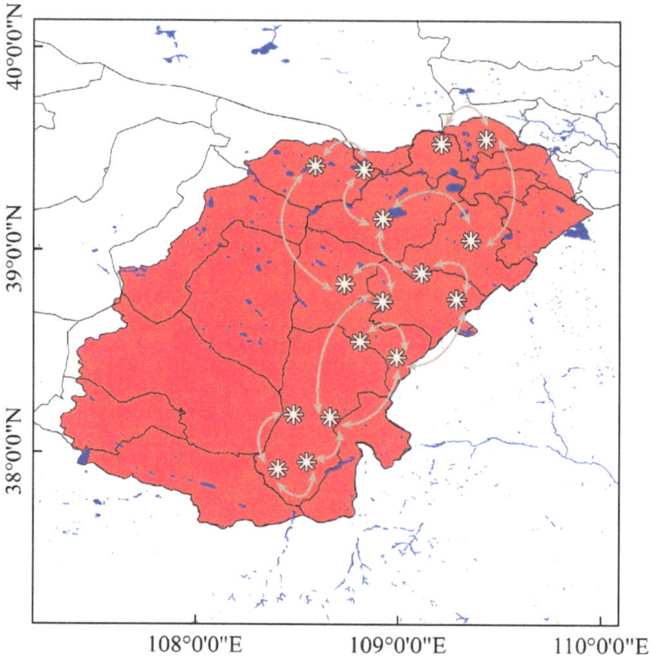

Figure 10. Multi-objective Optimization Loss Function: Capturing weight relationships and mutual influences between objectives.

3.5. WOA–Informer: Precise Long-Term Dynamic Sequence Forecasting

We picked the vegetation-drought index data spanning from 1 January 2010, to 18 April 2024, including 174 months, for multi-objective regional prediction. The training set comprises data from 1 January 2010, to 18 April 2023, while the test set includes data from 19 April 2023, to 22 April 2024. To evaluate the WOA–Informer model for the multi-objective parameter-optimization issue, minimal hyper-parameter adjustment was conducted, resulting in final hyper-parameter settings of a learning rate of 0.001, 20 iterations, and a batch size of 128. The WOA–Informer model's prediction results for the test set indicate the following error metrics: MAE $\leq$ 0.05, MSE $\leq$ 0.001, MSPE $\leq$ 0.01, and MAPE < 5% (Figure 11). The research demonstrates that the WOA–Informer model has great accuracy in regional long-term time-series forecasting, with errors remaining constant and exhibiting little fluctuation, making it particularly appropriate for extensive dry regions. A reduced learning rate with fewer iterations successfully mitigates overfitting and strengthens the model's generalization capability, while an increased batch size expedites model training and further bolsters model stability. The multi-objective prediction outcomes for extended time series indicate that regional multi-objective prediction may attain high accuracy and demonstrate significant applicability and universality.

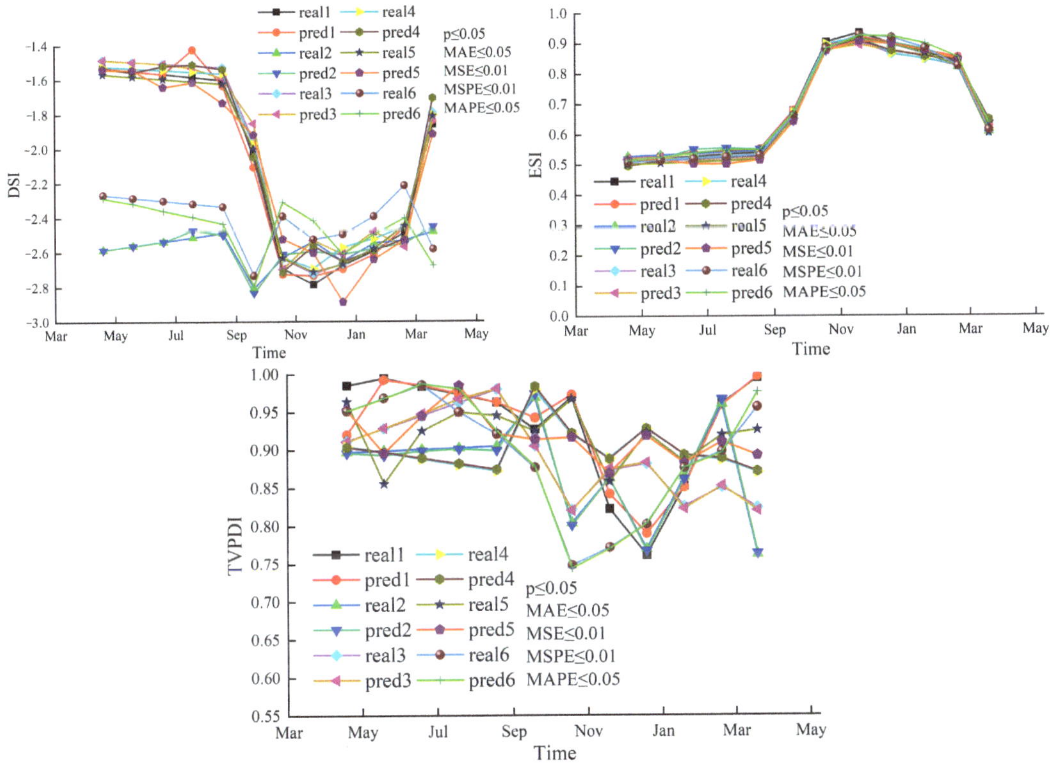

Figure 11. Forecast comparison of DSI, ESI, and TVPDI (April 2023–April 2024). Multi-objective prediction optimization for six sites: real (real value) 1–6 vs. pred (predicted value) 1–6.

4. Discussion

4.1. The Significance of Integrating WOA and Informer Models in Long-Term Time-Series Forecasting

This study integrates WOA with the Informer model, which incorporates a sparse self-attention mechanism, to introduce a novel modeling approach for regional multi-objective forecasting of the vegetation-drought index. We employ the Bionic Population Intelligent Optimization technique to enhance the hyperparameters of the Informer model. WOA effectively identifies the ideal hyperparameter combination, and when integrated with the *ProbSparse* attention mechanism, it accurately predicts lengthy time series in multi-objective forecasting [47]. The findings indicate that, in contrast to deep-learning model-optimization methods utilizing modal-decomposition techniques, WOA demonstrates superior adaptability in managing nonlinear, irregular, and complex data, and it is proficient in addressing single-objective, multi-objective, and constrained optimization challenges [48]. Furthermore, WOA possesses robust global search capabilities, successfully circumventing local optimization pitfalls, whereas modal-decomposition techniques frequently succumb to local optimization challenges when addressing complex non-convex or multi-peak structures [49]. When the training data are enough, WOA can replicate the "bubble net feeding" behavior of whales, together with appropriate configurations of population size and iteration count, yielding more optimal optimization outcomes [50]. Brodzicki, A. et al. demonstrated that the WOA-optimization algorithm can expedite the coupling process in high-dimensional search spaces by employing distributed or parallel computation via the population search mechanism [51]. Additionally, Ji, C. et al. discovered that WOA can dynamically modify its search strategy through the adaptive hunting behavior of whales, allowing it to respond to variations in the objective function over extended time-series data, thereby enhancing optimization efficacy [52]. The WOA–Informer model introduced in this paper exhibits robust regional predictive proficiency in the long-term multi-objective forecasting job. The model offers a more precise instrument for predicting the vegetation-aridity index in arid and semi-arid regions, addressing the information deficit in global vegetation predictions for these areas [53]. The model's training duration escalates with an increase in the number of regional multi-targets. The WOA–Informer model can enhance prediction accuracy by fine-tuning the pre-trained model.

4.2. Influence of Prolonged Regional Drought-Index Forecasts

Thus far, limited research has been undertaken on the selection of deep-learning models for the implementation of long time series to forecast the dynamic information of the vegetation-drought index. In 2021, Sanjoy Chakraborty et al. integrated the SHADE and WOA algorithms to address global optimization challenges within evolutionary optimization techniques, assessing the efficacy of the new algorithms in resolving real-world issues through two unconstrained and four constrained engineering design problems [54]. In 2020, Eslam M. Hassib et al. evaluated the efficacy of novel algorithms for addressing real-world challenges by employing BRNN-based models alongside the WOA-optimization algorithm to analyze large datasets [55]. They compared GWO-MLP with Particle Swarm Optimization (PSO-MLP), Genetic Algorithm (GA-MLP), Ant Colony Optimization (ACO-MLP), Evolutionary Strategies (ES-MLP), and Population-Based Incremental Learning (PBIL-MLP), demonstrating that the combination of WOA and BRNN achieves commendable accuracy and a high rate of local optimal avoidance [56]. The conventional Transformer model exhibits a computational complexity of $O(L^2)$ per layer and may provide predictions from extensive time-series datasets. However, this typically entails substantial computational expense and memory usage [57]. The Transformer model incorporates a self-attentive mechanism. Nonetheless, its performance deteriorates due to sequence overfields [58]. Consequently, for regional forecasting utilizing extensive datasets, it is imperative to assess the model's generalizability during pre-training and furnish a precise instrument for predicting the drought dynamics of vegetation in the designated arid and semi-arid domains by amalgamating regional predictions and establishing the weighting relationships among the targets within this area.

4.3. Influence of Multi-Objective Optimization on WOA–Informer Model Efficacy

This study demonstrates that enhancing the performance of the WOA–Informer model is achievable by augmenting the number of regional multi-targets. Our analysis revealed that the performance of the WOA–Informer model significantly improved while training due to the use of extensive time series of multi-target data, particularly when the number of monitoring points surpassed 10. Nonetheless, when the quantity of monitoring points attains 16, the model's performance appears to approach its maximum development potential. This indicates the existence of a multi-objective threshold for WOA–Informer to execute a global search throughout the modeling process. This may result from the model's susceptibility to local optima due to data complexity and high dimensionality [59]. Moreover, we discovered that WOA is susceptible to parameter configurations, and excessive tuning of these parameters may impact its performance and convergence rate. Consequently, determining whether excessive data input or a moderate amount suffices while utilizing the WOA–Informer model is a critical consideration for modeling efficiency [60]. Upon comparing the performance of the Transformer, Informer, and WOA–Informer models in the context of multi-objective optimization, it is evident that the prediction accuracy of WOA–Informer models surpasses that of the other models significantly [61]. The comprehensive acquisition of extensive time-series data allows the WOA–Informer model to more effectively manage the equilibrium among several objectives [62]. Furthermore, the WOA–Informer model enhances computing performance, reduces the number of parameters, and decreases model complexity, hence streamlining the parameter-tweaking process of the Informer model [63]. The study revealed that a reduction in monitoring points leads to a 30% to 42% decline in the performance of the WOA–Informer model. Nonetheless, it still outperforms the Transformer and Informer models despite this degradation [64]. Nonetheless, when sufficient monitoring points are present, the performance disparity between the WOA–Informer and Informer models is markedly diminished, mostly due to the *ProbSparse* attention mechanism and Hierarchical Time Embedding utilized in the Informer model [65]. This demonstrates that WOA is proficient in enhancing the Informer model, particularly for regional monitoring purposes. The integration of WOA and Informer can proficiently address the prediction issues associated with drought prone vegetation areas and efficiently tackle the challenge of forecasting lengthy time-series data.

4.4. Limitations and Prospects

The primary objective of this study is to tackle the critical challenges in multi-objective region prediction, particularly the generalizability and parameter complexity of deep-learning models managing extensive time-series data. While our suggested WOA–Informer model exhibits a marginally reduced training speed relative to the Informer model, its parameter-optimization procedure is more straightforward and can more rapidly identify the best global solution [66]. Subsequent studies may enhance this training velocity. For instance, migration learning can be employed to explore the spatial and physical determinants of human activities, climate change, and regional forecasting to identify the principal elements influencing the vegetation-drought index [67]. The incorporation of reinforcement learning can improve the generalization capacity of deep-learning models in extended time series, hence addressing continuous decision-making challenges. WOA necessitates numerous iterations throughout its search process, and in the context of intricate high-dimensional, non-convex problems, it may encounter early convergence, resulting in the failure to identify the global optimum [68]. The WOA–Informer model necessitates great data quality and is particularly susceptible to outliers, which can compromise the accuracy of its predictions. The model has inadequate sensitivity in managing extremely non-stationary data, hindering its ability to accurately capture the intricate variations within the data. Another problem is the optimization of hyperparameters, such as population size, maximum iterations, number of attention heads, window size, and learning rate, which generally necessitates considerable time and domain expertise. While the WOA–Informer model has considerable benefits in areas like water resource management and environmental time-series forecasting, its utilization in real-time or online prediction contexts is constrained by substantial computational resource requirements. The implementation of the differential evolution mechanism can effectively circumvent local optima by directing the search trajectory based on the disparities between solutions, hence offering a more resilient modeling strategy for regional forecasting. These enhancements and expansions will considerably augment the efficacy and applicability of our system in predicting the regional vegetation-drought index.

5. Conclusions

This paper presents a high-performance, regional, multi-target prediction model that integrates WOA with the Informer model to accurately forecast the vegetation-aridity index over an extended time series in arid and semi-arid locations. We assessed the model's performance at 16 aerial photography test locations in the hinterland of the Mu Us Sandy Land region in the Inner Mongolia Autonomous Region. Following the pre-training of the research area, we implemented the proposed WOA–Informer model in the target region, primarily utilizing the meta-heuristic optimization technique to augment its global search efficacy and significantly diminish the computational cost associated with the conventional self-attention mechanism. Experimental results indicate that the model's prediction accuracy markedly surpasses that of the conventional Transformer model, while the complexity of the parameter adjustment is much diminished in comparison to the Informer model. The research demonstrates that the WOA-optimization technique significantly enhances the performance of the Informer model. In comparison to the Informer model alone, the coupled model demonstrates superior performance in addressing the intricate nonlinear, regional, multi-objective prediction of the vegetation-drought index, exhibiting effective solution capabilities. This indicates that our suggested model can effectively address the challenging issue of vegetation monitoring and prediction in arid and semi-arid regions while also offering an optimum approach for attaining precise regional predictions globally in the future.

Author Contributions: The contributions of C.T., H.H., and Y.W. involved in designing the manuscript; C.T., H.H., and H.Z. carried out this experiment; H.H., H.Z., and J.L. analyzed the data and wrote the manuscript. All authors have read and agreed to the published version of the manuscript.

Funding: This research was funded by Ordos Water Science and Technology Project (MK20210222); Ordos Irrigation Experiment Station 2024 mission (MKGP2024ZC046); Key Project of "Science and Technology to Prosper Mongolia" Action (2021EEDSCXSFQZD010).

Data Availability Statement: The data presented in this study are available on request from the corresponding author. The data are not publicly available due to privacy.

Conflicts of Interest: The authors declare no conflict of interest.

References

1. Jha, S.; Das, J.; Sharma, A.; Hazra, B.; Goyal, M.K. Probabilistic evaluation of vegetation drought likelihood and its implications to resilience across India. *Glob. Planet. Chang.* **2019**, *176*, 23–35.
2. Rousta, I.; Olafsson, H.; Moniruzzaman, M.; Zhang, H.; Liou, Y.A.; Mushore, T.D.; Gupta, A. Impacts of drought on vegetation assessed by vegetation indices and meteorological factors in Afghanistan. *Remote Sens.* **2020**, *12*, 2433. [CrossRef]
3. Ding, Y.; Xu, J.; Wang, X.; Peng, X.; Cai, H. Spatial and temporal effects of drought on Chinese vegetation under different coverage levels. *Sci. Total Environ.* **2020**, *716*, 137166. [PubMed]
4. Han, H.; Bai, J.; Yan, J.; Yang, H.; Ma, G. A combined drought monitoring index based on multi-sensor remote sensing data and machine learning. *Geocarto Int.* **2021**, *36*, 1161–1177.
5. Piri, J.; Abdolahipour, M.; Keshtegar, B. Advanced machine learning model for prediction of drought indices using hybrid SVR-RSM. *Water Resour. Manag.* **2023**, *37*, 683–712.
6. Roy, B.; Sagan, V.; Haireti, A.; Newcomb, M.; Tuberosa, R.; LeBauer, D.; Shakoor, N. Early Detection of Drought Stress in Durum Wheat Using Hyperspectral Imaging and Photosystem Sensing. *Remote Sens.* **2023**, *16*, 155. [CrossRef]
7. Feng, P.; Wang, B.; Li Liu, D.; Yu, Q. Machine learning-based integration of remotely-sensed drought factors can improve the estimation of agricultural drought in South-Eastern Australia. *Agric. Syst.* **2019**, *173*, 303–316.
8. Aghelpour, P.; Bahrami-Pichaghchi, H.; Varshavian, V. Hydrological drought forecasting using multi-scalar streamflow drought index, stochastic models and machine learning approaches, in northern Iran. *Stoch. Environ. Res. Risk Assess.* **2021**, *35*, 1615–1635. [CrossRef]
9. Kafy, A.A.; Bakshi, A.; Saha, M.; Al Faisal, A.; Almulhim, A.I.; Rahaman, Z.A.; Mohammad, P. Assessment and prediction of index based agricultural drought vulnerability using machine learning algorithms. *Sci. Total Environ.* **2023**, *867*, 161394.
10. Tyagi, S.; Zhang, X.; Saraswat, D.; Sahany, S.; Mishra, S.K.; Niyogi, D. Flash drought: Review of concept, prediction and the potential for machine learning, deep learning methods. *Earth's Future* **2022**, *10*, e2022EF002723.
11. Zhou, H.; Zhang, S.; Peng, J.; Zhang, S.; Li, J.; Xiong, H.; Zhang, W. Informer: Beyond efficient transformer for long sequence time-series forecasting. In Proceedings of the AAAI Conference on Artificial Intelligence, Virtually, 2–9 February 2021; Volume 35, pp. 11106–11115.
12. Zhang, J. Gradient descent based optimization algorithms for deep learning models training. *arXiv* **2019**, arXiv:1903.03614.
13. Li, M.W.; Xu, D.Y.; Geng, J.; Hong, W.C. A ship motion forecasting approach based on empirical mode decomposition method hybrid deep learning network and quantum butterfly optimization algorithm. *Nonlinear Dyn.* **2022**, *107*, 2447–2467.
14. Jadhav, A.S.; Patil, P.B.; Biradar, S. Optimal feature selection-based diabetic retinopathy detection using improved rider optimization algorithm enabled with deep learning. *Evol. Intell.* **2021**, *14*, 1431–1448. [CrossRef]
15. Shrestha, A.; Mahmood, A. Review of deep learning algorithms and architectures. *IEEE Access* **2019**, *7*, 53040–53065.
16. Mirjalili, S.; Mirjalili, S. Genetic algorithm. In *Evolutionary Algorithms and Neural Networks: Theory and Applications*; Springer: Berlin/Heidelberg, Germany, 2019; pp. 43–55.
17. Slowik, A.; Kwasnicka, H. Evolutionary algorithms and their applications to engineering problems. *Neural Comput. Appl.* **2020**, *32*, 12363–12379.
18. Sharma AB HI SH, E.K.; Sharma, A.; Choudhary, S.; Pachauri, R.K.; Shrivastava, A.; Kumar, D. A review on artificial bee colony and it's engineering applications. *J. Crit. Rev.* **2020**, *7*, 4097–4107.
19. Shami, T.M.; El-Saleh, A.A.; Alswaitti, M.; Al-Tashi, Q.; Summakieh, M.A.; Mirjalili, S. Particle swarm optimization: A comprehensive survey. *IEEE Access* **2022**, *10*, 10031–10061.

20. Kennedy, J.; Eberhart, R. Particle swarm optimization. In Proceedings of the ICNN'95-International Conference on Neural Networks, Perth, WA, Australia, 27 November–1 December 1995; Volume 4, pp. 1942–1948.
21. Wang, D.; Tan, D.; Liu, L. Particle swarm optimization algorithm: An overview. *Soft Comput.* **2018**, *22*, 387–408.
22. Dorigo, M.; Birattari, M.; Stutzle, T. Ant colony optimization. *IEEE Comput. Intell. Mag.* **2006**, *1*, 28–39.
23. Couceiro, M.; Ghamisi, P.; Couceiro, M.; Ghamisi, P. *Particle Swarm Optimization*; Springer International Publishing: Berlin/Heidelberg, Germany, 2016; pp. 1–10.
24. Blum, C. Ant colony optimization: Introduction and recent trends. *Phys. Life Rev.* **2005**, *2*, 353–373.
25. Fidanova, S.; Fidanova, S. Ant colony optimization. In *Ant Colony Optimization and Applications*; Springer: Berlin/Heidelberg, Germany, 2021; pp. 3–8.
26. Mirjalili, S.; Lewis, A. The whale optimization algorithm. *Adv. Eng. Softw.* **2016**, *95*, 51–67.
27. Fan, Q.; Yu, F.; Xuan, M. Transformer fault diagnosis method based on improved whale optimization algorithm to optimize support vector machine. *Energy Rep.* **2021**, *7*, 856–866.
28. Hou, H.; Li, R.; Zheng, H.; Tong, C.; Wang, J.; Lu, H.; Wang, G.; Qin, Z.; Wang, W. Regional NDVI Attribution Analysis and Trend Prediction Based on the Informer Model: A Case Study of the Maowusu Sandland. *Agronomy* **2023**, *13*, 2882. [CrossRef]
29. Zheng, Y.; Dong, L.; Xia, Q.; Liang, C.; Wang, L.; Shao, Y. Effects of revegetation on climate in the Mu Us Sandy Land of China. *Sci. Total Environ.* **2020**, *739*, 139958. [PubMed]
30. Ji, X.; Yang, J.; Liu, J.; Du, X.; Zhang, W.; Liu, J.; Li, G.; Guo, J. Analysis of Spatial-Temporal Changes and Driving Forces of Desertification in the Mu Us Sandy Land from 1991 to 2021. *Sustainability* **2023**, *15*, 10399. [CrossRef]
31. Moravec, D.; Komárek, J.; López-Cuervo Medina, S.; Molina, I. Effect of atmospheric corrections on NDVI: Intercomparability of Landsat 8, Sentinel-2, and UAV sensors. *Remote Sens.* **2021**, *13*, 3550. [CrossRef]
32. Bian, Z.; Roujean, J.L.; Fan, T.; Dong, Y.; Hu, T.; Cao, B.; Li, H.; Du, Y.; Xiao, Q.; Liu, Q. An angular normalization method for temperature vegetation dryness index (TVDI) in monitoring agricultural drought. *Remote Sens. Environ.* **2023**, *284*, 113330.
33. Khan, R.; Gilani, H. Global drought monitoring with drought severity index (DSI) using Google Earth Engine. *Theor. Appl. Climatol.* **2021**, *146*, 411–427. [CrossRef]
34. Yoon, D.H.; Nam, W.H.; Lee, H.J.; Hong, E.M.; Kim, T. Drought hazard assessment using MODIS-based Evaporative Stress Index (ESI) and ROC analysis. *J. Korean Soc. Agric. Eng.* **2020**, *62*, 51–61.
35. Rana, N.; Latiff MS, A.; Abdulhamid SI, M.; Chiroma, H. Whale optimization algorithm: A systematic review of contemporary applications, modifications and developments. *Neural Comput. Appl.* **2020**, *32*, 16245–16277.
36. Yang, W.; Xia, K.; Fan, S.; Wang, L.; Li, T.; Zhang, J.; Feng, Y. A multi-strategy whale optimization algorithm and its application. *Eng. Appl. Artif. Intell.* **2022**, *108*, 104558. [CrossRef]
37. Liu, L.; Zhang, R. Multistrategy improved whale optimization algorithm and its application. *Comput. Intell. Neurosci.* **2022**, *2022*, 3418269. [CrossRef] [PubMed]
38. Hemasian-Etefagh, F.; Safi-Esfahani, F. Group-based whale optimization algorithm. *Soft Comput.* **2020**, *24*, 3647–3673. [CrossRef]
39. Hussien, A.G.; Hassanien, A.E.; Houssein, E.H.; Amin, M.; Azar, A.T. New binary whale optimization algorithm for discrete optimization problems. *Eng. Optim.* **2020**, *52*, 945–959. [CrossRef]
40. Deng, H.; Liu, L.; Fang, J.; Qu, B.; Huang, Q. A novel improved whale optimization algorithm for optimization problems with multi-strategy and hybrid algorithm. *Math. Comput. Simul.* **2023**, *205*, 794–817. [CrossRef]
41. Yan, Z.; Zhang, J.; Zeng, J.; Tang, J. Nature-inspired approach: An enhanced whale optimization algorithm for global optimization. *Math. Comput. Simul.* **2021**, *185*, 17–46. [CrossRef]
42. Chakraborty, S.; Saha, A.K.; Sharma, S.; Chakraborty, R.; Debnath, S. A hybrid whale optimization algorithm for global optimization. *J. Ambient. Intell. Humaniz. Comput.* **2023**, *14*, 431–467. [CrossRef]
43. Wang, H.K.; Song, K.; Cheng, Y. A hybrid forecasting model based on CNN and informer for short-term wind power. *Front. Energy Res.* **2022**, *9*, 788320. [CrossRef]
44. Zheng, H.; Hou, H.; Li, R.; Tong, C. Trend Prediction of Vegetation and Drought by Informer Model Based on STL-EMD Decomposition of Ha Cai Tou Dang Water Source Area in the Maowusu Sandland. *Agronomy* **2024**, *14*, 708. [CrossRef]
45. Wei, H.; Wang, W.S.; Kao, X.X. A novel approach to ultra-short-term wind power prediction based on feature engineering and informer. *Energy Rep.* **2023**, *9*, 1236–1250. [CrossRef]
46. Jiang, C.; Zhu, Q. Evaluating the most significant input parameters for forecasting global solar radiation of different sequences based on Informer. *Appl. Energy* **2023**, *348*, 121544. [CrossRef]
47. Xinxin, W.; Xiaopan, S.; Xueyi, A.; Shijia, L. Short-term wind speed forecasting based on a hybrid model of ICEEMDAN, MFE, LSTM and informer. *PLoS ONE* **2023**, *18*, e0289161. [CrossRef] [PubMed]
48. Zhou, J.; Zhu, S.; Qiu, Y.; Armaghani, D.J.; Zhou, A.; Yong, W. Predicting tunnel squeezing using support vector machine optimized by whale optimization algorithm. *Acta Geotech.* **2022**, *17*, 1343–1366. [CrossRef]
49. Abd Elaziz, M.; Lu, S.; He, S. A multi-leader whale optimization algorithm for global optimization and image segmentation. *Expert Syst. Appl.* **2021**, *175*, 114841. [CrossRef]
50. Hussain, N.; Khan, M.A.; Kadry, S.; Tariq, U.; Mostafa, R.R.; Choi, J.I.; Nam, Y. Intelligent deep learning and improved whale optimization algorithm based framework for object recognition. *Hum. Cent. Comput. Inf. Sci.* **2021**, *11*, 2021.
51. Brodzicki, A.; Piekarski, M.; Jaworek-Korjakowska, J. The whale optimization algorithm approach for deep neural networks. *Sensors* **2021**, *21*, 8003. [CrossRef]

52. Ji, C.; Zhang, C.; Hua, L.; Ma, H.; Nazir, M.S.; Peng, T. A multi-scale evolutionary deep learning model based on CEEMDAN improved whale optimization algorithm, regularized extreme learning machine and LSTM for AQI prediction. *Environ. Res.* **2022**, *215*, 114228. [CrossRef]
53. Hu, Q.; Hu, H.X.; Lin, Z.Z.; Chen, Z.H.; Zhang, Y. A decision-making method for reservoir operation schemes based on deep learning and whale optimization algorithm. *Front. Plant Sci.* **2023**, *14*, 1102855. [CrossRef]
54. Chakraborty, S.; Sharma, S.; Saha, A.K.; Chakraborty, S. SHADE–WOA: A metaheuristic algorithm for global optimization. *Appl. Soft Comput.* **2021**, *113*, 107866. [CrossRef]
55. Hassib, E.M.; El-Desouky, A.I.; Labib, L.M.; El-Kenawy, E.S.M. WOA+ BRNN: An imbalanced big data classification framework using Whale optimization and deep neural network. *Soft Comput.* **2020**, *24*, 5573–5592. [CrossRef]
56. Yang, P.; Wang, T.; Yang, H.; Meng, C.; Zhang, H.; Cheng, L. The performance of electronic current transformer fault diagnosis model: Using an improved whale optimization algorithm and RBF neural network. *Electronics* **2023**, *12*, 1066. [CrossRef]
57. Toren, M. Optimization of transformer parameters at distribution and power levels with hybrid Grey wolf-whale optimization algorithm. *Eng. Sci. Technol. Int. J.* **2023**, *43*, 101439.
58. Ibrahim, A.; El-kenawy ES, M.; Khodadadi, N.; Eid, M.M.; Abdelhamid, A.A. Guided whale optimization algorithm (guided WOA) with its application. In *Handbook of Whale Optimization Algorithm*; Academic Press: Cambridge, MA, USA, 2024; pp. 243–251.
59. Ye, H.; Zhu, Q.; Zhang, X. Short-Term Load Forecasting for Residential Buildings Based on Multivariate Variational Mode Decomposition and Temporal Fusion Transformer. *Energies* **2024**, *17*, 3061. [CrossRef]
60. Zhou, Y.; Yang, X.; Tao, L.; Yang, L. Transformer fault diagnosis model based on improved gray wolf optimizer and probabilistic neural network. *Energies* **2021**, *14*, 3029. [CrossRef]
61. Wu, Z.; Zhang, Z.; Zheng, L.; Yan, T.; Tang, C. The Denoising Method for Transformer Partial Discharge Based on the Whale VMD Algorithm Combined with Adaptive Filtering and Wavelet Thresholding. *Sensors* **2023**, *23*, 8085. [CrossRef]
62. Guan, S.; Yang, H.; Wu, T. Transformer fault diagnosis method based on TLR-ADASYN balanced dataset. *Sci. Rep.* **2023**, *13*, 23010.
63. Dai, X.; Yi, K.; Wang, F.; Cai, C.; Tang, W. Bearing fault diagnosis based on POA-VMD with GADF-Swin Transformer transfer learning network. *Measurement* **2024**, *238*, 115328.
64. Wang, B.; Zhao, H.; Wang, X.; Lyu, G.; Chen, K.; Xu, J.; Cui, G.; Zhong, L.; Yu, L.; Huang, H.; et al. Bamboo classification based on GEDI, time-series Sentinel-2 images and whale-optimized, dual-channel DenseNet: A case study in Zhejiang province, China. *ISPRS J. Photogramm. Remote Sens.* **2024**, *209*, 312–323. [CrossRef]
65. Liu, C.; Fan, H.; Jiang, Y.; Ma, R.; Song, S. Gully erosion susceptibility assessment based on machine learning-A case study of watersheds in Tuquan County in the black soil region of Northeast China. *Catena* **2023**, *222*, 106798.
66. Xue, Z.; Yi, X.; Feng, W.; Kong, L.; Wu, M. Prediction and mapping of soil thickness in alpine canyon regions based on whale optimization algorithm optimized random forest: A case study of Baihetan Reservoir area in China. *Comput. Geosci.* **2024**, *191*, 105667. [CrossRef]
67. Chen, S.; Huang, J.; Wang, P.; Tang, X.; Zhang, Z. A coupled model to improve river water quality prediction towards addressing non-stationarity and data limitation. *Water Res.* **2024**, *248*, 120895. [CrossRef]
68. Yan, S.; Wu, L.; Fan, J.; Zhang, F.; Zou, Y.; Wu, Y. A novel hybrid WOA-XGB model for estimating daily reference evapotranspiration using local and external meteorological data: Applications in arid and humid regions of China. *Agric. Water Manag.* **2021**, *244*, 106594. [CrossRef]

land

Article

Comprehensive Assessment of Drought Susceptibility Using Predictive Modeling, Climate Change Projections, and Land Use Dynamics for Sustainable Management

Jinping Liu [1,2,3], Mingzhe Li [1], Renzhi Li [4,*], Masoud Jafari Shalamzari [5], Yanqun Ren [1] and Esmaeil Silakhori [6]

1 College of Surveying and Geo-Informatics, North China University of Water Resources and Electric Power, Zhengzhou 450046, China
2 Key Laboratory of Mine Spatio-Temporal Information and Ecological Restoration, Ministry of Natural Resources, Jiaozuo 454003, China
3 Hydraulics and Geotechnics Section, KU Leuven, Kasteelpark Arenberg 40, BE-3001 Leuven, Belgium
4 National Institute of Natural Hazards, Ministry of Emergency Management of the People's Republic of China, Beijing 100085, China
5 Department of Environment, Tabas Branch, Tabas 9791735618, Iran
6 Department of Arid Zone Management, Gorgan University of Agricultural Sciences and Natural Resources (GUASNR), Gorgan 4913815739, Iran
* Correspondence: renzhili@ninhm.ac.cn

Abstract: This study assessed the drought susceptibility in Golestan Province, Northeastern Iran, using land use change modeling and climate projections from the CMIP6 framework, under three Shared Socioeconomic Pathways (SSP1-2.6, SSP2-4.5, and SSP5-8.5) for 2030–2050. The development of current (2022) and future drought susceptibility maps was based on agrometeorological sample points and 14 environmental factors—such as land use, precipitation, mean temperature, soil moisture, and remote sensing-driven vegetation indices—used as inputs into a machine learning model, maximum entropy. The model showed a very robust predictive capacity, with AUCs for the training and test data of 0.929 and 0.910, thus certifying the model's reliability. The current analysis identified major hotspots in Gomishan and Aqqala, where 66.12% and 36.12% of their areas, respectively, exhibited "very high" susceptibility. Projections under the SSP scenarios, particularly SSP5-8.5, indicate that the risk of drought will be the most severe in Maraveh Tappeh, where 72.09% of the area exhibits a "very high" risk. The results revealed that Golestan Province is at a crossroads. Rising temperatures, exceeding 35 °C in summer, combined with declining rainfall, intensify agricultural and hydrological droughts. These aggravated risks are compounded with land use transitions from rangelands to bare land, mostly in Aqqala and Gomishan, besides urban expansion in Bandar-e Torkman and Bandar Gaz, all of which face less groundwater recharge and increased surface runoff. Golestan's drought vulnerability has both local and regional impacts, with its increased susceptibility affecting neighboring communities and ecosystems. Trade, migration, and ecological stresses linked to declining water resources may emerge as critical challenges, requiring regional collaboration for mitigation. Targeted interventions prioritizing sustainable land use practices, regional cooperation, and collaborative strategies are essential to address and mitigate these cascading risks and safeguard vulnerable communities.

Keywords: agrometeorological drought; maximum entropy; drought vulnerability; drought risk; drought management

https://doi.org/10.3390/land14020337

1. Introduction

Drought is defined as a prolonged period of deficient precipitation and related climatic conditions that significantly affect agrometeorological factors, such as soil moisture and crop yields, leading to adverse impacts on agricultural systems and water availability [1]. Drought is a complex natural disaster with multiple facets, characterized by prolonged water deficiency, which has major effects on ecosystems, agricultural systems, and human societies [2]. It has critically affected the water availability, agricultural productivity, and livelihoods of millions of people [3]. The United Nations Convention to Combat Desertification (UNCCD) has reported a global economic loss due to drought of approximately USD 124 billion from 2000 to 2019, with the last 10 years witnessing an increased frequency and severity [4]. Droughts have a significant societal impact in addition to financial costs, including food insecurity, malnutrition, migration, and conflict over scarce resources; therefore, it represents a significant concern regarding sustainable development [5]. In the case of Iran, an arid and semi-arid country, drought represents one of the most significant environmental challenges. It has critically affected the water availability, agricultural productivity, and livelihoods of millions of people. Iran has witnessed a number of recent serious droughts, with the drought from 1999 to 2001 causing an estimated USD 3.3 billion in economic losses, affecting over 37 million people [6]. Agricultural production in Iran relies heavily on irrigation, to the point that it is estimated that agriculture requires, on average, over 90 percent of Iran's water resources [7]. This heavy reliance on farming renders Iran particularly vulnerable to precipitation and temperature variability, where the results of drought exacerbate water scarcity and increase the extraction of groundwater, land subsidence, and desertification. Additionally, the socioeconomic repercussions of drought include the loss of income from agricultural operations, leading to rural–urban migration, placing an additional burden on the already stretched socioeconomic system [8]. In Northern Iran, Golestan Province, which experiences comparatively higher precipitation levels and milder temperatures than the arid and semi-arid southern regions, still faces significant drought challenges, leading to repeated reductions in crop production, diminished water availability, and increased socioeconomic stress.

Recent advances in climate science have emphasized the role of human-induced climate change in modifying the hydrological cycle and increasing the frequency and severity of droughts in many regions of the globe [9]. Previous global studies have applied different climate models and emission scenarios to project future drought conditions and reported unprecedented drought risks for the Mediterranean area, Central America, and parts of the Middle East and other regions of Africa and Asia, including Iran [10,11]. These global drought studies use global circulation models (GCMs) and regional climate models (RCMs) to simulate changes in key variables like the temperature, precipitation, and other climate variables, and they are used as inputs to assess the impact of the projected climate on the future drought dynamics under various greenhouse gas emission scenarios [12]. However, integrating LULC changes with climate projections remains a novel approach in Iran. While LULC studies in Iran have explored its impact on hydrological processes and droughts, no comprehensive research has integrated both climate and LULC dynamics to assess future drought risks under changing conditions, as is considered in our study. In the case of Iran, a plethora of studies have been conducted to analyze the link between climate change and drought. These studies have found significant changes in the temperature and precipitation patterns over the last few decades in different climatic zones of Iran [13,14]. Climate studies lack localized drought indicators, which are crucial for decision makers to address future vulnerabilities [15]. In Iran, LULC changes, such as deforestation, urbanization, and agricultural expansion, significantly affect hydrological processes and exacerbate drought conditions [16]. Studies have shown that these LULC changes can influence key factors like evapotranspiration, soil moisture, and surface runoff, further intensifying the drought risk [16–20]. For example, deforestation in some regions of Iran has led to increased vulnerability to drought [21,22], while urbanization has contributed to a reduced water retention capacity in the soil [23–25]. Still, to the authors' knowledge, there are no studies on the impact of LULC change on the future drought risks in Iran under different climate change scenarios.

While much promise is found in the literature related to drought and climate change, several underpinning issues remain. The first is that there have been few, if any, comprehensive studies that fully integrate the latest climate projections with dynamics associated with land use change to assess the drought susceptibility moving forward into the future. Most studies have assessed either the impacts of climate change or land use changes in isolation, neglecting the interconnected nature of these factors, which could significantly shape the drought conditions [26]. Secondly, many of the existing studies utilize earlier climate models, such as the Climate Model Intercomparison Project Phase 5 (CMIP5) [27,28], which uses climate models that have a lower spatial resolution compared to those from CMIP6 [29]. CMIP6 enhances climate simulations by incorporating higher-resolution models and a broader range of socioeconomic pathways, providing deeper insights into future drought risks and opportunities. This advancement is particularly relevant for Iran, where previous climate studies have often relied on lower-resolution models (e.g., CMIP5), resulting in generalized climate projections. The higher spatial resolution and diverse socioeconomic pathways provided by CMIP6 offer more region-specific and nuanced insights, which are critical in addressing the unique drought challenges faced by Iran, where agriculture and water management are heavily impacted by such changes. Our study therefore aims to fill this gap through projections of the future precipitation and temperature derived from CMIP6 climate models and using a Markov chain model for land use projection in Golestan Province, Iran. The integration of CMIP6 projections with LULC dynamics is particularly innovative for Iran. Previous studies in Iran and other semi-arid regions have typically focused either on climate projections alone [30–33], LULC impacts on drought [34–36], or older climate models such as CMIP5 [37–40]. By combining high-resolution climate projections (CMIP6) with future LULC changes, our approach allows for a more accurate and localized assessment of the drought risk, addressing gaps in the current research, where the interaction between the climate and land use has been overlooked in this region. For example, recent studies have similarly noted the importance of coupling LULC dynamics with climate change projections for a comprehensive understanding of the future drought conditions [41–44]. Together with a machine learning model (MaxEnt) for drought susceptibility assessment and projection, with the use of several drought-controlling predictors, including the current and future projected land use and climate variables, we aim to enable better spatial literacy in understanding how climate change and land use dynamics will influence the future drought risks. In particular, we consider and interpret the current and three unique future scenarios representing optimistic, intermediate, and pessimistic conditions, allowing us to offer a range of possible futures and options to consider with respect to drought management.

The selection of the MaxEnt model for drought susceptibility analysis in this study is justified by its ability to handle complex, non-linear relationships between environmental variables and drought occurrences, making it particularly well suited for this context. MaxEnt has been extensively validated in ecological and environmental studies, demonstrating robust performance in modeling susceptibility and distribution under uncertain conditions [45]. Compared to other machine learning approaches like random forest or support vector machines, MaxEnt requires only presence data, which is advantageous when absence data are unavailable or unreliable. Additionally, its feature regularization techniques minimize overfitting, ensuring reliable predictions even with diverse and interrelated drought-controlling factors [45,46]. Despite its established utility and success in studies of drought [47–49] and other natural hazards, such as landslides and floods, and multi-hazard susceptibility analysis [50–53], its application under diverse future climate and land use change scenarios remains underexplored. By integrating a comprehensive range of drought-controlling factors and employing the Standardized Drought Condition Index (SDCI) as a reliable proxy for agrometeorological drought, our study demonstrates this model's ability to achieve high predictive accuracy [46]. This methodological approach underscores the model's potential to advance drought management strategies in vulnerable regions. In addition, the use of the most recent CMIP6 climate projections in our study should also be noted. The CMIP6 projections represent a major enhancement over previous climate projections with respect to both the spatial resolution and complex Earth system processes [29]. Coupling projections with land use change helps to incorporate a more systemic approach to the future drought risk and avoids problems with static land use assumptions, as frequently used in past studies [54]. This transition from relying solely on climate change projections to integrating both climate and land use changes is essential in improving drought management and policy planning [55]. Our study provides actionable outputs that can be directly applied by decision makers in Golestan Province. By identifying the areas that are most susceptible to drought under varying future conditions, the findings offer a basis for targeted intervention. For instance, regions projected to experience severe drought could benefit from the immediate implementation of water-saving irrigation practices, the selection of drought-resistant crops, or support for reforestation initiatives. These actions are not merely suggested but are supported by a clear understanding of future vulnerabilities, allowing local governments to prioritize and allocate resources effectively [56]. Additionally, the scenario-based approach adopted in our study enables policymakers to test and refine different strategies, ensuring that resource allocation aligns with the most likely future conditions, thereby minimizing uncertainties in long-term planning [57].

In light of the above, the current study seeks to offer an integrated assessment of the future drought risks in Golestan Province, considering climate projections and associated land use changes, using sophisticated modeling techniques. Furthermore, the aim is to fill gaps in existing studies and offer actionable science to inform management and policy, in order to facilitate the development of more resilient and sustainable approaches to the management of drought in Iran and potentially in other similarly arid or semi-arid regions with comparable climatic and land use conditions, such as parts of Central Asia, the Middle East, and North Africa.

2. Materials and Methods

The methodological framework adopted in this research is depicted in Figure 1 and discussed further below.

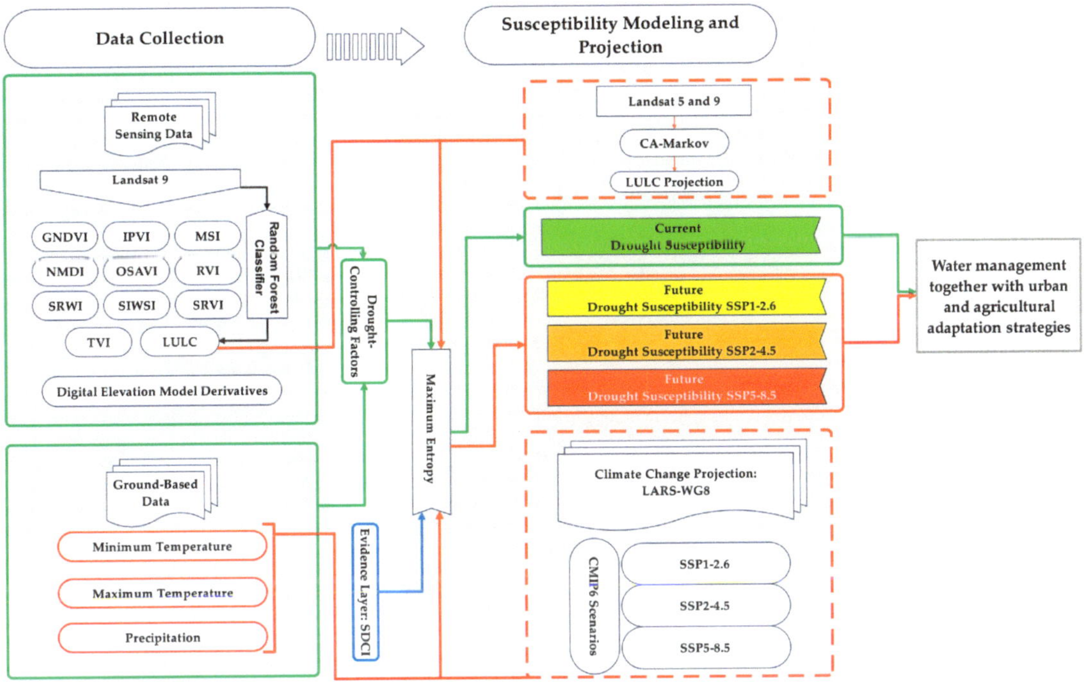

Figure 1. Flowchart showing the procedures adopted in this study (GNDVI: Green Normalized Difference Vegetation Index; IPVI: Infrared Percentage Vegetation Index; MSI: Moisture Stress Index; NMDI: Normalized Multi-Band Drought Index; OSAVI: Optimized Soil-Adjusted Vegetation Index; RVI: Ratio-Based Vegetation Index; SRWI: Simple Ratio Water Index; SIWSI: Soil Moisture Stress Index; SRVI: Stress-Related Vegetation Index; TVI: Transformed Vegetation Index; LULC: Land Use and Land Cover; SDCI: Standardized Drought Condition Index; CA-Markov: Cellular Automata-Markov Chain; SSP: Shared Socioeconomic Pathways; LARS-WG: Long Ashton Research Station Weather Generator).

2.1. Study Area

Golestan Province, Northeastern Iran, spans 20,367 km^2. Geographically, its landscape exhibits a variety of topographical attributes, including the coast along the Caspian Sea in the north, plains in the middle, and the Alborz Mountains in the southern part (Figure 2). Due to the contrasting geographical features, the elevation varies significantly throughout the province, with elevation of −27 m a.s.l on the northern coastal areas and over 3000 m in the southern mountainous areas. Since the elevation range is considerable, the climatic conditions are also diverse, ranging from humid subtropical in the coastal and plain regions to temperate and semi-arid in the mountainous regions. Golestan Province receives average annual precipitation of approximately 550 mm, where the northern plains have a total annual precipitation amount of less than 200 mm, while the southern and central areas, especially the highlands, receive up to 1000 mm of total annual rainfall. Golestan experiences cold winters (−10 °C) and hot summers (>35 °C). These extreme climate conditions make the region highly prone to both drought and flooding, as two extremes with recognized impacts on water resource availability [58].

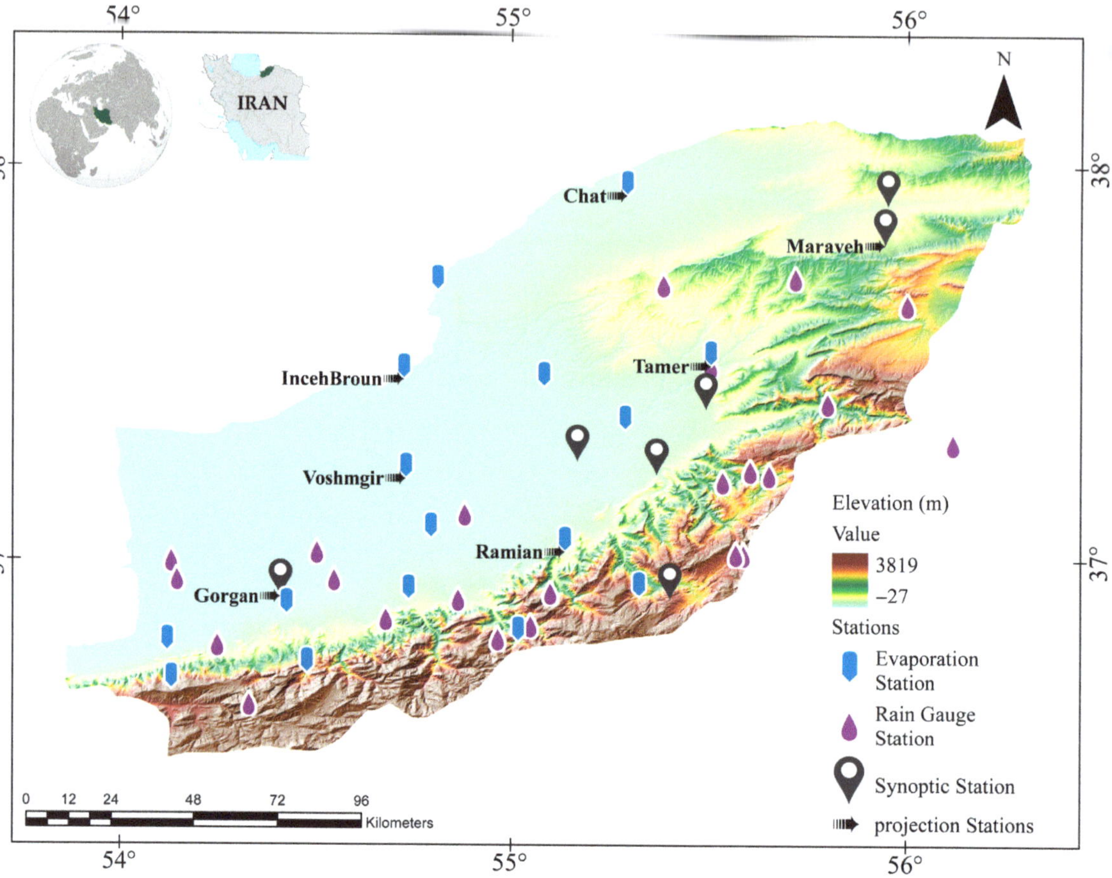

Figure 2. Location of the study area in Iran and the distribution of various meteorological stations used to map precipitation and temperature (the labeled projection stations refer to the synoptic and evaporation stations employed for the downscaling of future climate change data).

LULC changes across Golestan Province have been observed in the last few decades as a result of the increased population, agricultural expansion, and urbanization. The major land use types in Golestan Province include agricultural land (35%), forests (20%), rangeland (30%), and urban areas (15%). Agriculture, concentrated in fertile plains, plays a critical role in the local economy, with key crops including wheat, barley, rice, and cotton, alongside fruits, vegetables, and livestock production. However, deforestation, land degradation, and urbanization have heightened the region's susceptibility to drought and water stress [59]. The reliance on agriculture makes the province highly vulnerable to drought, leading to economic losses, food shortages, and socioeconomic instability. Recent droughts have strained agriculture, the water supply, and livelihoods, fueling resource conflicts [8].

The province has developed irrigation canals, reservoirs, and water transfer systems to sustain agriculture. However, these infrastructures are under growing pressure due to population growth, climate change, and the over-extraction of water resources. Inadequate water resource management and limited drought response practices further exacerbate these vulnerabilities. The severe droughts in 1999–2001 and 2007–2009 caused considerable economic damage, highlighting the need for effective drought management and adaptation strategies [7]. The population is unevenly distributed, with a higher density in the northern and central parts, where agricultural activities are concentrated and the drought impacts are the most severe due to reduced agricultural outputs and water stress.

2.2. Data Sources

The data utilized in this study were sourced from various platforms and used to evaluate the drought susceptibility in Golestan Province under both the current and future conditions. Remote sensing indices, derived from Landsat 9, offered essential insights into the vegetation health, soil moisture, and environmental stress. A land use map for the study area's current state was also created using Landsat 9 data. These tools were supplemented with ground-based data, such as the temperature, precipitation, and other meteorological variables from local measurement stations. To develop the agrometeorological evidence layer, this study combined the Vegetation Condition Index (VCI) and Temperature Condition Index (TCI), resulting in the Standardized Drought Condition Index (SDCI). The SDCI evidence layer, alongside the remote sensing indices and ground-based climatic data, was used to generate the drought susceptibility map for the present conditions. Additionally, future land use and climate projections were derived using the LARS-WG8 tool, which provided high-resolution climate data for the years 2030 to 2050 under various emission scenarios. Collectively, these datasets enabled the assessment and prediction of the drought susceptibility under changing climate and land use scenarios for the future.

2.2.1. Drought-Controlling Factors

In order to evaluate the susceptibility to agrometeorological drought in Golestan Province, we exploited 14 drought-controlling predictors, which included remote sensing indices and climate data derived from ground measurements. The selected indices were chosen due to their sensitivity to the root causes of drought, ensuring that we captured different climatic and environmental dimensions of drought. In particular, they were chosen based on their demonstrated ability to capture the key dimensions of drought—vegetation health, soil moisture, and climatic stressors. Remote sensing indices, such as the Green Normalized Difference Vegetation Index (GNDVI) and Soil Moisture Stress Index (SIWSI), are particularly effective in assessing vegetation health and water stress over large spatial scales. Ground-based data, including the temperature and precipitation, provide localized, high-resolution climatic inputs that are critical for drought modeling. Together, these predictors ensured the comprehensive representation of the complex interactions driving drought conditions. Relevant studies (e.g., [60–63]) demonstrate that combining these factors enhances the reliability and accuracy of drought susceptibility models. These factors were used to produce the current (2022) and future drought susceptibility models (2030–2050) using the MaxEnt method. The year 2022 was selected to reflect the current conditions, being the year that a drought strongly impacted the province, while the two following years (2023 and 2024) were characterized as wet years.

Remote Sensing Indices

Remote sensing indices were derived from Landsat 9 Level-2 imagery from 21 January 2022 to 1 January 2023, considering only scenes with less than 5% cloud cover. The data provider (USGS) had already applied atmospheric, radiometric, and geometric corrections, including orthorectification. Additional preprocessing steps were performed to prepare the data for analysis. Specifically, pixel values were normalized to a range of 0 to 1 by dividing them by 10,000, and a cloud and cloud shadow masking process was implemented to improve the dataset's quality and accuracy. The study area was covered by three paths (161, 162, and 163) and three rows (33, 34, and 35), resulting in 74 extracted scenes. These scenes were then incorporated to generate both a land use map and corresponding vegetation indices. Remote sensing allows for continuous, large-scale landscape monitoring and provides high-resolution data for the detection of drought at large scales. Remote sensing indices provide key information about vegetation health, soil moisture, and environmental stress (Table 1).

Table 1. Detailed descriptions and formulas of the remote sensing indices used as drought-controlling factors.

Index Name	Description	Formula	Reference
Green Normalized Difference Vegetation Index (GNDVI)	GNDVI is an index used to evaluate the health and greenness of vegetation. It primarily measures the chlorophyll content in leaves, which decreases during drought conditions.	$GNDVI = \frac{NIR-Green}{NIR+Green}$	[64]
Infrared Percentage Vegetation Index (IPVI)	IPVI is a simple vegetation index focusing on the percentage of infrared reflectance. Vegetation health can be inferred from the balance between infrared and visible light reflected by plant surfaces.	$IPVI = \frac{NIR}{NIR+Red}$	[65]
Moisture Stress Index (MSI)	MSI is a reflectance measurement, sensitive to increases in leaf water content. It is applicable for densely vegetated areas and is more sensitive at the canopy level rather than the leaf level.	$MSI = \frac{SWIR2}{NIR}$	[66]
Normalized Multi-Band Drought Index (NMDI)	NMDI is used to monitor the water content in vegetation and soil, making it highly relevant in detecting droughts. It leverages both near-infrared (NIR) and shortwave infrared (SWIR) bands.	$NMDI = \frac{NIR-(SWIR1-SWIR2)}{NIR+(SWIR1+SWIR2)}$	[67]
Optimized Soil-Adjusted Vegetation Index (OSAVI)	OSAVI is a modification of NDVI, designed to reduce soil brightness influences. It works well in environments with minimal vegetation cover, helping to accurately assess vegetation health.	$OSAVI = \frac{(NIR-Red)}{(NIR+Red+0.16)}$	[68]
Ratio-Based Vegetation Index (RVI)	RVI is a straightforward vegetation index comparing near-infrared and red reflectance, providing a ratio that correlates with vegetation health.	$RVI = \frac{NIR}{Red}$	[69]
Simple Ratio Water Index (SRWI)	SRWI is designed to detect moisture levels in vegetation and soil, particularly in water-stressed environments.	$SRWI = \frac{NIR}{SWIR1}$	[69]
Soil Moisture Stress Index (SIWSI)	SIWSI measures water stress by combining information from the near-infrared and shortwave infrared bands.	$SIWSI = \frac{(SWIR1-NIR)}{(SWIR1+NIR)}$	[70]
Stress-Related Vegetation Index (SRVI)	SRVI is sensitive to stress in vegetation and can detect early signs of drought stress before visible symptoms appear.	$SRVI = MIR + \left(\frac{Red}{NIR}\right)$	[71,72]
Transformed Vegetation Index (TVI)	TVI is a normalized vegetation index that helps to reduce atmospheric effects, making it more robust under variable conditions.	$TVI = \frac{(SWIR1-Red)}{(SWIR1+Red)}$	[73]

Land Use

In this study, we employed Landsat 9 satellite imagery and a supervised classification approach to produce a land cover map. To achieve this, Landsat 9 images were first preprocessed by applying temporal filters for each season of 2022, capturing the relevant data for spring, summer, fall, and winter. These temporal filters allowed us to select seasonal data for each period, ensuring that the images were relevant for seasonal land cover changes. Subsequently, various vegetation indices, including the Normalized Difference Vegetation Index (NDVI) and Enhanced Vegetation Index (EVI), were calculated for each image using the appropriate bands of Landsat 9, such as the NIR, RED, and BLUE bands, to provide insights into the vegetation health and land cover dynamics over time. The seasonal indices were subsequently integrated into the dataset and utilized as input features for a supervised random forest classifier, a method that has demonstrated strong efficacy for LULC classification [74–76]. This classifier was trained on labeled ground truth samples that were randomly split into training and testing datasets. By applying the random forest algorithm, we were able to classify the land cover types in the study area based on the spectral information from the Landsat 9 imagery and the derived vegetation indices. In addition to the optical data, we also incorporated other relevant features, such as terrain data, including the elevation and slope, obtained from a digital elevation model (DEM), to enhance the classification results. The digital elevation data used in this study were sourced from the NASADEM dataset, available on the Google Earth Engine platform. NASADEM represents a reprocessed version of the Shuttle Radar Topography Mission (SRTM) data, offering enhanced accuracy by integrating auxiliary datasets, including ASTER GDEM, ICESat GLAS, and PRISM. This dataset provided crucial topographic information, such as the elevation and slope, which were derived and incorporated into the analysis to improve the accuracy of the classification and land use mapping processes.

In this work, all reference samples were split into training and testing sets (70% for training, 30% for validation) based on a randomly assigned column in the Google Earth Engine (GEE) environment. A total of 3226 reference samples were used, with 30% allocated for testing and 70% for training. The random forest classifier was then trained on the training subset to generate spectral signatures for each land cover class. Next, the trained model was validated on the remaining test set, allowing the calculation of error metrics through a confusion matrix. This matrix yielded the overall accuracy and the Kappa coefficient. This approach enabled a reliable and scalable classification performance assessment by relying on GEE's robust data handling and efficient parallel processing capabilities.

Ground-Based Data

In addition to the remote sensing indices, we utilized ground-based datasets including the minimum temperature (Tmin), maximum temperature (Tmax), and precipitation, which were based on data from synoptic, rainfall, and evaporation stations across Golestan Province, to provide important climatic variables for a more thorough examination of drought in the area. Ground-based data were acquired from the Regional Water Company of Golestan.

2.2.2. Evidence Layer

The Standardized Drought Condition Index (SDCI) was adopted to extract 500 drought occurrence points, which served as the evidence layer for MaxEnt modeling. The SDCI integrates remote sensing and ground data to capture spatial drought patterns. Notably, the SDCI index effectively represents meteorological and agricultural drought conditions in the region when exploring the impacts of drought. The SDCI integrates three vegetation and meteorological indices, listed in Table 2. These drought occurrence points represent historical drought events in the region and serve as a critical dataset for drought susceptibility mapping.

Table 2. Detailed descriptions and formulas of the SDCI and its constituents (TCI, VCI, and PCI).

Index Name	Description	Formula	Reference
Standardized Drought Condition Index (SDCI)	The SDCI is an integrated index that combines multiple drought-related factors to provide a comprehensive assessment of drought conditions.	$SDCI = (0.25 \times TCI) + (0.25 \times VCI) + (0.50 \times PCI)$	[77]
Temperature Condition Index (TCI)	TCI assesses vegetation stress due to temperature extremes, helping to identify areas impacted by heat stress or cooling deficits.	$TCI = \frac{(LST_{max} - LST_i)}{(LST_{max} - LST_{min})}$	[78]
Vegetation Condition Index (VCI)	VCI evaluates vegetation health and stress using satellite-derived Normalized Difference Vegetation Index (NDVI) values.	$VCI = \frac{(NDVI_i - NDVI_{min})}{(NDVI_{max} - NDVI_{min})}$	[79]
Precipitation Condition Index (PCI)	PCI measures precipitation anomalies to identify areas experiencing drought or excess rainfall.	$PCI = \frac{(TRMM - TRMM_{min})}{(TRMM_{max} - TRMM_{min})}$	[80]

2.2.3. Future Land Use and Climate Data

In our endeavor to project drought conditions for the timeline spanning from 2030 to 2050, we harnessed land use and climate change projections, explained as follows.

- Land Use Projection: In this study, we utilized Land Change Modeler (LCM) to project future land use changes, employing both the cellular automata–Markov chain model and machine learning techniques for more accurate predictions. Land Change Modeler is a robust tool for the analysis of historical land use patterns and projection of future changes, which are essential in understanding how land use evolves over time. To create the input maps for this projection, we used Landsat satellite imagery from the years 2000 and 2022. These images were first preprocessed and classified into different land use categories using the random forest classification model. This machine learning model, known for its accuracy and efficiency, was trained on labeled data, allowing us to classify the satellite imagery into relevant categories, such as urban areas, agricultural land, forests, and water bodies. Once the land use maps for the years 2000 and 2022 were generated, we applied the Markov chain model to simulate transitions between different land use types over time. This model helps to quantify the likelihood of one land use category changing into another based on observed historical trends. Using these tools, we projected land use maps for the period between 2030 and 2050.
- Climate Data: Data from the CMIP6 report were used under three shared socioeconomic pathways (SSPs).
    1. SSP1-2.6: Low emissions, representing a future with stringent climate mitigation;
    2. SSP2-4.5: Moderate emissions, being the middle path; and
    3. SSP5-8.5: High emissions, for rapid warming and extreme climate impacts [11].

The Long Ashton Research Station Weather Generator (LARS-WG8), as the most recent downscaling tool, was used to downscale the future climate change data at the station level. In this study, an ensemble model was constructed using an Raisanen approach, drawing on outputs from five models participating in the Coupled Model Intercomparison Project Phase 6 (CMIP6) [81,82] (Table 3). Daily minimum temperature (Tmin), maximum temperature (Tmax), and precipitation data were collected from seven well-distributed synoptic and evaporation stations across the Golestan Province, each having 30 years of consistent data from 1985 to 2015, and were input into LARS-WG8. The Tmin, Tmax, and precipitation projection data will serve as critical inputs for future drought susceptibility modeling.

Table 3. Description of the CMIP6 models used in this study.

Model	Modeling Center	Resolution (Atmosphere) Latitude × Longitude	Reference
CNRM-CM6-1	Centre National de Recherches Meteorologiques/Centre Europeen de Recherche et Formation Avancees en Calcul Scientifique	1.4° × 1.4°	[83]
GFDL-ESM4	Geophysical Fluid Dynamics Laboratory (GFDL), USA	1.25° × 1.0°	[84]
HadGEM3-GC31-LL	Meteorological Office Hadley Centre, UK	1.88° × 1.25°	[85]
MPI-ESM1-2-LR	Meteorological Research Institute, Japan	1.9° × 1.9°	[86]
MRI-ESM2.0	Meteorological Research Institute, Japan	1.0° × 1.0°	[87]

2.3. *Data Analysis and Processing*

2.3.1. Current Conditions (2022)

For the current drought susceptibility analysis, we used the maximum entropy (MaxEnt) model. The 14 controlling factors, together with the 500 drought evidence points, were employed to train and validate the model, ensuring that it accurately reflected the current agrometeorological drought patterns in Golestan Province. MaxEnt uses a probability distribution to predict areas most susceptible to drought, with the evidence points serving as the foundation for model calibration. By running the model under the current climatic conditions (2022), we were able to generate a baseline map of the drought susceptibility for the region.

2.3.2. Future Scenarios (2030–2050)

Using the MaxEnt model, we projected the drought susceptibility under four scenarios: the current conditions (2022 data as baseline) and three future climate pathways—SSP1-2.6 (optimistic), SSP2-4.5 (moderate), and SSP5-8.5 (severe). Each future scenario incorporated 2030–2050 climate data and projected land use changes, while other thematic layers, like the vegetation indices, remained constant. These scenarios illustrated potential climate-driven shifts in the drought risk.

2.3.3. Drought Susceptibility Modeling Using Maximum Entropy

The MaxEnt model, a machine learning algorithm for presence-only data, is widely used in environmental studies to predict areas susceptible to specific conditions, such as drought. It maximizes the entropy to estimate the least-biased probability distribution that meets environmental constraints, utilizing factors like the temperature, precipitation, and land use [88,89]. MaxEnt compares environmental conditions at known presence locations (e.g., drought-affected areas) with the entire study area to calculate susceptibility probabilities [89,90]. Rooted in information theory, the model assumes minimal bias by selecting distributions with the maximum entropy, ensuring that the predictions align with the provided data while avoiding unsupported assumptions [91]. For this study, MaxEnt processed 500 SDCI points alongside environmental variables to produce probabilistic maps of the drought susceptibility under the current and future climate scenarios, offering critical insights into Golestan Province's drought risk [46,89,92].

2.3.4. Model Validation and Performance Metrics

To assess the MaxEnt model's performance, we calculated the area under the curve (AUC) metric of the ROC curve. The ROC plots the false positive rate (FPR) on the x-axis and the true positive rate (TPR) on the y-axis. The AUC of the ROC curve indicates the accuracy of the model's prediction. The ROC values can be categorized into four descriptive accuracy classes: (1) 0.6–0.7, indicating poor accuracy; (2) 0.7–0.8, indicating fair accuracy; (3) 0.8–0.9, indicating good accuracy; and (4) 0.9–1.0, indicating excellent accuracy. Therefore, AUC values closer to 1 will indicate higher model accuracy in predicting the drought susceptibility [93–95].

3. Results

3.1. Monthly Climate Change Data

Figure 3 displays the monthly projected climate change data for the precipitation and maximum and minimum temperatures for two stations. These projections include changes under SSP2-4.5 and SSP5-8.5 over the historical baseline.

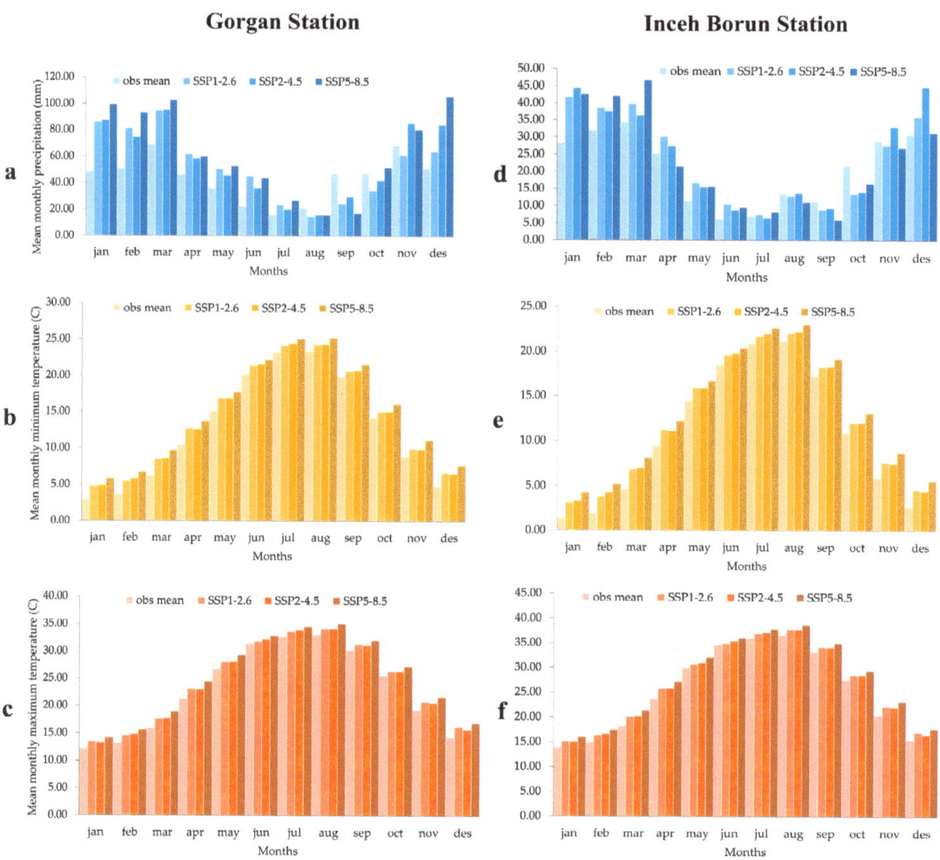

Figure 3. Monthly projected climate change data for Gorgan, including (**a**) mean monthly precipitation (mm), (**b**) mean monthly minimum temperature (°C), and (**c**) mean monthly maximum temperature (°C), and for Inceh Borun station, including (**d**) mean monthly precipitation (mm), (**e**) mean monthly minimum temperature (°C), and (**f**) mean monthly maximum temperature (°C).

The temperatures rise across all scenarios, with notable increases in the minimum and maximum values. Under SSP5-8.5, the most pessimistic scenario, the minimum temperature increases range from +1.07 °C (Maraveh in June) to +3.44 °C (Inceh Borun in March). The maximum temperature increases under SSP5-8.5 are also substantial, with values reaching +3.38 °C (Inceh Borun in April), +3.15 °C (Gorgan in April), and +2.95 °C (Maraveh in November). In the optimistic scenario, SSP1-2.6, the minimum temperature increases are more modest, generally ranging from +0.36 °C (Maraveh in July) to +2.25 °C (Chat in December). The maximum temperature under SSP1-2.6 also increases, but the peaks are typically below +2.0 °C, with exceptions like Gorgan in December (+1.90 °C).

The precipitation varies significantly by station, month, and scenario. Under SSP5-8.5, the precipitation increases significantly in some months, such as +57.24 mm in December at Ramian and +54.76 mm in December at Gorgan. Conversely, the same scenario predicts sharp declines in other months, such as −30.52 mm in September at Gorgan and −26.78 mm in September at Ramian. SSP1-2.6 generally results in more stable changes, although some notable variations occur, including a +19.14 mm increase in January at Ramian and a −8.37 mm decrease in October at Inceh Borun. The standard deviation for precipitation changes is considerably higher than for the temperature, indicating greater variability, with values as high as 18.66 mm at Ramian under SSP5-8.5 and 23.22 mm at Gorgan under the same scenario. The temperature's standard deviations are comparatively lower, often below 0.6 °C, reflecting more consistent trends.

Stations like Gorgan and Ramian show significant susceptibility to extreme precipitation changes, with the maximum values exceeding +50 mm in some months and negative extremes of −23.35 mm or more in others. Tamer and Maraveh, by contrast, show smaller fluctuations in precipitation, with fewer extreme positive or negative values. However, even these stations exhibit clear warming trends across all scenarios, with the minimum temperature increases ranging from +0.36 °C to +2.58 °C and the maximum temperature increases from +0.54 °C to +2.95 °C.

3.2. Annual Climate Change Data

Figure 4 presents the annual projected climate change data for the precipitation and maximum and minimum temperatures across all seven stations used for downscaling. Figure 5 summarizes the percentage changes in the examined variables under the future climate change scenarios compared to the current conditions (observed mean values).

Figure 5 uses a color-coded format to facilitate a graphical spatiotemporal comparison across various stations under different scenarios. As illustrated in Figures 4 and 5, the increases in the mean annual precipitation across all synoptic stations are consistent among climate scenarios SSP1-2.6, SSP2-4.5, and SSP5-8.5. For the Gorgan station, the precipitation under SSP5-8.5 shows the highest increase, rising by 224.20 mm, from 525.56 mm to 749.76 mm. The Ramian station experiences an increase of 147.86 mm under the same scenario, from 688.78 mm to 836.64 mm. In Voshmgir, the precipitation rises from 320.68 mm to 375.03 mm, reflecting an increase of 54.35 mm, while, in Inceh Borun, the precipitation increases by 26.37 mm, from 251.47 mm to 277.84 mm.

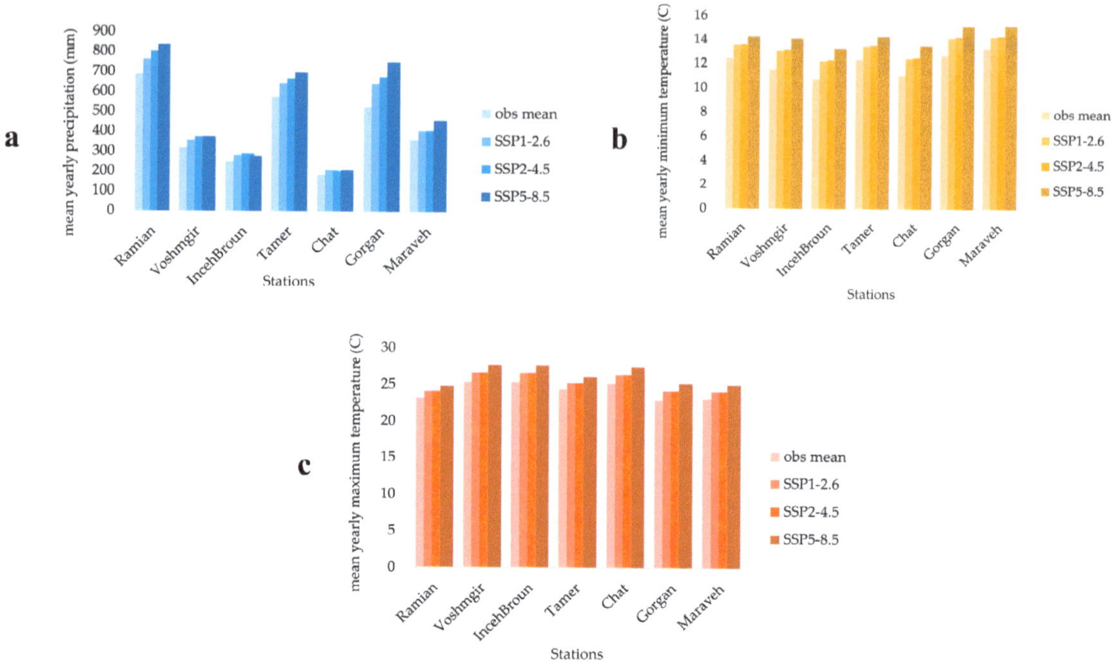

Figure 4. Mean annual projected climate change data for (**a**) precipitation (mm), (**b**) minimum temperature (°C), and (**c**) maximum temperature (°C) across stations.

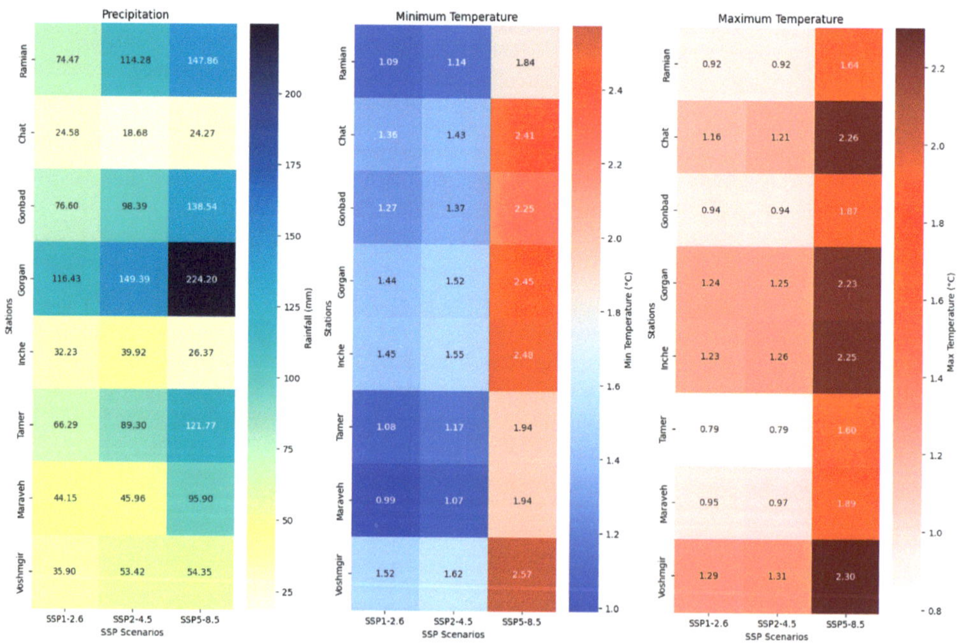

Figure 5. Annual changes in precipitation and maximum and minimum temperature under future climate change scenarios compared to the current conditions (the light to dark color palette indicates the range from low to amplified percentage changes).

All scenarios project a steady rise in the minimum temperatures, with the largest rises being found with SSP5-8.5. For example, in Ramian, the minimum temperature increases by 1.84 °C, from 12.45 °C to 14.29 °C, under SSP5-8.5. Similarly, in Maraveh Tappeh, the temperature rises by 1.94 °C, from 13.29 °C to 15.23 °C, while Tamer sees an increase of 1.94 °C, from 12.36 °C to 14.31 °C. All stations exhibit a similar pattern of increase for the maximum temperatures, with greater changes observed under the SSP5-8.5 scenario. Voshmgir and Inceh Broun show the highest projected maximum temperatures, reaching 27.69 °C and 27.68 °C, respectively, under SSP5-8.5, reflecting increases of 2.30 °C and 2.25 °C compared to the current conditions. Gorgan sees a rise in the maximum temperature from 22.99 °C to 25.22 °C, an increase of 2.23 °C, under SSP5-8.5, while Ramian experiences a rise of 1.64 °C, from 23.19 °C to 24.83 °C. Figure 6 illustrates the spatial distribution maps of the examined climatic variables under the current and future climate change scenarios.

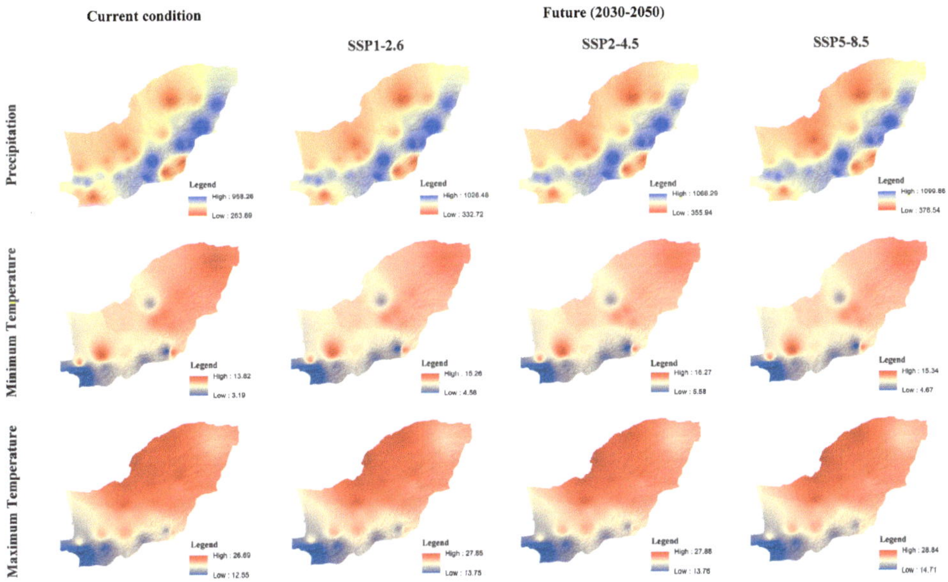

Figure 6. Spatial distribution maps of precipitation and minimum and maximum temperatures under current and future climate change scenarios.

3.3. Analysis of Land Use Change and Its Impact on Drought Conditions

The accuracy assessment of the LULC classification results indicated a high level of performance. The overall accuracy was 0.98, demonstrating that the classifier correctly identified 98% of the test samples. Furthermore, the Kappa coefficient was calculated to be 0.98, indicating the strong agreement between the classified map and the reference data. When trained with the selected spectral and textural features, these results suggest that the random forest classifier provided highly accurate and reliable land cover classification for the study area. The overall accuracy and Kappa coefficient were well above the acceptable threshold, confirming the robustness of the classification model. Accordingly, the LULC map of Golestan Province was classified into seven distinct categories. Forests, predominantly part of the Hyrcanian forests, are located in the northern foothills of the Alborz Mountains and the southern Caspian coastal plains, characterized by dense vegetation consisting of ancient broadleaf species such as oak (*Quercus castaneifolia*), beech (*Fagus orientalis*), hornbeam (*Carpinus betulus*), and maple (*Acer velutinum*), playing a vital role in the ecological balance and biodiversity. Rangelands, primarily found in the eastern and southeastern parts of the province, consist of natural grasslands and shrublands that are essential for grazing livestock and the preservation of native flora and fauna. Agricultural lands dominate the fertile lowlands of the central and western plains, actively cultivated for crops, orchards, and plantations, forming the cornerstone of the region's economy. Bare lands, with minimal vegetation, are scattered across eroded zones, particularly in the northern plains, often resulting from deforestation or natural processes. Water bodies, such as rivers, lakes, and reservoirs, are primarily concentrated in the central plains, with significant examples being the Gorgan River and the Golestan Dam. Wetlands, such as the Gomishan Wetland near the Caspian Sea, are defined by water-saturated soils and support aquatic plants and wildlife, serving as natural water filters and habitats. Built-up areas, including cities such as Gorgan, Bandar-e Torkman, and Gonbad-e Kavus, encompass urban and rural settlements, industrial zones, and infrastructure, reflecting the extent of human habitation and activity. Understanding these categories and their spatial distribution is essential in managing the land resources and planning sustainable development in the province.

Table 4 and Figure 7 presents a summary of the land use in 2022, as well as the projected changes for 2030–2050, revealing significant shifts in land categories like forests, rangeland, agricultural land, and bare land. These changes give important insights into how the drought conditions, both agricultural and hydrological, can be worsened. Figure 8 presents spatial maps of the land use classes for the years 2022 and 2030–2050. Next, we will discuss how each land use change case relates to drought's impacts and the generalized environmental repercussions.

Table 4. Areal changes in land use classes from the current conditions (2022) to future projections (2023–2050). Units are ha.

Class	Land Use	2022 (ha)	2030–2050 (ha)	Percent Change	Areal Change (ha)
1	Forest	339,409	338,110	−0.06	−1299
2	Rangeland	708,347	651,052	−2.83	−57,295
3	Agricultural	805,055	747,267	−2.85	−57,788
4	Bare land	123,473	229,675	5.24	106,202
5	Water body	7210	11,920	0.23	4710
6	Wetland	8818	7878	−0.05	−940
7	Built-up	35,801	42,221	0.32	6420

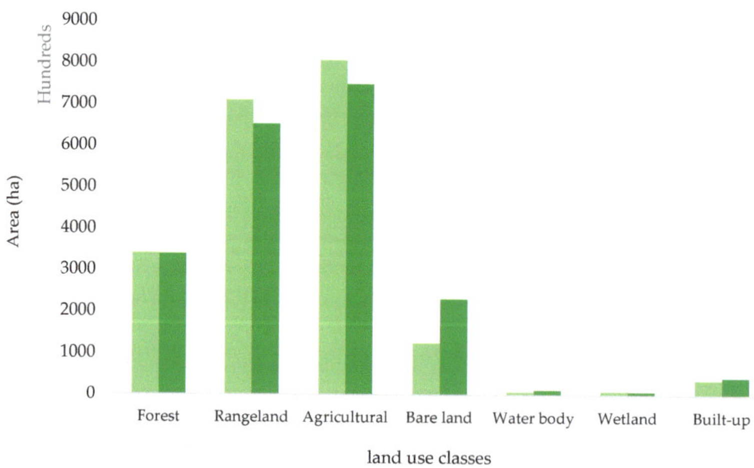

Figure 7. Schematic comparison of projected areal shifts in land use classes from 2022 to 2030–2050.

In 2022, the total forest area in the region was 339,409 ha, which is projected to decrease by 0.06% (1299 ha) by 2030–2050, with this area being converted to rangeland and built-up spaces. The total rangeland area in 2022 was 708,347 ha, and it is expected to decline by 2.83% (57,295 ha) by 2030–2050, with much of this land being converted into bare land. Agricultural land, which covered 805,055 ha in 2022, will decrease by 2.85% (57,788 ha), primarily transitioning into bare land and built-up areas. The bare land area in 2022 was 123,473 ha and is expected to increase significantly by 5.24% (106,202 ha) by 2030–2050. Water bodies, which accounted for 7210 ha in 2022, are projected to increase by 0.23% (4710 ha) by 2030–2050. Wetland areas, which covered 8818 ha in 2022, will decrease by 0.05% (940 ha) by 2030–2050, largely transitioning into water bodies. Built-up areas, which totaled 35,801 ha in 2022, will see a small increase of 0.32% (6420 ha) by 2030–2050 due to urbanization.

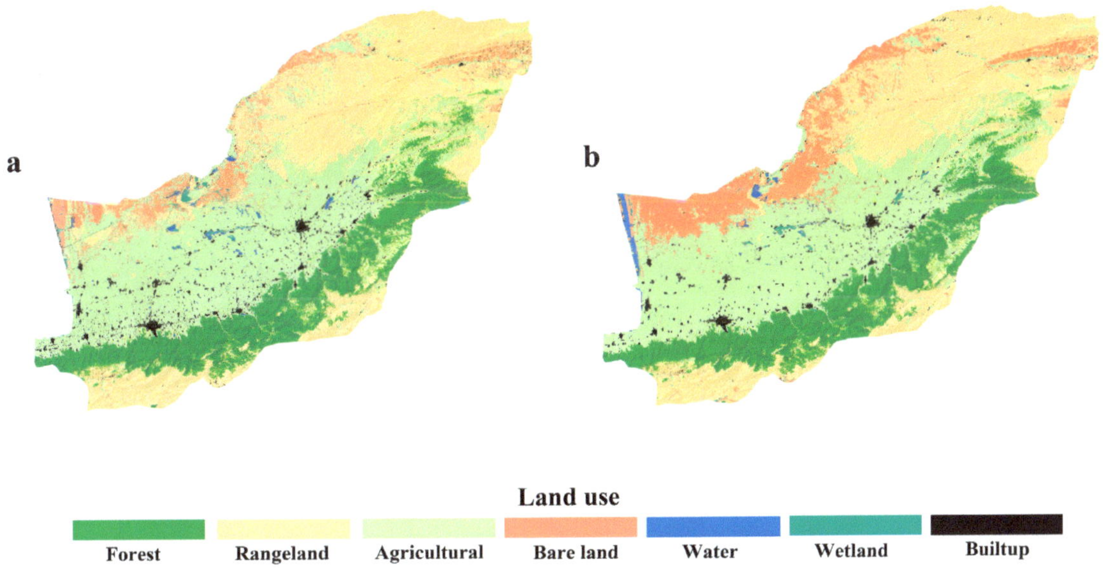

Figure 8. Spatial maps of land use classes for the years 2022 (**a**) and 2030–2050 (**b**).

3.4. Localized Analysis of Land Use Change in Golestan Province (2022 vs. 2030–2050)

Significant land use changes are projected across Golestan Province by 2030–2050 (Table 5). In sum, Ali Abad showed slight forest cover increases (45.79% to 46.13%) and minimal rangeland losses, while Aqqala experienced sharp rangeland and agricultural land declines, with bare land rising dramatically (13.68% to 26.13%). Azadshahr, Bandar Gaz, and Bandar-e Torkman exhibited minor shifts, including small reductions in rangeland and slight increases in built-up areas. Galikesh and Kalaleh saw moderate rangeland declines and agricultural land increases, while Gomishan and Gonbad-e Kavus faced substantial rangeland and agricultural land losses, with significant bare land expansion. Gorgan, Kordkuy, Maraveh Tappeh, Minudasht, and Ramiyan recorded minor changes, including slight increases in forest or agricultural land and small reductions in rangeland. More details are provided in Table 5.

Table 5. Land use change projections and the suggested management strategies for different counties of Golestan Province (Alternating row colors were applied to improve visual separation and readability).

County	Land Use Change		Status
Ali Abad	Forest	Slight increase (+0.34%) from 45.79% to 46.13%	Ali-Abad's forest area is well preserved, with a slight increase projected, which should help to buffer hydrological drought by maintaining water retention and reducing surface runoff. However, the increase in bare land (+0.30%) is concerning, as this trend could exacerbate agricultural drought by reducing the productive land available for farming. The rise in agricultural land may reflect increased irrigation needs, which could place pressure on local water resources. With Ali Abad located centrally, near the forested and agricultural heart of the province, its stable forest cover will mitigate some drought effects, but land degradation (bare land) must be addressed to avoid worsening conditions.
	Rangeland	Decrease (−1.62%)	
	Agricultural	Increase (+0.60%)	
	Bare Land	Significant rise from 0.07% to 0.37%	
Aqqala	Rangeland	Substantial decline from 7.21% to 2.32%	Aqqala faces severe degradation, with rangeland loss (−4.89%) and an alarming increase in bare land (+12.45%). Aqqala's location in the northern plains, which is prone to drought and aridification, exacerbates this trend. The significant decrease in agricultural land reflects the impact of increasing drought severity, as water shortages likely make farming unsustainable. The bare land expansion directly correlates with worsening agricultural and hydrological drought, and climate projections from nearby stations likely indicate drier future conditions, requiring urgent intervention to halt desertification.
	Agricultural	Decrease (−7.13%)	
	Bare Land	Significant rise from 13.68% to 26.13%	
Azadshahr	Rangeland	Decrease (−1.09%)	Azadshahr's moderate rangeland decrease and slight agricultural increase suggest a transition in land use towards farming, but the rise in bare land signals potential soil degradation. This county, located in the southeast, is prone to drought stress due to climate variability. The increase in bare land may be driven by reduced rainfall and rising temperatures projected for nearby synoptic stations, pushing the land towards more desert-like conditions. The conservation of the remaining rangelands will be crucial to maintain moisture in the soil and prevent the further expansion of degraded areas.
	Agricultural	Slight increase (+0.62%)	
	Bare Land	Increase from 0.28% to 0.44%	

Table 5. *Cont.*

County	Land Use Change		Status
Bandar Gaz	Forest	Slight decrease (−0.28%)	Urbanization in Bandar Gaz, located along the Caspian Sea, is notable, with a +0.60% increase in built-up areas. This rise, along with the small reduction in forest and agricultural lands, could worsen hydrological drought due to increased impermeable surfaces (urban areas) that reduce groundwater recharge. Although forest loss is minimal, the rising urban footprint suggests that the water demand and management issues could become more severe as climate change leads to warmer temperatures and more erratic rainfall patterns.
	Agricultural	Minimal decrease (−0.50%)	
	Built-up	Rise from 5.57% to 6.17%	
Bandar-e Torkman	Agricultural	Slight decrease (−0.63%)	Bandar-e Torkman, another Caspian coastal area, shows a steady expansion in urban areas (+0.73%) and a small increase in bare land (+0.03%), while agricultural land shrinks slightly. Although the changes are not dramatic, this county's strong dependence on agriculture makes even minor land use shifts critical. With projections of rising temperatures and altered precipitation patterns in the coastal zones, agricultural drought could intensify, particularly if water resources are strained by expanding urbanization. Efforts to improve urban water management and protect agricultural land from overuse will be vital.
	Bare Land	Small increase (+0.03%)	
	Built-up	Increase from 6.28% to 7.01%	
Galikesh	Forest	Slight increase (+0.17%)	Galikesh, situated near the Alborz Mountains, retains a large forested area, which sees a small increase (+0.17%), helping to stabilize the local hydrological cycle. However, the reduction in rangeland (−2.60%) and the increase in agricultural land (+2.21%) point to intensifying land use pressures, likely driven by an increasing demand for farming. Given Galikesh's mountainous geography, changes in climate (precipitation and temperature) will have a pronounced effect on the agricultural drought risk, especially if rainfall patterns shift and temperatures rise.
	Rangeland	Sharp decline (−2.60%)	
	Agricultural	Increase (+2.21%)	

Table 5. *Cont.*

County	Land Use Change		Status
Gomishan	Rangeland	Sharp decline (−7.84%)	Gomishan experiences one of the most dramatic shifts, with a large increase in bare land (+13.24%) and a steep reduction in rangeland (−7.84%). This county, located in the low-lying coastal areas of the province, is vulnerable to land degradation and salinization due to its proximity to the Caspian Sea. The increase in water bodies (+5.11%) may reflect rising water levels or changes in water management, but it is not enough to offset the effects of widespread land degradation. These trends suggest a growing risk of desertification and hydrological drought due to increased bare land exposure.
	Bare Land	Large increase (+13.24%)	
	Water Bodies	Significant rise (+5.11%)	
Gonbad-e Kavus	Rangeland	Decrease (−2.04%)	Gonbad-e Kavus, a central agricultural hub, shows one of the largest bare land increases (+11.44%), coupled with a sharp decline in agricultural land (−9.27%). These changes indicate a strong correlation between increasing agricultural drought and the degradation of farmland into bare land, which exacerbates water scarcity issues. Rangeland also sees a notable decline (−2.04%), further highlighting the challenges faced by this region. The high degree of land degradation points to rising temperatures and reduced rainfall contributing to worsening drought conditions.
	Agricultural	Large decline (−9.27%)	
	Bare Land	Significant rise from 9.96% to 21.40%	
Gorgan	Rangeland	Decrease (−2.55%)	As the provincial capital, Gorgan shows a balance between urban expansion and land use changes. The rise in bare land (+0.54%) and reduction in rangeland (−2.55%) suggest increasing pressure on the available land, while agriculture expands slightly. The city's location near the Alborz foothills suggests that the increase in bare land could lead to hydrological drought, as urbanization and reduced rangeland disrupt natural water retention mechanisms.
	Agricultural	Increase (+1.48%)	
	Bare Land	Increase (+0.54%)	
Kalaleh	Moderate rangeland decline (−2.97%), with an increase in agricultural land (+2.86%)		This shift may lead to a heightened drought risk as land degradation continues.

Table 5. *Cont.*

County	Land Use Change	Status
Kordkuy	Small but important rise in bare land (+0.39%)	It indicates growing degradation, with slight reductions in forest and rangeland.
Maraveh Tappeh	Significant bare land expansion (+2.62%)	It signals increasing vulnerability to drought, particularly as rangelands shrink.
Minudasht and Ramiyan	Small increases in forest and agricultural land but some declines in rangeland	These counties show moderate stability. Minimal bare land expansion helps to maintain local drought resilience.

3.5. Key Environmental Drivers and Evaluation of Drought Susceptibility Using MaxEnt: Insights from Jackknife and AUC Analysis

The jackknife test of the maximum entropy (MaxEnt) model evaluated the predictive significance of 14 environmental variables regarding the agrometeorological drought susceptibility within Golestan Province (Figure 9). The analysis highlighted precipitation, the maximum temperature, and land use as the most influential predictors of drought, based on the lengths of their blue bars in the test. Precipitation exhibited a blue bar value of 0.771, followed by the maximum temperature at 0.788 and land use at 0.755. These factors demonstrated the strongest capabilities to independently predict the Standardized Drought Condition Index (SDCI). The green bar values, which reflect the decline in the AUC when a variable is excluded from the model, further emphasized the critical role of these variables. For example, precipitation's green bar value was 0.9, while the maximum temperature and land use had values of 0.898 and 0.878, respectively. The exclusion of these variables significantly reduced the model's performance, indicating their unique contributions to drought prediction.

Variables like the Soil Moisture Stress Index (SIWSI) and Simple Ratio Water Index (SRWI) also contributed notably, with blue bar values of 0.779 and 0.78, respectively. Although their individual predictive capacities were moderate, their inclusion in the model enhanced the accuracy, as evidenced by their consistent green bar values of 0.91. Conversely, vegetation indices such as the Green Normalized Difference Vegetation Index (GNDVI) and Infrared Percentage Vegetation Index (IPVI) showed lower individual predictive power, with blue bar values of 0.755 and 0.749, respectively, despite maintaining strong overall model performance.

According to the ROC curve, the MaxEnt model demonstrated reliable performance for drought susceptibility assessment (Figure 10). The generated AUC values of the ROC curves for the training and test data were 0.929 and 0.910. The generated ROC curve of the training data indicated that the MaxEnt model achieved a good discrimination capacity between drought and non-drought occurrences in the training points, evidenced by the generated AUC value of 0.929.

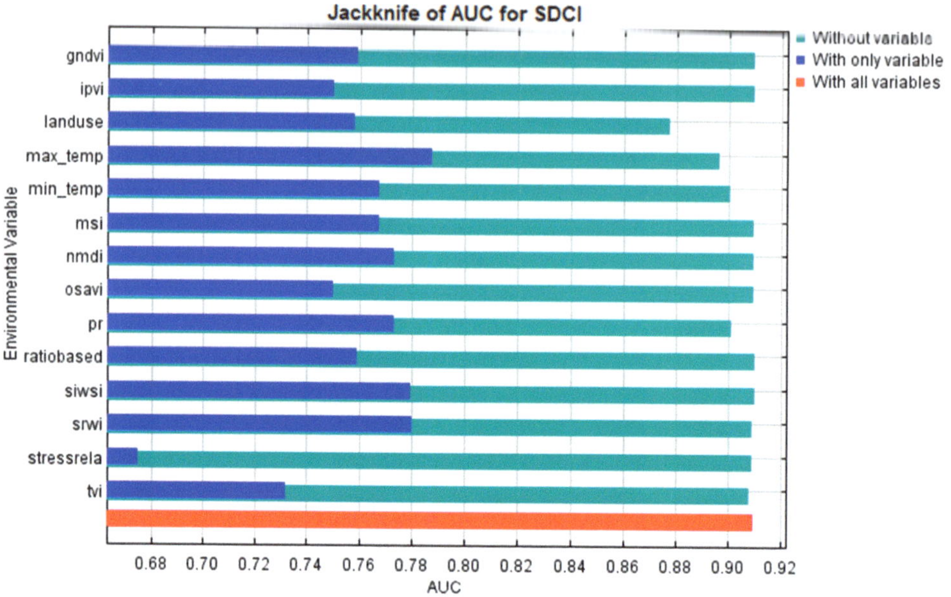

Figure 9. Jackknife test results for the MaxEnt model, showing the contributions of various environmental factors to drought susceptibility modeling.

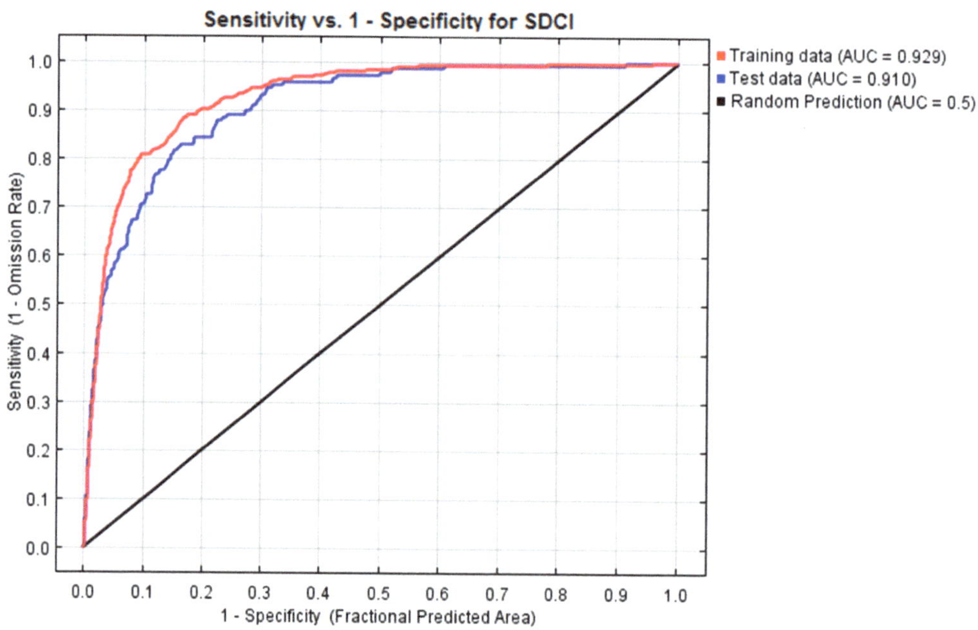

Figure 10. ROC curves from the MaxEnt model for the training and test stages of drought susceptibility modeling.

3.6. Integrated Analysis of Drought Susceptibility Classes and Their Link to Climate Change and Land Use Dynamics

The drought susceptibility maps were classified into five susceptibility categories (i.e., very low, low, moderate, high, and very high) using Jenks's natural breaks classification method [96], given the skewed, light-tailed distribution of the susceptibility frequency histogram [97,98]. Building upon the modeling of the 14 drought-controlling factors and 500 SDCI sample points (Figure 11), the drought susceptibility maps of Golestan Province reveal significant heterogeneity across the counties (Figure 12). In Gomishan, for instance, 66.12% of the county exhibits "very high" susceptibility, indicating severe threats to agricultural and hydrological resources. In contrast, Bandar Gaz shows 97.68% of its area categorized as "very low," which is a significant achievement for this region. The projected data for the future scenarios—SSP1-2.6, SSP2-4.5, and SSP5-8.5—indicate a troubling fate for the region. Under SSP1-2.6, Bandar-e Torkman shows a shift, with 89.25% of its area moving to the "very high" class. For SSP2-4.5, in Aqqala and Gomishan, most areas remain "highly" and "very highly" susceptible. Under SSP5-8.5, Aqqala is projected to have large areas of "moderate" (48.10%) and "high" (31.25%) susceptibility. The land use patterns show significant changes, particularly the conversion of rangeland to bare land in the counties of Aqqala and Gomishan. These changes correspond with the increased maximum temperatures and decreased summer rainfall. Gonbad-e Kavus shows the progressive degradation of agricultural land, linked to increasing temperatures, resulting in water shortages. Urban expansion in Bandar Gaz and Bandar-e Torkman has led to increased surface runoff and decreased groundwater recharge.

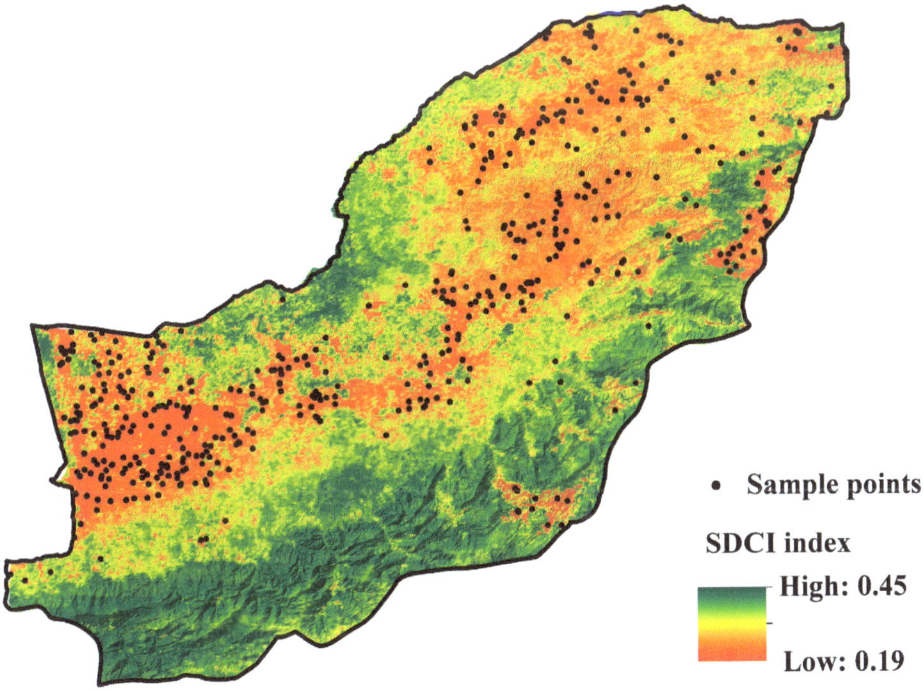

Figure 11. The SDCI map of Golestan Province, along with 500 sample points, which served as evidence of agrometeorological drought.

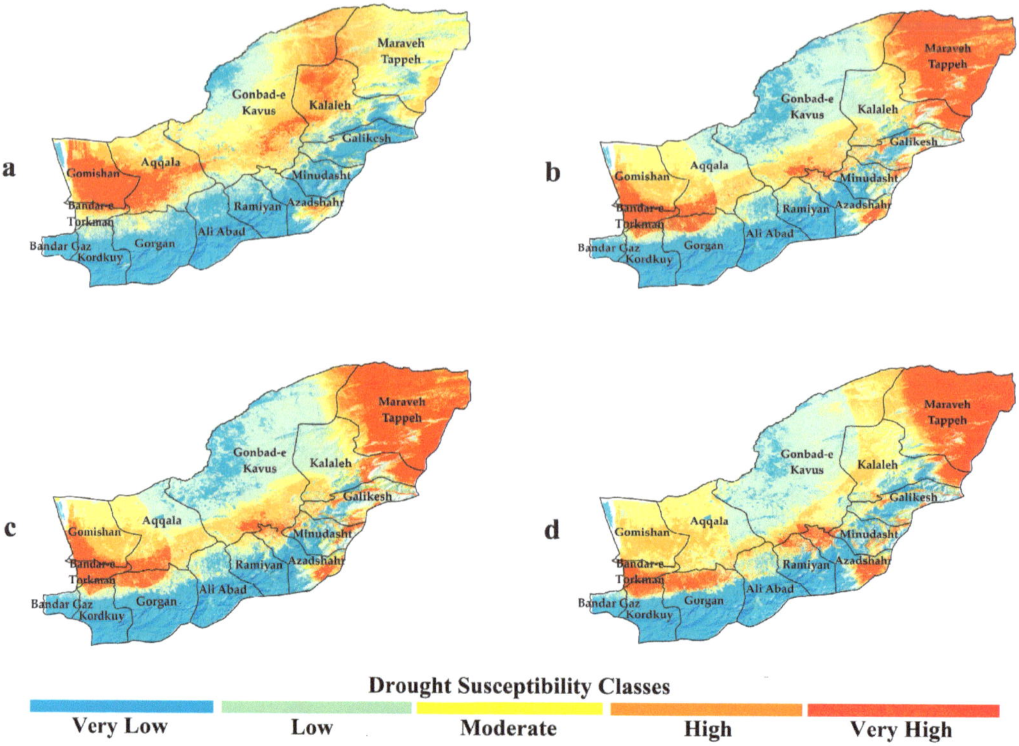

Figure 12. Drought susceptibility maps of Golestan Province for the current (**a**), SSP1-2.6 (**b**), SSP2-4.5 (**c**), and SSP5-8.5 (**d**) scenarios.

4. Discussion

4.1. Monthly Climate Change Data

The consistent warming projected across all scenarios underscores the inevitability of climate change impacts on Golestan Province, with SSP5-8.5 highlighting the most severe outcomes. Temperature increases, particularly in the summer months, could exacerbate the drought conditions by accelerating the evaporation rates and reducing the soil moisture availability. The projected temperature rises (+3.15 °C in April, +1.85 °C in July at Gorgan) under SSP5-8.5 may heighten heat stress, reducing crop yields and increasing the water demand. Similarly, the persistent rise in the minimum temperatures, such as +3.44 °C at Inceh Borun in March, suggests warmer nights, which may impact plant growth and lead to increased energy demands for cooling.

The high variability in precipitation, particularly under SSP5-8.5, poses additional challenges for water resource management. Stations like Gorgan and Ramian, with extreme precipitation swings such as −30.52 mm in September and +54.76 mm in December, highlight the dual risk of severe droughts and floods. These fluctuations are likely to strain the existing water infrastructure, including reservoirs, irrigation systems, and flood defenses, necessitating substantial investments in adaptive measures. In contrast, stations like Tamer and Maraveh show less extreme variability but still experience notable changes, such as precipitation decreases of −7.83 mm in September (Tamer under SSP5-8.5) and increases of +22.18 mm in December (Tamer under SSP1-2.6), indicating that even relatively stable regions are not immune to the effects of climate change.

Changes in land use, combined with these climatic shifts, are likely to exacerbate the drought susceptibility. The interplay of increased temperatures and erratic precipitation can reduce groundwater recharge, especially in areas experiencing urban expansion or transitions from rangeland to bare land. This is particularly concerning for regions where urbanization is likely to compound the effects of reduced rainfall in certain months. Moreover, higher precipitation variability can lead to flash flooding, as projected under SSP5-8.5 in stations like Ramian and Gorgan during December, potentially damaging infrastructure and ecosystems.

A coordinated climate adaptation and land use strategy is essential. Water storage solutions (e.g., reservoirs, rainwater harvesting) are vital in balancing rainfall extremes and mitigating drought. Crop selection and irrigation techniques should be adjusted to account for the warmer and drier conditions expected in the summer months, while flood-prone areas may require upgraded drainage and flood control systems. Monitoring systems are also essential in tracking climatic changes and informing adaptive management, especially in stations like Gorgan and Ramian, which exhibit significant variability.

The results emphasize the urgent need for scenario-based planning to mitigate the adverse effects of climate change. While SSP1-2.6 offers a more optimistic outlook with manageable temperature and precipitation changes, even these scenarios present challenges that require proactive measures. Under SSP5-8.5, the stakes are significantly higher, necessitating robust, coordinated action to protect both natural ecosystems and human communities in Golestan Province.

4.2. Annual Climate Change Data

The increase in annual rainfall foretells changes in regional atmospheric dynamics, conditioned, most likely, by changes in circulation patterns and moisture availability concerning global warming. Stations such as Gorgan, lying relatively close to the Caspian Sea, may be subject to more moisture influx. Although the increases in Voshmgir and Inceh Borun are significant, they are relatively smaller compared to Gorgan and Ramian due to the smaller amounts of observed precipitation at lower altitudes. Such an increase in precipitation also signals rising future challenges like increasing flood risks, particularly in areas with insufficient drainage systems and water management infrastructure.

The annual minimum temperature increase may alter local microclimates, affecting the length and intensity of the growing season for crops. An option that raises the minimum temperatures could diminish the risk of frost but could also disrupt the ecosystem dynamics. When the minimum temperatures rise, less diurnal temperature variation occurs—this influences plant growth, evapotranspiration rates, and the ecosystem's capacity to resist heat stress. The rise in the maximum temperatures presents two challenges: higher temperatures may lengthen the growing season in cooler regions but also increase the likelihood of drought if the water availability (from precipitation) cannot be maintained with higher temperatures. In combination with the expected increases in precipitation, this may yield more erratic weather, with more storms or drought periods becoming more frequent or intense. Even small increases in the maximum temperature can increase heat stress in human populations and natural systems, with more frequent and intense heatwaves.

4.3. Combined Monthly and Annual Climate Chnage

To evaluate whether any seasonal or long-term patterns might emerge, the mean annual and monthly results regarding the precipitation, minimum temperature, and maximum temperature across the synoptic stations were compared. An unequivocal and demonstrable increase in both precipitation and temperature is evident under all SSP scenario types, with the magnitude of the change increasing progressively from SSP1-2.6

through to SSP5-8.5. For example, in Gorgan, the annual precipitation is projected to increase from 642 mm under SSP1-2.6 to 749.7 mm under SSP5-8.5, a clear indicator of intensified atmospheric moisture availability. Similarly, the minimum temperatures show a rise of up to +3.44 °C in Inceh Borun in March under SSP5-8.5, while the maximum temperatures in Gorgan could rise by as much as +3.15 °C in April under the same scenario. These increases point to a pronounced trend towards a hotter and wetter climate, particularly under severe climate change scenarios such as SSP5-8.5.

The data highlight distinct seasonal patterns superimposed on these long-term trends. The winter and early spring months (e.g., January to April) are projected to experience the most significant increases in precipitation across stations, aligning with global observations in temperate regions, where the precipitation peaks during these seasons. ese months, which raises concerns about flood risks in areas with insufficient drainage infrastructure, such as parts of the Voshmgir and Ramian districts. For instance, the December precipitation at Ramian could rise by +57.24 mm under SSP5-8.5, compared to the current levels. Conversely, the summer months (e.g., June to August) show mixed precipitation trends, with some stations experiencing reductions (e.g., −30.52 mm in September at Gorgan under SSP5-8.5), potentially exacerbating drought conditions.

Seasonal contrasts in the temperature changes are similarly pronounced. The summertime maximum temperatures are projected to rise sharply, with Gorgan, for example, seeing increases of up to +3.15 °C in April under SSP5-8.5. The rise in the minimum temperatures across all seasons further suggests reduced frost occurrence, which could benefit agriculture by extending the growing season. However, the accompanying reduction in diurnal temperature variation and elevated heat stress may negatively affect crop yields, particularly during critical summer growth stages, when higher evapotranspiration rates may outpace the water availability. For instance, Maraveh shows minimum temperature increases of +2.58 °C in November under SSP5-8.5, signaling the potential for higher nighttime heat stress.

This pattern of accentuated seasonal contrasts suggests that the winter and early spring months will become wetter, posing increased flood risks, while the summers may become drier and hotter, raising the likelihood of prolonged heatwaves and drought conditions. Stations like Ramian and Voshmgir may see amplified intensities and frequencies of rainfall events during wetter months, while regions like Gorgan and Inceh Borun could face more extreme summer heat, exacerbating water scarcity and agricultural stress. These patterns are consistent with global climate projections that indicate more erratic weather, including intensified wet and dry periods.

These combined monthly and annual trends point towards an aggravated seasonal cycle characterized by heightened contrasts between wetter winters and drier summers. This dynamic may benefit crops by offering a longer growing season due to warmer winters but impose significant challenges during the summer months, when water deficits and heat stress could impair yields. The projected increases in annual and seasonal rainfall suggest opportunities for water storage and harvesting; however, adaptive measures will be necessary to mitigate the flood risks during wetter months and address water scarcity during prolonged dry spells. Effective strategies will require integrated water management, agricultural adaptation (e.g., drought-resistant crops and irrigation efficiencies), and disaster risk reduction to cope with the dual threats of intensified floods and droughts.

The findings emphasize the need for a multifaceted approach to climate adaptation, balancing the opportunities offered by a longer growing season with the challenges posed by increased seasonal variability. This unique combination of warmer temperatures and fluctuating precipitation patterns reinforces the urgency of developing resilient infrastructure and systems to safeguard the livelihoods and ecosystems of Golestan Province.

4.4. Analysis of Land Use Change and Its Impact on Drought Conditions

The land use analysis reveals a clear trend of land degradation, with significant shifts that will exacerbate both agricultural and hydrological droughts. The small decrease in the forest area of 1299 ha (0.06%) may seem insignificant but has critical implications for drought mitigation. Forests, which play an essential role in water retention, groundwater recharge, and preventing soil erosion, will lose their capacity to buffer drought conditions as they are converted into rangeland and built-up areas. The reduction in forests in this manner diminishes the land's ability to absorb moisture, thus contributing to higher rates of evapotranspiration and reinforcing drought conditions. The 2.83% decline in rangeland (57,300 ha) is also concerning. As rangelands are converted to bare land, their role in moisture retention is significantly diminished, leading to higher soil erosion rates and exacerbating the onset of hydrological droughts. This loss of rangeland may also result in overgrazing, further reducing the soil moisture and heightening the vulnerability to drought. Similarly, the 2.85% reduction in agricultural land (57,800 ha) intensifies the risk of agricultural drought. As agricultural lands shrink and are replaced by bare land and built-up areas, the remaining fertile land faces increased water stress, potentially forcing farmers to utilize less productive soils that are more prone to degradation. This shrinkage in cultivable land also places additional strain on water resources, exacerbating water scarcity and the depletion of local aquifers, thus further aggravating agricultural droughts. The 5.24% rise in bare land (106,202 ha) is particularly alarming, as bare land significantly reduces the land's capacity to retain water. Increased bare land leads to more runoff, less water infiltration, and higher evaporation rates, all of which amplify hydrological drought. The transformation of agricultural and rangeland areas into bare land further compounds these effects, reducing productivity and pushing the region towards desertification.

On the other hand, there is a small increase in water bodies (0.23% or 4710 ha), which could potentially serve as a buffer against drought, offering additional resources for irrigation. However, this gain is limited in scope and cannot counterbalance the broader negative effects of land use change. The slight reduction in wetland areas (0.05% or 940 ha) is also noteworthy, as wetlands play a crucial role in groundwater recharge and flood control. The conversion of wetlands into water bodies reduces their natural buffering capacity, which could further exacerbate hydrological droughts in the region. Finally, the 0.32% increase in built-up areas (6420 ha) driven by urbanization leads to the conversion of permeable land into impervious surfaces. This transformation reduces groundwater recharge and increases surface runoff, both of which contribute to hydrological drought. Urbanization also exacerbates the urban heat island effect, increasing the evaporation rates and reducing the water availability for agriculture, further intensifying the drought conditions. In sum, the land use changes observed in this study point to a growing challenge in mitigating drought conditions in the region. The increase in bare land and the reductions in forest, rangeland, and agricultural areas undermine the landscape's ability to retain moisture and support vegetation. While the growth of water bodies offers some potential relief, it is insufficient to offset the broader land degradation trends. These findings underscore the urgent need for targeted land management strategies and climate-smart agricultural practices to mitigate the impacts of climate change and land degradation on the drought susceptibility in the region.

4.5. Localized Analysis of Land Use Change in Golestan Province (2022 vs. 2030–2050)

The localized land use changes observed in Golestan Province highlight significant trends that could amplify the drought risks, both agricultural and hydrological. The widespread transformation of rangeland into bare land, especially in Aqqala, Gomishan, and Gonbad-e Kavus, is particularly concerning. The increase in bare land, by 12.45% in

Aqqala, 13.24% in Gomishan, and 11.44% in Gonbad-e Kavus, represents a loss of vital vegetation cover, which normally helps to retain moisture and prevent desertification. This transformation leads to a reduced ability to absorb and retain water, exacerbating both agricultural and hydrological droughts. Bare land increases the evaporation rates and accelerates soil erosion, making recovery from drought conditions more difficult. Additionally, the sharp decrease in agricultural land in Aqqala (−7.13%) and Gonbad-e Kavus (−9.27%) suggests that farming is becoming unsustainable in these areas, primarily due to water scarcity and land degradation. As agricultural land is converted to barren soil, the land becomes more vulnerable to further degradation, creating a feedback loop of soil erosion and reduced water availability, thereby heightening agricultural droughts. Urbanization in coastal areas, such as Bandar Gaz and Bandar-e Torkman, introduces another layer of stress. The increases in built-up areas of +0.60% and +0.73%, respectively, although modest, signify a growing trend towards impermeable surfaces that hinder water infiltration and increase surface runoff. This urban expansion contributes to the urban heat island effect, which intensifies evaporation and exacerbates water stress, further contributing to hydrological drought.

On the other hand, the slight increase in forest cover in counties like Galikesh (+0.17%) and Ali Abad (+0.34%) offers hope for resilience against drought. Forests play a critical role in sustaining microclimates, reducing surface runoff, and enhancing groundwater recharge. Their preservation helps to stabilize the soil, prevent erosion, and maintain the moisture levels in the landscape. These forests not only mitigate drought effects but also contribute to carbon sequestration and local climate regulation, supporting the region's long-term resilience to climate change and land degradation. However, the slight reduction in forest cover in Bandar Gaz and Kordkuy is concerning, as these areas are already vulnerable to urban encroachment and the expansion of barren land. Addressing the loss of forested areas and preventing further land degradation should be a priority in these counties to maintain their capacity to buffer drought conditions. Overall, the land-use changes across Golestan Province underscore the growing vulnerability of the region to drought, driven by a combination of land degradation, urban expansion, and shifts in agricultural practices. Strategies focusing on preserving forested areas, improving water management, and reversing land degradation are essential in mitigating the adverse impacts of drought in these counties.

4.6. Climate Change Projections and Their Interaction with Land Use Changes

Overall, the synoptic stations across Golestan Province project marked changes in the temperature and precipitation under the future climate scenarios (SSP1-2.6, SSP2-4.5, and SSP5-8.5). In general, the minimum and maximum temperatures are projected to increase under all scenarios, leading to increased evapotranspiration and deepening water stress, particularly in agricultural areas. Winter precipitation is expected to increase for all scenarios, pointing to a wetter winter and a greater probability of flooding, which is problematic under conditions of deepening drought and desertification. Reduced precipitation under future scenarios is projected during the growing season for most synoptic stations; the results from Gorgan, Ramian, and Gonbad-e Kavus are particularly alarming. These trends are consistent with the land use observations. The expansion of bare land in Aqqala and Gomishan, for example, may be directly related to consecutive increases in the temperature, leading to increasing soil moisture deficits. This result is corroborated by the reduction in the rangeland area in these counties, suggesting declining water resources and signaling a potential water shortage for vegetative regrowth. Urban expansion in Bandar Gaz and Bandar-e Torkman is likely to be a driver of water runoff and potential water shortage in these densely built areas; this pattern may also be related to the potential risk of extreme precipitation events and the increasing likelihood of water shortages under repetitive drought cycles. Rising temperatures increase the likelihood of increasing water runoff in deforested areas, but also increase the urban water needs in drier months.

The combined evidence of climate and land use changes indicates that Golestan Province is at a crossroads as it is rapidly transforming under the pressures of land degradation, urban expansion, and climate change, triggering mounting threats of future drought and potential desertification events. The present expansion of bare land, particularly in Aqqala and Gonbad-e Kavus, suggests the importance of developing and implementing conservation measures to prevent the further exacerbation of this condition. This should be prioritized for rangeland management to prevent or recover from desertification. In Ali Abad and Galikesh, it appears to be most crucial to provide some forest protection from immediate drought; the forests must be maintained and a mechanism established to preserve forests that are currently declining. The parts of the province that are experiencing forest loss, including Bandar Gaz, would benefit from reforestation, which has potentially positive influences on the local water cycle and drought likelihood. Sustainability-oriented urban expansion planning is tasked with integrating green infrastructure, like permeable pavements and living roofs, in urbanized counties. The goal of this is to minimize surface runoff and groundwater depletion and to control for urban heat islands or, at least, enhance the reliable supply of water to growing cities. As climate change accelerates, a crucial role will be played by adaptation options in relation to agriculture, particularly in places like Gonbad-e Kavus and Kalaleh, marked by significant reductions in productive land. In this way, food security will be sustained against escalating climatic change.

The combination of alterations in land use and climatic pressures has multiplied the risks of agricultural and hydrological drought in Golestan Province. With an extensive increase in bare land and a decrease in rangelands and agricultural land, Golestan faces increasing challenges in provisioning and supporting ecosystem services. County-wide trends reflect greater pressure on the environment, arising from climate change as well as unsustainable approaches to land use management. Targeted interventions focused on land degradation, supporting efforts towards forest conservation, and promoting sustainable agriculture are crucial to mitigate these growing risks. The future of the province lies in its capacity to cope with these emerging changes: integrated land use planning and climate-resilient development strategies will guarantee the long-term sustainability of natural and human systems under a climate characterized by increasingly strong droughts.

4.7. Key Environmental Drivers and Evaluation of Drought Susceptibility Using MaxEnt: Insights from Jackknife and AUC Analysis

The results underscore the dominant role of climate-related factors, particularly precipitation and the maximum temperature, in shaping the drought susceptibility within Golestan Province. As the primary driver, precipitation directly influences the water availability, with reduced rainfall exacerbating meteorological and agricultural drought. The maximum temperature acts synergistically, intensifying the evapotranspiration rates and depleting the soil moisture, thereby contributing to the drought severity. Land use emerges as another critical determinant, reflecting how anthropogenic alterations—such as agricultural expansion or rangeland degradation—affect the region's vulnerability to drought. The observed green bar value of 0.878 for land use signifies its indispensable role in improving the model's accuracy. Strategic land management practices, aimed at mitigating surface runoff and enhancing groundwater recharge, are essential to reducing drought's impacts. While vegetation indices, including the GNDVI and IPVI, exhibit lower standalone predictive capabilities, their integration into the MaxEnt model bolsters the overall performance. These indices provide indirect measures of vegetation health and stress, acting as supplementary inputs that refine the spatial characterization of drought.

Similarly, moisture-specific variables such as the SIWSI and SRWI enhance the model's reliability by offering detailed insights into the soil and water conditions. The findings highlight the need for targeted interventions to address the underlying drivers of drought. Adaptive strategies, such as optimizing irrigation practices, promoting afforestation, and implementing sustainable land use policies, can mitigate the risks in hotspots like Gomishan and Aqqala. Furthermore, proactive measures to combat rising temperatures and declining precipitation—projected under climate change scenarios—are crucial in safeguarding the region's agricultural productivity and water security. By integrating robust environmental data and predictive modeling, this study provides a comprehensive framework for the understanding and management of drought dynamics, emphasizing the interplay between climatic variables and land use in shaping regional susceptibility. These insights form a foundation for evidence-based planning and resilience-building efforts in Golestan Province.

The high AUC value in the ROC curve suggests that the areas with high vulnerability to drought were well discriminated from areas with low susceptibility, meaning that the model can be considered to be applicable when predicting the drought susceptibility using the training data. Meanwhile, the generated AUC value of 0.910 indicated that the model's performance was good for the test data within the same area, thus possessing good generalization abilities for independent data, meaning that we can assess the robustness of the model in predicting drought at different spatial scales. The overfitting of the model has been effectively avoided, as evidenced by the close alignment of the AUC training and AUC test curves, indicating that the developed model can be applied to new areas to assess the drought susceptibility. We also strongly attribute this to the fact that the model has a strong generalization capacity and is not only applicable to the specific study area. The excellent performance observed during model validation demonstrates its practical application value for various fields, such as scientific research, large-scale public disaster management, and various stakeholders. This is due to the MaxEnt model's ability to handle high-dimensional data and capture non-linear relationships, providing large space for the model's application. The predictive power of the MaxEnt setup in our study allows it to be easily integrated into practical decision-making tools for drought-prone areas such as Golestan Province, including the development of early warning systems and resource allocation optimization.

4.8. Integrated Analysis of Drought Susceptibility Classes and Their Link to Climate Change and Land Use Dynamics

The transition of hotspot areas from one scenario to another for Golestan Province highlights the strong need for timely interventions. For example, the Aqqala hotspot area will experience an alarming shift, with 48.10% of its area exhibiting "moderate" and 31.25% exhibiting "high" susceptibility under the SSP5-8.5 scenario. Moreover, Gomishan maintains high levels, with 66.12% of its territory falling within the "very high" category, underscoring this area as a hotspot that requires specific attention. It is imperative to address these transitions by applying comprehensive management strategies that are tailored to these hotspot areas and their respective challenges. Both Aqqala and Gomishan require the immediate adoption of more sustainable agricultural practices, such as the introduction of drought-resistant plant varieties and improved irrigation efficiency to maximize the potential water use. Additionally, improved soil conservation practices can decrease the drought severity by increasing water retention and reducing erosion. In the SSP5-8.5 scenario, Maraveh Tappeh is projected to have about 72.09% of its area experiencing "very high" susceptibility to drought. Temperature rises, combined with reduced summer precipitation, can exacerbate the depletion of water for agricultural and ecosystem-related needs. Management strategies urgently need to be employed to manage the risks related to drought in Maraveh Tappeh. The isolation of high-drought-risk regions through drought-resistant agriculture would help to maintain productivity in the face of natural disasters. Enhanced water management through rainwater harvesting systems for additional irrigation during dry seasons would be prudent.

The involvement of the community is crucial in the area of Maraveh Tappeh, as the greatest benefits could be obtained by local stakeholders. This involvement could provide valuable insights into sustainable practices that are acceptable from both cultural and ecological perspectives. Local initiatives for soil conservation and vegetation restoration would significantly improve the resilience of the land to dry conditions. Thus, Maraveh Tappeh would benefit from adopting these interventions, as it would prepare the area for the expected drought conditions and make it more environmentally sustainable. Lastly, continuous monitoring and community engagement will be crucial. Enabling local communities to participate in decision-making processes regarding water resource management and raising their awareness of sustainable practices could greatly boost the adaptation capacity of the region. Therefore, interventions within these hotspots are of the utmost importance to ensure that Golestan Province achieves greater resilience against the increasing threats of climate change and land use.

4.9. Comparison with Other Works

The findings of this study align closely with a substantial body of global and regional research, while also contributing novel insights into the interplay of climate change and land use dynamics. Our projections of increased precipitation during winter and early spring, coupled with rising temperatures under all SSP scenarios, resonate with the global trends identified by Dai (2013), who highlighted the amplification of drought risks due to reduced warm-season precipitation and elevated evapotranspiration under global warming [9]. Similarly, Kirtman et al. (2013) emphasized heightened seasonal variability in the temperature and precipitation, which supports our observations of wetter winters and hotter, drier summers in Golestan Province [12]. At a regional level, studies such as Soltani et al. (2012) and Asakereh et al. (2023) similarly documented increasing precipitation variability and temperature trends across Iran, which correspond closely with our findings [6,14]. Moreover, Madani (2014) underscored Iran's water management challenges, a perspective that complements our analysis of the dual threats of intensified winter rainfall and heightened summer drought conditions [7]. Our examination of land use changes, particularly the conversion of rangeland to bare land in Aqqala and Gomishan, aligns with the findings of Rahmati et al. (2019), who demonstrated the exacerbation of drought conditions through reduced groundwater recharge and increased surface runoff [17]. Furthermore, our integration of CMIP6 climate scenarios with land use projections represents a methodological advancement. While Wu et al. (2022) similarly coupled land use and climate models, their focus on habitat quality contrasts with our emphasis on drought susceptibility, demonstrating the broader applicability of such integrated frameworks [54].

Methodologically, our use of the MaxEnt model for drought susceptibility assessment is consistent with its widespread application in environmental modeling, as described by Phillips et al. (2006) and Elith et al. (2011) [45,46]. The model's strong predictive performance in this study, with AUC values of 0.929 for training and 0.910 for testing, confirms its robustness in assessing complex environmental risks such as drought, comparable to its success in species distribution modeling. This reliability further underscores the strength of our predictive framework. In comparison to earlier studies, such as Zhao and Dai (2017), which addressed uncertainties in global drought projections, our approach provides a localized and integrated perspective by combining climate change scenarios with land use dynamics [26]. While most prior research in Iran, such as that of Soltani et al. (2012)

and Morid et al. (2006), has focused on either climatic variables or land use changes in isolation, our study bridges these domains to offer a more comprehensive assessment of the future drought risks in Golestan Province [6,15]. This integration of spatiotemporal climate projections with land use modeling not only validates existing findings but also extends them by addressing critical gaps in understanding the multifaceted drivers of drought susceptibility.

In summary, the results of this study are well supported by previous works, yet they contribute important new insights into the interactions between climate and land use changes. This synthesis of global trends and localized impacts advances the discourse on drought risk assessment and highlights the pressing need for integrated management strategies tailored to regional vulnerabilities.

5. Conclusions

This study assessed Golestan's drought risks, emphasizing climate change and land use shifts. By applying the MaxEnt model, we were able to produce high-resolution drought susceptibility maps, which identified current drought hotspots like Gomishan and Aqqala, and we forecasted severe future drought conditions under various climate scenarios. The model effectively predicts droughts, providing insights into regional challenges, including land degradation, water scarcity, urban expansion, climate change, and the loss of critical ecosystems, all of which contribute to the growing vulnerability of Golestan Province to severe drought conditions. The study emphasizes the critical role of climate change and the land use dynamics in exacerbating the drought risk. In particular, the conversion of rangelands to bare land in areas like Aqqala and Gomishan, combined with rising temperatures and declining summer rainfall, is expected to aggravate the drought conditions. Urbanization, particularly in areas like Bandar-e Torkman and Bandar Gaz, further exacerbates the situation by increasing surface runoff and decreasing groundwater recharge, placing more pressure on the already strained water resources. In response to these findings, this study advocates for urgent intervention in the most vulnerable regions, particularly Gomishan, Aqqala, Bandar-e Torkman, and Bandar Gaz, where land degradation, water scarcity, urban expansion, and climate change are the most pronounced. Sustainable water strategies, including efficient irrigation and rainwater harvesting, are crucial. Additionally, land restoration initiatives, including the rehabilitation of degraded rangelands and reforestation, are critical in improving soil moisture retention and reversing the trend of land degradation. Urban planning must also incorporate green infrastructure, such as permeable pavements and green roofs, to effectively manage stormwater and enhance groundwater recharge.

In conclusion, our study underscores the need for integrated regional policies that align local drought management with broader strategies for water resource resilience. Collaborative frameworks between neighboring regions could help to mitigate shared challenges arising from ecological degradation, resource scarcity, and climate-induced migration. Given the alarming trends observed in Golestan Province, future research should focus on the development of adaptive land management strategies tailored to the mitigation of drought risks. Future studies could explore the potential for climate-smart agriculture, innovative irrigation techniques, and real-time monitoring systems to improve the water use efficiency, as well as the roles of afforestation and reforestation in enhancing ecosystem resilience. Additionally, evaluating the long-term impacts of urbanization and land use changes on water resources and soil quality will be essential in designing integrated water management strategies. It is also crucial to explore community-based participatory approaches to drought management, ensuring that local populations are actively involved in the design and implementation of sustainable practices. Future studies should explore the interactions between social, hydrological, and agrometeorological droughts to provide a more holistic understanding of the drought dynamics in Golestan Province. Incorporating social aspects, such as the impacts on local communities and livelihoods, alongside hydrological changes and water availability, will enable a comprehensive approach to drought management.

Author Contributions: Conceptualization, J.L. and M.J.S.; methodology, J.L.; software, J.L. and E.S.; validation, M.J.S. and E.S.; formal analysis, J.L.; investigation, M.J.S. and E.S.; resources, M.J.S.; data curation, M.J.S.; writing—original draft preparation, J.L. and M.J.S.; writing—review and editing, J.L., M.L., R.L., M.J.S., Y.R. and E.S.; visualization, J.L. and M.L.; supervision, R.L. and M.J.S.; project administration, J.L. and M.J.S.; funding acquisition, J.L. and R.L. All authors have read and agreed to the published version of the manuscript.

Funding: This research was jointly funded by the National Key Research and Development Project (Grant No. 2021YFB3901203), the Key Laboratory of Mine Spatio-Temporal Information and Ecological Restoration, MNR (No. KLM202301), Henan Provincial Science and Technology Research (No. 242102320017), and the Henan Province Joint Fund Project of Science and Technology (No. 222103810097).

Data Availability Statement: The data presented in this study are available upon request from the corresponding author.

Conflicts of Interest: The authors declare no conflicts of interest.

References

1. Wilhite, D.A.; Glantz, M.H. Understanding: The Drought Phenomenon: The Role of Definitions. *Water Int.* **1985**, *10*, 111–120. [CrossRef]
2. Orimoloye, I.R.; Belle, J.A.; Orimoloye, Y.M.; Olusola, A.O.; Ololade, O.O. Drought: A Common Environmental Disaster. *Atmosphere* **2022**, *13*, 111. [CrossRef]
3. Wilhite, D.A.; Buchanan-Smith, M. Drought as Hazard: Understanding the Natural and Social Context. *Drought Water Cris. Sci. Technol. Manag. Issues* **2005**, *3*, 29.
4. Byron-Cox, R. From Desertification to Land Degradation Neutrality: The UNCCD and the Development of Legal Instruments for Protection of Soils. In *Legal Instruments for Sustainable Soil Management in Africa*; Springer: Berlin/Heidelberg, Germany, 2020; pp. 1–13.
5. Wilhite, D.A.; Sivakumar, M.V.K.; Pulwarty, R. Managing Drought Risk in a Changing Climate: The Role of National Drought Policy. *Weather. Clim. Extrem.* **2014**, *3*, 4–13. [CrossRef]
6. Soltani, S.; Saboohi, R.; Yaghmaei, L. Rainfall and Rainy Days Trend in Iran. *Clim. Change* **2012**, *110*, 187–213. [CrossRef]
7. Madani, K. Water Management in Iran: What Is Causing the Looming Crisis? *J. Environ. Stud. Sci.* **2014**, *4*, 315–328. [CrossRef]
8. Karimi, S.; Rahemi, M.; Rostami, A.A.; Sedaghat, S. Drought Effects on Growth, Water Content and Osmoprotectants in Four Olive Cultivars with Different Drought Tolerance. *Int. J. Fruit Sci.* **2018**, *18*, 254–267. [CrossRef]
9. Dai, A. Increasing Drought under Global Warming in Observations and Models. *Nat. Clim. Chang.* **2013**, *3*, 52–58. [CrossRef]
10. Yasuhara, M.; Cronin, T.M.; Peter, B.; Okahashi, H.; Linsley, B.K.; Yang, H.; Wang, Y.; Liu, Z.; Work Group II; Weaver, A.J.; et al. IPCC 2014. Impacts, Adaptation and Vulnerability—Europe. *Clim. Dyn.* **2014**, *26*, 661–668. [CrossRef]
11. Masson-Delmotte, V.; Zhai, P.; Pirani, A.; Connors, S.L.; Péan, C.; Berger, S.; Caud, N.; Chen, Y.; Goldfarb, L.; Gomis, M.I. Climate Change 2021: The Physical Science Basis. *Contrib. Work. Group I Sixth Assess. Rep. Intergov. Panel Clim. Change* **2021**, *2*, 2391.
12. Kirtman, B.; Power, S.B.; Adedoyin, A.J.; Boer, G.J.; Bojariu, R.; Camilloni, I.; Doblas-Reyes, F.; Fiore, A.M.; Kimoto, M.; Meehl, G. Near-Term Climate Change: Projections and Predictability. In *Climate Change 2013: The Physical Science Basis. IPCC Working Group I Contribution to AR5*; Cambridge University Press: Cambridge, UK, 2013.
13. Silakhori, E.; Dahmardeh Ghaleno, M.R.; Meshram, S.G.; Alvandi, E. To Assess the Impacts of Climate Change on Runoff in Golestan Province, Iran. *Nat. Hazards* **2022**, *112*, 281–300. [CrossRef]
14. Asakereh, H.; Masoodian, S.A.; Tarkarani, F.; Zandkarimi, S. Decadal Change of Iran's Precipitation Regime. *Pure Appl. Geophys.* **2023**, *180*, 4275–4293. [CrossRef]
15. Morid, S.; Smakhtin, V.; Moghaddasi, M. Comparison of Seven Meteorological Indices for Drought Monitoring in Iran. *Int. J. Climatol.* **2006**, *26*, 971–985. [CrossRef]
16. Ghaffari, G.; Keesstra, S.; Ghodousi, J.; Ahmadi, H. SWAT-simulated hydrological impact of land-use change in the Zanjanrood basin, Northwest Iran. *Hydrol. Process. Int. J.* **2010**, *24*, 892–903. [CrossRef]

17. Rahmati, O.; Samadi, M.; Shahabi, H.; Azareh, A.; Rafiei-Sardooi, E.; Alilou, H.; Melesse, A.M.; Pradhan, B.; Chapi, K.; Shirzadi, A. SWPT: An Automated GIS-Based Tool for Prioritization of Sub-Watersheds Based on Morphometric and Topo-Hydrological Factors. *Geosci. Front.* **2019**, *10*, 2167–2175. [CrossRef]
18. Bhimala, K.R.; Rakesh, V.; Prasad, K.R.; Mohapatra, G.N. Identification of Vegetation Responses to Soil Moisture, Rainfall, and LULC over Different Meteorological Subdivisions in India Using Remote Sensing Data. *Theor. Appl. Climatol.* **2020**, *142*, 987–1001. [CrossRef]
19. Kouassi, J.-L.; Gyau, A.; Diby, L.; Bene, Y.; Kouamé, C. Assessing Land Use and Land Cover Change and Farmers' Perceptions of Deforestation and Land Degradation in South-West Côte d'Ivoire, West Africa. *Land* **2021**, *10*, 429. [CrossRef]
20. Silakhori, E.; Maleki, S.; Pourghasemi, H.R. Spatiotemporal Variations in Land Use of Mahvelat Plain in Iran Using Google Earth Engine from 2011 to 2030. In *Advanced Tools for Studying Soil Erosion Processes*; Elsevier: Amsterdam, The Netherlands, 2024; pp 413–433.
21. Emadodin, I.; Reinsch, T.; Taube, F. Drought and Desertification in Iran. *Hydrology* **2019**, *6*, 66. [CrossRef]
22. Avazpour, N.; Faramarzi, M.; Omidipour, R.; Mehdizadeh, H. Monitoring the Drought Effects on Vegetation Changes Using Satellite Imagery (Case Study: Ilam Catchment). *Geogr. Environ. Sustain.* **2021**, *11*, 125–143.
23. Khaledian, Y.; Kiani, F.; Ebrahimi, S. The Effect of Land Use Change on Soil and Water Quality in Northern Iran. *J. Mt. Sci.* **2012**, *9*, 798–816. [CrossRef]
24. Zare, M.; Samani, A.A.N.; Mohammady, M. The Impact of Land Use Change on Runoff Generation in an Urbanizing Watershed in the North of Iran. *Environ. Earth Sci.* **2016**, *75*, 1–20. [CrossRef]
25. Keshtkar, M.; Mokhtari, Z.; Sayahnia, R. Impact of Land Use/land Cover Change on Soil Retention Service: A Case of Agricultural-Urbanized Landscape in Northern Iran. *J. Landsc. Ecol.* **2022**, *15*, 34–58. [CrossRef]
26. Zhao, T.; Dai, A. Uncertainties in Historical Changes and Future Projections of Drought. Part II: Model-Simulated Historical and Future Drought Changes. *Clim. Change* **2017**, *144*, 535–548. [CrossRef]
27. Taylor, K.E.; Stouffer, R.J.; Meehl, G.A. An Overview of CMIP5 and the Experiment Design. *Bull. Am. Meteorol. Soc.* **2012**, *93*, 485–498. [CrossRef]
28. IPCC. *Climate Change. The Physical Science Basis. Contribution of Working Group I to the Fourth Assessment Report of the Intergovernmental Panel on Climate Change*; Cambridge University Press: Cambridge, UK, 2007; pp. 747–785.
29. Eyring, V.; Bony, S.; Meehl, G.A.; Senior, C.A.; Stevens, B.; Stouffer, R.J.; Taylor, K.E. Overview of the Coupled Model Intercomparison Project Phase 6 (CMIP6) Experimental Design and Organization. *Geosci. Model Dev.* **2016**, *9*, 1937–1958. [CrossRef]
30. Zhai, J.; Mondal, S.K.; Fischer, T.; Wang, Y.; Su, B.; Huang, J.; Tao, H.; Wang, G.; Ullah, W.; Uddin, M.J. Future Drought Characteristics through a Multi-Model Ensemble from CMIP6 over South Asia. *Atmos. Res.* **2020**, *246*, 105111. [CrossRef]
31. Li, H.; Li, Z.; Chen, Y.; Liu, Y.; Hu, Y.; Sun, F.; Kayumba, P.M. Projected Meteorological Drought over Asian Drylands under Different CMIP6 Scenarios. *Remote Sens.* **2021**, *13*, 4409. [CrossRef]
32. Yang, L.; Tian, J.; Fu, Y.; Zhu, B.; He, X.; Gao, M.; Odamtten, M.T.; Kong, R.; Zhang, Z. Will the Arid and Semi-Arid Regions of Northwest China Become Warmer and Wetter Based on CMIP6 Models? *Hydrol. Res.* **2022**, *53*, 29–50. [CrossRef]
33. Yahaya, I.; Zhou, J.; Jiang, S.; Xu, R.H.; Huang, J.; Su, B.; Jing, C.; Qiu, F.; Jiang, T. Depiction of Drought Over Africa in the Light of Changing Climate from CMIP6 Models. *Earth Syst. Environ.* **2024**, *8*, 1011–1035. [CrossRef]
34. Sierra-Soler, A.; Adamowski, J.; Qi, Z.; Saadat, H.; Pingale, S. High Accuracy Land Use Land Cover (LULC) Maps for Detecting Agricultural Drought Effects in Rainfed Agro-Ecosystems in Central Mexico. *J. Water Land Dev.* **2015**, *26*, 19–35. [CrossRef]
35. Akinyemi, F.O. Vegetation Trends, Drought Severity and Land Use-Land Cover Change during the Growing Season in Semi-Arid Contexts. *Remote Sens.* **2021**, *13*, 836. [CrossRef]
36. Patel, S.K.; Sharma, A.; Barla, A.; Singh, G.S. Traditional Farming in the Indian Himalayan Region: Preserving Agro-Biodiversity and Future Food Security. In *The Resilience of Traditional Knowledge Systems for a Sustainable Future. Advances in Asian Human-Environmental Research*; Springer: Cham, Switzerland, 2024; pp. 27–57.
37. Gondim, R.; Silveira, C.; de Souza Filho, F.; Vasconcelos, F.; Cid, D. Climate Change Impacts on Water Demand and Availability Using CMIP5 Models in the Jaguaribe Basin, Semi-Arid Brazil. *Environ. Earth Sci.* **2018**, *77*, 550. [CrossRef]
38. Xu, K.; Wu, C.; Zhang, C.; Hu, B.X. Uncertainty Assessment of Drought Characteristics Projections in Humid Subtropical Basins in China Based on Multiple CMIP5 Models and Different Index Definitions. *J. Hydrol.* **2021**, *600*, 126502. [CrossRef]
39. Alemu, M.G.; Wubneh, M.A.; Worku, T.A.; Womber, Z.R.; Chanie, K.M. Comparison of CMIP5 Models for Drought Predictions and Trend Analysis over Mojo Catchment, Awash Basin, Ethiopia. *Sci. Afr.* **2023**, *22*, e01891. [CrossRef]
40. El-Rawy, M.; Batelaan, O.; Al-Arifi, N.; Alotaibi, A.; Abdalla, F.; Gabr, M.E. Climate Change Impacts on Water Resources in Arid and Semi-Arid Regions: A Case Study in Saudi Arabia. *Water* **2023**, *15*, 606. [CrossRef]
41. Dixit, S.; Atla, B.M.; Jayakumar, K. V Evolution and Drought Hazard Mapping of Future Meteorological and Hydrological Droughts Using CMIP6 Model. *Stoch. Environ. Res. Risk Assess.* **2022**, *36*, 3857–3874. [CrossRef]

42. Ty, T.V.; Lavane, K.; Nguyen, P.C.; Downes, N.K.; Nam, N.D.G.; Minh, H.V.T.; Kumar, P. Assessment of Relationship between Climate Change, Drought, and Land Use and Land Cover Changes in a Semi-Mountainous Area of the Vietnamese Mekong Delta. *Land* **2022**, *11*, 2175. [CrossRef]
43. Haider, S.; Masood, M.U.; Rashid, M.; Alshehri, F.; Pande, C.B.; Katipoğlu, O.M.; Costache, R. Simulation of the Potential Impacts of Projected Climate and Land Use Change on Runoff under CMIP6 Scenarios. *Water* **2023**, *15*, 3421. [CrossRef]
44. Mohseni, U.; Agnihotri, P.G.; Pande, C.B.; Durin, B. Understanding the Climate Change and Land Use Impact on Streamflow in the Present and Future under CMIP6 Climate Scenarios for the Parvara Mula Basin, India. *Water* **2023**, *15*, 1753. [CrossRef]
45. Phillips, S.J.; Anderson, R.P.; Schapire, R.E. Maximum Entropy Modeling of Species Geographic Distributions. *Ecol. Modell.* **2006**, *190*, 231–259. [CrossRef]
46. Elith, J.; Phillips, S.J.; Hastie, T.; Dudík, M.; Chee, Y.E.; Yates, C.J. A Statistical Explanation of MaxEnt for Ecologists. *Divers. Distrib.* **2011**, *17*, 43–57. [CrossRef]
47. Pérez Navarro, M.Á.; Sapes, G.; Batllori, E.; Serra-Diaz, J.M.; Esteve, M.A.; Lloret, F. Climatic Suitability Derived from Species Distribution Models Captures Community Responses to an Extreme Drought Episode. *Ecosystems* **2019**, *22*, 77–90. [CrossRef]
48. Ghazipour, F.; Mahjouri, N. A Multi-Model Data Fusion Methodology for Seasonal Drought Forecasting under Uncertainty: Application of Bayesian Maximum Entropy. *J. Environ. Manag.* **2022**, *304*, 114245. [CrossRef] [PubMed]
49. Eghtedarnezhad, M.; Malekinezhad, H.; Rafiei, E. Investigating the Effectiveness of the Maximum Entropy Model in Measuring the Potential of Areas Prone to Hydrological Drought of Underground Water Resources (Case Study: Jiroft Plain). *Iran. J. Ecohydrol.* **2024**, *11*, 105–124.
50. Javidan, N.; Kavian, A.; Pourghasemi, H.R.; Conoscenti, C.; Jafarian, Z.; Rodrigo-Comino, J. Evaluation of Multi-Hazard Map Produced Using MaxEnt Machine Learning Technique. *Sci. Rep.* **2021**, *11*, 6496. [CrossRef] [PubMed]
51. Soltanpour, H.; Serrhini, K.; Gill, J.C.; Fuchs, S.; Mohadjer, S. Multi-Hazard Susceptibility Mapping in the Karst Context Using a Machine-Learning Method (MaxEnt). *EGUsphere* **2024**, *2024*, 1–39.
52. Orefice, S.; Innocenti, C. Regional Assessment of Coastal Landslide Susceptibility in Liguria, Northern Italy, Using MaxEnt. *Nat. Hazards* **2024**, 1–27. [CrossRef]
53. Qasimi, A.B.; Isazade, V.; Berndtsson, R. Flood Susceptibility Prediction Using MaxEnt and Frequency Ratio Modeling for Kokcha River in Afghanistan. *Nat. Hazards* **2024**, *120*, 1367–1394. [CrossRef]
54. Wu, J.; Luo, J.; Zhang, H.; Qin, S.; Yu, M. Projections of Land Use Change and Habitat Quality Assessment by Coupling Climate Change and Development Patterns. *Sci. Total Environ.* **2022**, *847*, 157491. [CrossRef]
55. Zougmoré, R.; Partey, S.; Ouédraogo, M.; Omitoyin, B.; Thomas, T.; Ayantunde, A.; Ericksen, P.; Said, M.; Jalloh, A. Toward Climate-Smart Agriculture in West Africa: A Review of Climate Change Impacts, Adaptation Strategies and Policy Developments for the Livestock, Fishery and Crop Production Sectors. *Agric. Food Secur.* **2016**, *5*, 1–16. [CrossRef]
56. Gohari, A.; Mirchi, A.; Madani, K. System Dynamics Evaluation of Climate Change Adaptation Strategies for Water Resources Management in Central Iran. *Water Resour. Manag.* **2017**, *31*, 1413–1434. [CrossRef]
57. Walker, W.E.; Haasnoot, M.; Kwakkel, J.H. Adapt or Perish: A Review of Planning Approaches for Adaptation under Deep Uncertainty. *Sustainability* **2013**, *5*, 955–979. [CrossRef]
58. Hoseini Rostami, F.; Jahani, M.A. Analysis of Service Quality Gap between Perceptions and Expectations of Service Recipients Using SERVQUAL Approach in Selected Hospitals in Golestan Province. *Iran. J. Health Sci.* **2018**, *6*, 58–67. [CrossRef]
59. Rahmati, M.; Mirás-Avalos, J.M.; Valsesia, P.; Davarynejad, G.H.; Bannayan, M.; Azizi, M.; Lescourret, F.; Génard, M.; Vercambre, G. Assessing the Effects of Water Stress on Peach Fruit Quality and Size Using the QualiTree Model. In Proceedings of the XXX International Horticultural Congress IHC2018: International Symposium on Cultivars, Rootstocks and Management Systems of 1281, Istanbul, Turkey, 12–16 August 2018; pp. 539–546.
60. Ji, L.; Peters, A.J. Assessing Vegetation Response to Drought in the Northern Great Plains Using Vegetation and Drought Indices. *Remote Sens. Environ.* **2003**, *87*, 85–98. [CrossRef]
61. Bajgiran, P.R.; Darvishsefat, A.A.; Khalili, A.; Makhdoum, M.F. Using AVHRR-Based Vegetation Indices for Drought Monitoring in the Northwest of Iran. *J. Arid Environ.* **2008**, *72*, 1086–1096. [CrossRef]
62. Rousta, I.; Olafsson, H.; Moniruzzaman, M.; Zhang, H.; Liou, Y.-A.; Mushore, T.D.; Gupta, A. Impacts of Drought on Vegetation Assessed by Vegetation Indices and Meteorological Factors in Afghanistan. *Remote Sens.* **2020**, *12*, 2433. [CrossRef]
63. Zhu, X.; Li, Q.; Guo, C. Evaluation of the Monitoring Capability of Various Vegetation Indices and Mainstream Satellite Band Settings for Grassland Drought. *Ecol. Inform.* **2024**, *82*, 102717. [CrossRef]
64. Gitelson, A.A.; Kaufman, Y.J.; Merzlyak, M.N. Use of a Green Channel in Remote Sensing of Global Vegetation from EOS-MODIS. *Remote Sens. Environ.* **1996**, *58*, 289–298. [CrossRef]
65. Crippen, R.E. Calculating the Vegetation Index Faster. *Remote Sens. Environ.* **1990**, *34*, 71–73. [CrossRef]
66. Hunt, E.R., Jr.; Rock, B.N. Detection of Changes in Leaf Water Content Using near-and Middle-Infrared Reflectances. *Remote Sens. Environ.* **1989**, *30*, 43–54.

67. Wang, L.; Qu, J.J. NMDI: A Normalized Multi-band Drought Index for Monitoring Soil and Vegetation Moisture with Satellite Remote Sensing. *Geophys. Res. Lett.* **2007**, *34*, L20405. [CrossRef]
68. Rondeaux, G.; Steven, M.; Baret, F. Optimization of Soil-Adjusted Vegetation Indices. *Remote Sens. Environ.* **1996**, *55*, 95–107. [CrossRef]
69. Zarco-Tejada, P.J.; Ustin, S.L. Modeling Canopy Water Content for Carbon Estimates from MODIS Data at Land EOS Validation Sites. In Proceedings of the IGARSS 2001. Scanning the Present and Resolving the Future. Proceedings. IEEE 2001 International Geoscience and Remote Sensing Symposium (Cat. No. 01CH37217), Sydney, NSW, Australia, 9–13 July 2001; Volume 1, pp. 342–344.
70. Fensholt, R.; Sandholt, I. Derivation of a Shortwave Infrared Water Stress Index from MODIS near-and Shortwave Infrared Data in a Semiarid Environment. *Remote Sens. Environ.* **2003**, *87*, 111–121. [CrossRef]
71. Jafari, R.; Lewis, M.M.; Ostendorf, B. Evaluation of Vegetation Indices for Assessing Vegetation Cover in Southern Arid Lands in South Australia. *Rangel. J.* **2007**, *29*, 39–49. [CrossRef]
72. Milner, N. *Advanced Spatial Regression Modeling of Corn and Soybean Yield with 30 Vegetation Indices Derived from a Sentinel-2 Time-Series*; Southern Illinois University at Edwardsville: Edwardsville, IL, USA, 2021; ISBN 9798780611400.
73. Alahacoon, N.; Edirisinghe, M. A Comprehensive Assessment of Remote Sensing and Traditional Based Drought Monitoring Indices at Global and Regional Scale. *Geomat. Nat. Hazards Risk* **2022**, *13*, 762–799. [CrossRef]
74. Gislason, P.O.; Benediktsson, J.A.; Sveinsson, J.R. Random Forests for Land Cover Classification. *Pattern Recognit. Lett.* **2006**, *27*, 294–300. [CrossRef]
75. Rodriguez-Galiano, V.F.; Ghimire, B.; Rogan, J.; Chica-Olmo, M.; Rigol-Sanchez, J.P. An Assessment of the Effectiveness of a Random Forest Classifier for Land-Cover Classification. *ISPRS J. Photogramm. Remote Sens.* **2012**, *67*, 93–104. [CrossRef]
76. Phan, T.N.; Kuch, V.; Lehnert, L.W. Land Cover Classification Using Google Earth Engine and Random Forest classifier—The Role of Image Composition. *Remote Sens.* **2020**, *12*, 2411. [CrossRef]
77. Rhee, J.; Im, J.; Carbone, G.J. Monitoring Agricultural Drought for Arid and Humid Regions Using Multi-Sensor Remote Sensing Data. *Remote Sens. Environ.* **2010**, *114*, 2875–2887. [CrossRef]
78. Kogan, F.N. Application of Vegetation Index and Brightness Temperature for Drought Detection. *Adv. Sp. Res.* **1995**, *15*, 91–100. [CrossRef]
79. Liu, W.T.; Kogan, F.N. Monitoring Regional Drought Using the Vegetation Condition Index. *Int. J. Remote Sens.* **1996**, *17*, 2761–2782. [CrossRef]
80. Du, L.; Tian, Q.; Yu, T.; Meng, Q.; Jancso, T.; Udvardy, P.; Huang, Y. A Comprehensive Drought Monitoring Method Integrating MODIS and TRMM Data. *Int. J. Appl. Earth Obs. Geoinf.* **2013**, *23*, 245–253. [CrossRef]
81. Räisänen, J.; Ruokolainen, L.; Ylhäisi, J. Weighting of Model Results for Improving Best Estimates of Climate Change. *Clim. Dyn.* **2010**, *35*, 407–422. [CrossRef]
82. Xu, Z.P.; Li, Y.P.; Huang, G.H.; Wang, S.G.; Liu, Y.R. A Multi-Scenario Ensemble Streamflow Forecast Method for Amu Darya River Basin under Considering Climate and Land-Use Changes. *J. Hydrol.* **2021**, *598*, 126276. [CrossRef]
83. Voldoire, A.; Saint-Martin, D.; Sénési, S.; Decharme, B.; Alias, A.; Chevallier, M.; Colin, J.; Guérémy, J.; Michou, M.; Moine, M. Evaluation of CMIP6 Deck Experiments with CNRM-CM6-1. *J. Adv. Model. Earth Syst.* **2019**, *11*, 2177–2213. [CrossRef]
84. John, J.G.; Blanton, C.; McHugh, C.; Radhakrishnan, A.; Rand, K.; Vahlenkamp, H.; Wilson, C.; Zadeh, N.T.; Dunne, J.P.; Dussin, R. *Noaa-Gfdl Gfdl-esm4 Model Output Prepared for cmip6 Scenariomip*; World Data Center for Climate: Hamburg, Germany, 2018.
85. Good, P. *MOHC HadGEM3-GC31-LL Model Output Prepared for CMIP6 ScenarioMIP*; World Data Center for Climate: Hamburg, Germany, 2019.
86. Jungclaus, J.; Mikolajewicz, U.; Kapsch, M.-L.; D'Agostino, R.; Wieners, K.-H.; Giorgetta, M.; Reick, C.; Esch, M.; Bittner, M.; Legutke, S. *MPI-M MPIESM1. 2-LR Model Output Prepared for CMIP6 PMIP*; World Data Center for Climate: Hamburg, Germany, 2019.
87. Yukimoto, S.; Koshiro, T.; Kawai, H.; Oshima, N.; Yoshida, K.; Urakawa, S.; Tsujino, H.; Deushi, M.; Tanaka, T.; Hosaka, M. *MRI MRI-ESM2. 0 Model Output Prepared for CMIP6 ScenarioMIP*; World Data Center for Climate: Hamburg, Germany, 2019.
88. Chen, W.; Pourghasemi, H.R.; Kornejady, A.; Zhang, N. Landslide Spatial Modeling: Introducing New Ensembles of ANN, MaxEnt, and SVM Machine Learning Techniques. *Geoderma* **2017**, *305*, 314–327. [CrossRef]
89. Kornejady, A.; Kouchaki, E.; Boustan, A.; Pourghasemi, H.R.; Sadeghinia, M.; Heidari, A. Multihazard Risk Analysis and Governance across a Provincial Capital in Northern Iran. In *Computers in Earth and Environmental Sciences*; Elsevier: Amsterdam, The Netherlands, 2022; pp. 655–673.
90. Seidenfeld, T. Entropy and Uncertainty. *Philos. Sci.* **1986**, *53*, 467–491. [CrossRef]
91. Karmeshu, A.; Pal, N.R. Uncertainty, Entropy and Maximum Entropy Principle—An Overview. In *Entropy Measures, Maximum Entropy Principle and Emerging Applications*; Springer: Berlin/Heidelberg, Germany, 2003; pp. 1–78.
92. Merow, C.; Smith, M.J.; Silander, J.A., Jr. A Practical Guide to MaxEnt for Modeling Species' Distributions: What It Does, and Why Inputs and Settings Matter. *Ecography* **2013**, *36*, 1058–1069. [CrossRef]

93. Verbakel, J.Y.; Steyerberg, E.W.; Uno, H.; De Cock, B.; Wynants, L.; Collins, G.S.; Van Calster, B. ROC Curves for Clinical Prediction Models Part 1. ROC Plots Showed No Added Value above the AUC When Evaluating the Performance of Clinical Prediction Models. *J. Clin. Epidemiol.* **2020**, *126*, 207–216. [CrossRef]
94. Hou, S.; Liu, Y.; Yang, Q. Real-Time Prediction of Rock Mass Classification Based on TBM Operation Big Data and Stacking Technique of Ensemble Learning. *J. Rock Mech. Geotech. Eng.* **2022**, *14*, 123–143. [CrossRef]
95. Zhou, Y.; Wu, W.; Wang, H.; Zhang, X.; Yang, C.; Liu, H. Identification of Soil Texture Classes under Vegetation Cover Based on Sentinel-2 Data with SVM and SHAP Techniques. *IEEE J. Sel. Top. Appl. Earth Obs. Remote Sens.* **2022**, *15*, 3758–3770. [CrossRef]
96. Jenks, Y.F. The data model concept in statistical mapping. *Int. Yearb. Cartogr.* **1967**, *7*, 186–190.
97. Ayalew, L.; Yamagishi, H. The application of GIS-based logistic regression for landslide susceptibility mapping in the Kakuda-Yahiko Mountains, Central Japan. *Geomorphology* **2005**, *65*, 15–31. [CrossRef]
98. Akgun, A. A comparison of landslide susceptibility maps produced by logistic regression, multi-criteria decision, and likelihood ratio methods: A case study at İzmir, Turkey. *Landslides* **2012**, *9*, 93–106. [CrossRef]

land

MDPI

Article

Assessing Climate and Land-Use Change Scenarios on Future Desertification in Northeast Iran: A Data Mining and Google Earth Engine-Based Approach

Weibo Yin [1], Qingfeng Hu [2,*], Jinping Liu [2,3], Peipei He [2], Dantong Zhu [2] and Abdolhossein Boali [4]

1 School of Civil Engineeing and Transportation, North China University of Water Resources and Electric Power, Zhengzhou 450046, China; yinweib@ncwu.edu.cn
2 College of Surveying and Geo-Informatics, North China University of Water Resources and Electric Power, Zhengzhou 450046, China
3 Hydraulics and Geotechnics Section, KU Leuven, Kasteelpark Arenberg 40, BE-3001 Leuven, Belgium
4 Department of Arid Zone Management, Gorgan University of Agricultural Sciences and Natural Resources, Gorgan 4913815739, Iran
* Correspondence: huqingfeng@ncwu.edu.cn

Abstract: Desertification poses a significant threat to dry and semi-arid regions worldwide, including Northeast Iran. This study investigates the impact of future climate and land-use changes on desertification in this region. Six remote sensing indices were selected to model desertification using four machine learning algorithms: Random Forest (RF), Support Vector Machine (SVM), Gradient Boosting Machine (GBM), and Generalized Linear Models (GLM). To enhance the model's reliability, an ensemble model was employed. Future climate and land-use scenarios were projected using the CNRM-CM6 model and Markov chain analysis, respectively. Results indicate that the RF and SVM models performed best in mapping current desertification patterns. The ensemble model highlights a 2% increase in decertified areas by 2040, primarily in the northwestern regions. The study underscores the importance of land-use change and climate change in driving desertification and emphasizes the need for sustainable land management practices and climate change adaptation strategies to mitigate future impacts.

Keywords: land degradation; desertification; prediction; modelling; ensemble model

1. Introduction

Desertification, a complex environmental issue, is accelerating globally, particularly in arid and semi-arid regions. This phenomenon, a process leading to land degradation and reduced productivity, poses a significant environmental challenge in arid and semi-arid regions worldwide [1]. Climate change and unsustainable land-use practices are major drivers of this phenomenon, leading to significant ecological, economic, and social consequences. Although various factors contribute to desertification, climate change and subsequent land-use changes are the primary drivers [2]. Climate change, characterized by long-term shifts in temperature, precipitation, wind patterns, and other climatic variables, is a major contributor to desertification [3]. This phenomenon has led to an expansion of desert areas, affecting over 250 million people directly and billions more indirectly [4]. Desertification can occur in all climates, accompanied by a decline in ecological and biological capacity [5]. Rising temperatures and altered precipitation patterns reduce agricultural and rangeland productivity, often prompting land-use changes such as the conversion of agricultural land to urban or industrial areas [6]. This interplay between climate change and land-use change creates a negative feedback loop, exacerbating desertification [7].

Land-use change, such as the conversion of forests and grasslands to agricultural or urban areas, exacerbates these effects by reducing vegetation cover, increasing soil erosion, and decreasing soil infiltration. These changes disrupt the local and regional water and energy balance, amplifying the impacts of climate change and accelerating desertification. For example, the reduction in vegetation cover caused by land-use change leads to increased surface temperatures and decreased soil moisture, further altering precipitation patterns and intensifying droughts. In essence, desertification, land-use change, and climate change are interconnected and mutually reinforcing processes. Addressing desertification requires a comprehensive and integrated approach that simultaneously addresses climate change mitigation, sustainable land management, and restoration of degraded ecosystems.

The Middle East, including Iran, is highly susceptible to desertification due to its naturally arid and semi-arid climate, rendering its ecosystems fragile and vulnerable to degradation [8]. Northeast Iran, with its particularly harsh climatic conditions, is especially vulnerable to this environmental threat. Human activities, such as unsustainable land use, have further intensified these pressures, accelerating the desertification process. In Iran, desertification is a pressing environmental issue with profound implications for the environment, economy, and society [9]. Therefore, assessing and predicting desertification in Iran is essential for developing effective mitigation and adaptation strategies to promote sustainable development in arid regions.

Numerous studies have been conducted globally to assess and map desertification, leading to the development of various models. However, desertification assessment models require extensive spatial and temporal data, and given that this phenomenon occurs in arid regions, extracting large-scale and long-term information is challenging [10]. Therefore, establishing a data extraction process based on remote sensing techniques for desertification assessment is essential [11]. Nonetheless, due to the complex nature of desertification, researchers believe that its assessment can be satisfactory only when reliable remote sensing indices are considered for this purpose [12]. To date, it has not been possible to determine such indices that can be used at various global, regional, national, and local scales. In most studies, three to five remote sensing indices have been used to assess the desertification hazard. However, given the multifaceted nature of this phenomenon, the obtained results may not be sufficiently accurate [8]. For instance, in many studies, the NDVI threshold is used to classify the degree of desertification [13,14]. Consequently, as this method considers only vegetation cover, the accuracy of classification results will decrease [15].

In this study, machine learning techniques were employed to model and project desertification. These methods are multi-index approaches that utilize remote sensing data to analyze changes in land cover, soil moisture, temperature, and other indicators that influence desertification. By applying advanced machine learning algorithms such as random forests, neural networks, and support vector machines, complex patterns of desertification can be identified, and highly accurate models can be developed to predict future trends [16]. The use of a single method may not yield satisfactory results. When models are correctly combined, they can produce more accurate results. Ensemble models, which combine multiple machine learning algorithms, have a significant advantage over individual models. These models can improve prediction accuracy by combining the results of different models. In the context of desertification assessment, the use of ensemble models can lead to more accurate predictions and better decision-making in natural resource management and mitigating the effects of desertification. In this study, an ensemble machine learning model in the SDM package in R software (R-4.4.1) was used to achieve a more accurate classification. Predicting future desertification is also of paramount importance as this phenomenon not only leads to a reduction in arable land and increased pressure on natural resources but also jeopardizes the food, economic, and social security of communities [17]. With climate change and increased human activities such as land-use change and unsustainable agriculture, the rate of desertification has accelerated [18]. Therefore, predicting desertification based on climate and land-use change scenarios is crucial as this approach allows for a more accurate analysis and evaluation of the interactions between these two key factors on desertification trends. Scenario modeling of these two factors can help identify high-hazard areas and develop effective management strategies to mitigate the impacts of desertification, as well as provide guidance for decision-makers in developing sustainable policies.

Golestan Province in northeastern Iran has faced land degradation, water and wind erosion, land-use change, and groundwater depletion over the past decade. Therefore, this study aimed to model and predict the risk of desertification based on climate and land-use changes in northeastern Iran. This research introduced the use of a combined model based on remote sensing indices as a suitable method for accurately assessing desertification. Additionally, producing future desertification maps for the study area based on climate and land-use change scenarios can contribute to better planning and management of natural resources, developing environmental protection policies, and adopting effective strategies to mitigate the negative impacts of desertification, thereby preventing environmental and human crises.

This study advances the field of desertification research by integrating advanced machine learning techniques with remote sensing data to accurately assess and predict future trends. By employing an ensemble modeling approach, we aim to improve the robustness and reliability of the results. Furthermore, the incorporation of climate and land-use change scenarios enables a comprehensive assessment of the potential impacts of these factors on desertification, facilitating the development of informed strategies for mitigation and adaptation. This research contributes to the existing body of knowledge by providing a comprehensive assessment of desertification in Northeast Iran and offering valuable insights into the potential impacts of future climate and land-use changes. The findings of this research can inform policymakers and decision-makers in developing effective strategies to mitigate the adverse effects of desertification and promote sustainable land management practices.

2. Materials and Methods

2.1. Study Area

The study area, spanning 20,367 square kilometers, is geographically located between 36°25′ and 38°08′ North latitude and 53°50′ and 56°18′ East longitude. This region is bordered by the Turkmen desert to the north, the Caspian Sea coastal plain to the west, and the Alborz mountain range to the south (Figure 1). The highest and lowest elevations in the region are approximately 3813 m above sea level (Shahkoh Mountain) and −32 m below sea level (Gomishoan Lagoon), respectively. The average annual precipitation, average evaporation, and average temperature of the region over a 30-year period (1993–2023) were estimated to be 530 mm, 1338 mm, and 17.3 °C, respectively. The De Martonne (1926) and Ivanov (1941) methods classified the region's climate as arid and steppe, respectively.

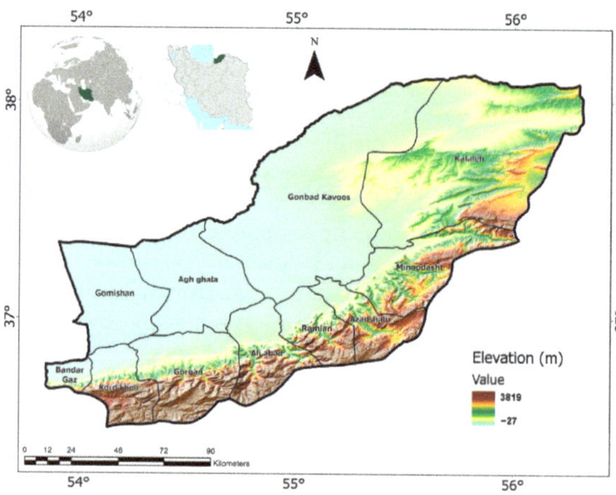

Figure 1. Geographical location of the study area in Iran and Golestan province.

2.2. Data Collection

In July 2021, 42 soil samples were collected from the study area to measure salinity and texture indices. To assess vegetation cover, one-meter square plots were systematically established along 100-m transects (3 transects per land use and 10 plots per transect) in the region. The percentage of vegetation cover in each plot was determined by expert opinion and extrapolated to the entire land use. Ten meteorological stations with a 30-year statistical period (1993–2023) were used to analyze climatic parameters such as temperature and precipitation. To evaluate groundwater depth, data from 75 wells (obtained from the Golestan Province Water Resources Organization) were used over a 30-year period (1993–2023). To investigate the impact of human factors on desertification, a land-use map was prepared. During field visits, training points were established in each land usage to prepare the land-use map.

This study employs two types of atmospheric data: observed and simulated. Observed data is sourced from Stations data, while simulated data was derived from the CNRM-CM6 global climate model. Developed by the Met Office Hadley Centre, this model was utilized for the CMIP6 climate simulations and is noteworthy for its integrated Earth system components. The research focuses on three future scenarios (SSP1-2.6, SSP2-4.5, and SSP5-8.5) simulated by CNRM-CM6 for the period of 2031–2050.

2.3. Remote Sensing Indices for Desertification Modeling

In this study, Google Earth Engine was utilized to generate remote sensing indices for modeling desertification in 2023. For this purpose, surface reflectance band 1 Landsat images were selected, which have undergone atmospheric and geometric corrections, as well as cross-calibration between different sensors to ensure the accuracy and consistency of the data used in the analysis. To develop the model, the most significant indices influencing desertification needed to be identified. Therefore, by taking inspiration from MEDALUS model indicators and reviewing comprehensive sources, the most important influencing indicators in the region's desertification were identified. In the next step, several remote sensing indicators were considered for each selected indicator. Subsequently, the index that exhibited the highest correlation with ground-truth data was selected for desertification modeling. The root mean square error (RMSE) and coefficient of determination (R^2) were used as criteria for selecting the best indices.

2.4. Desertification Modeling Using Machine Learning Algorithms

Selecting an Appropriate Number of Training Samples is the first step in implementing machine learning algorithms. Various approaches have been proposed for selecting the number of training samples, with no single method universally accepted. Therefore, to reduce spatial autocorrelation, 100 samples were randomly selected using visual interpretation of Google Earth images and field visits for implementing machine learning algorithms. Fifty samples were taken from areas where desertification had occurred, and 50 samples were taken from areas where desertification had not occurred. Ultimately, 70% of these samples were randomly selected as the training group, and the remaining 30% were selected as the testing group for modeling.

Four machine learning algorithms were employed to model desertification using the SDM package in R: Support Vector Machine (SVM) [1,2], Gradient Boosting Machine (GBM) [3,4], Generalized Linear Model (GLM) [5], and Random Forest (RF) [6,7]. Each algorithm was run randomly three times to ensure that the classification was not influenced by the sampling point distribution. To evaluate the accuracy of the machine learning models, three indices were used: Kappa coefficient, Receiver Operating Characteristic (ROC) curve, and True Skill Statistic (TSS). The Kappa coefficient is a measure of model accuracy, ranging from -1 (lowest accuracy) to 1 (highest accuracy) [8]. The ROC curve, plotted as the true positive rate against the false positive rate, is a graphical tool for evaluating model performance, and its area under the curve (AUC) is a measure of model accuracy, ranging from 0.5 (random) to 1 (perfect). Previous studies have categorized AUC values into several performance levels, including excellent (0.9–1), very good (0.8–0.9), good (0.7–0.8), moderate (0.6–0.7), and poor (0.5–0.6) [9,10]. The TSS also evaluates model accuracy by considering sensitivity and specificity at a specific threshold, ranging from 0 to 1, with higher values indicating better model accuracy. These indices, by providing complementary information, allow for a more comprehensive evaluation of the performance of machine learning models [11].

2.5. Ensemble Modeling for Desertification Prediction

Ensemble models, which combine the strengths of multiple base models, offer improved accuracy and reliability in predictions [12]. By aggregating results from various models and mitigating noise and biases, ensemble models enhance predictive accuracy and reduce error probabilities [13]. Moreover, ensemble methods provide greater flexibility in handling complex datasets, leading to more informed decision-making and effective management strategies in desertification and other environmental studies [14]. In this study, after evaluating the performance of four machine learning models, a weighted ensemble model was implemented using the SDM package in R to develop an optimal model for desertification modeling. Subsequently, to predict desertification in 2044 under static conditions, two scenarios were considered: land-use change and precipitation. Future land-use change was predicted using a Markov chain model, while precipitation was estimated using a LARS model.

2.6. CA–Markov Model

The CA–Markov model integrates cellular automata, Markov chains, and multi-objective land allocation (MOLA) to forecast future land-use changes [15]. Initially, a Markov chain model is employed to compute the probability of land-use class transitions, represented as a transition probability matrix, based on changes observed between time T0 and T1. While the Markov chain model provides temporal probabilities, it lacks spatial information regarding the geographic location of land uses. To address this limitation, a CA–Markov model is utilized, which incorporates spatial adjacency and user-defined spatial distribution of transition probabilities into the Markov chain framework [16]. In this study, land-use maps from 2004 to 2023 were used to predict the 2042 map. To assess the accuracy of the CA–Markov model, a 2023 map was predicted using the 2004 and 2014 maps, and the results were compared with a reference map created through supervised classification.

2.7. Statistical Downscaling

This study investigated the influence of climate change on desertification in the Golestan province. Three climate change scenarios (SSP1-2.6, SSP2-4.5, and SSP5-8.5) from the CNRM-CM6 model were used to predict scenarios for the period of 2031–2050. To apply these large-scale climate projections to the local watershed, the LARS-WG8 statistical model was employed for downscaling. A primary challenge in regional climate change impact studies is the mismatch between the large scale of climate models and the smaller scale of actual impact areas. This discrepancy, arising from variations in topography and climatic conditions, limits the direct application of model results to specific locations. To bridge this gap, downscaling techniques were employed [17]. This research utilized the statistical downscaling method, specifically the LARS-WG8 model, to generate regional climate scenarios.

3. Results

3.1. The Selection of Remote Sensing Indicators

For desertification modeling, six indicators based on remote sensing were selected (Table 1). The literature review revealed that the TGSI is the most appropriate index for evaluating soil surface texture. An increase in soil surface grain size is a clear indicator of land degradation [18,19]. When surface soil particles become coarser, this change is considered a warning sign of the onset or progression of destructive processes such as erosion and desertification. This indicates that the TGSI index can be an effective tool for monitoring and predicting soil degradation in desert and semi-desert areas. The NDSI, VSSI, and SI indices were used to assess soil salinity. The NDSI index, with a low root mean square error (RMSE = 5.21) and a suitable coefficient of determination (R^2 = 0.71), was selected for modeling. By analyzing data from the Chrips, PERSIANN, GPM, and TRMM satellites, it was determined that Chrips satellite data had the highest correlation with the synoptic station data in the region and, thus, was chosen for modeling (Table 1). Five indices NDVI, DVI, EVI, and SAVI were considered for evaluating vegetation cover. The NDVI index, with a root mean square error (RMSE = 2.41) and a suitable coefficient of determination (R^2 = 0.83), had the highest correlation with ground data (percentage of vegetation cover in plots). The land-use change map was used as a remote sensing indicator to examine the role of human factors in desertification. The accuracy of the land-use change map was validated using ground truth data collected during field visits and Google Earth imagery (Table 1). The WEHI model has three indicators; wind speed, soil moisture, and bare soil. The high correlation coefficients between the WEHI model, which is capable of remote sensing, and the ground evidence map show the high capability of this model in estimating wind erosion (Table 1 and Figure 2).

Table 1. Indicators based on the degree of correlation with ground surface data.

Numbers	Indicators	Remote Sensing Indicators	Ground Surface Data	Correlation Rate Between Satellite Indicators and Ground Surface Data		Reference
				R^2	RMSE	
1	Soil texture	TGSI	Soil surface profile	0.82	0.87	[18,19]
2	EC	NDSI		0.71	5.21	
3	Rainfall	Chrips	synoptic stations	0.65	141.33	[20]
4	Groundwater depth	-	The water depth of the wells	0.79	4.01	[21]
5	Vegetation percent	NDVI	Vegetation percentage in plots	0.83	2.41	[22]
6	Wind erosion	WEHI model = /NDMI) wind speed * MBI(	Map of ground evidence	0.86	0.89	[23]
7	Land use	Land-use changes	Map of ground evidence	0.71	0.64	[24]

3.2. Desertification Modeling

After selecting the remote sensing indices, desertification modeling was conducted in 2023 using four machine learning models (Figure 3). The results of all four models indicated severe land degradation in the northern regions of the study area, with scattered patches of high degradation intensity also observed in the southern and northeastern areas. To evaluate the modeling results, statistical parameters including the Kappa coefficient, Receiver Operating Characteristic (ROC) curve, and True Skill Statistic (TSS) were used (Table 2). Based on the results, the RF and SVM models outperformed the GLM and GBM models. Overall, the RF model was identified as the best performer with AUC = 0.91, TSS = 0.88, and Kappa = 0.90, while the GLM model was the weakest with AUC = 0.79, TSS = 0.77, and Kappa = 0.65.

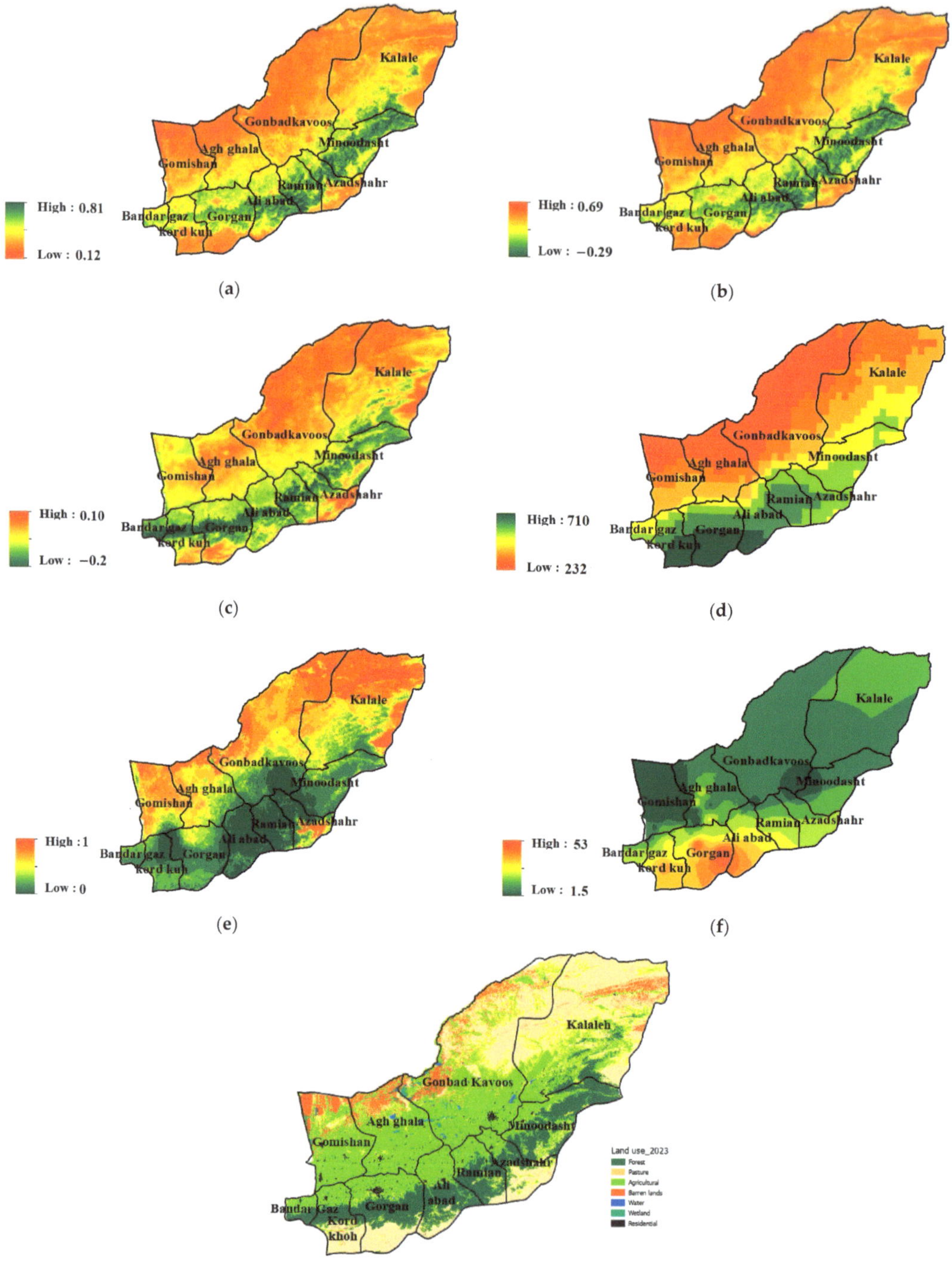

Figure 2. Remote sensing indicators for desertification modeling, (**a**): NDVI, (**b**): NDSI, (**c**): TGSI, (**d**): Chrips, (**e**): WEHI, (**f**): Groundwater and (**g**): Land use.

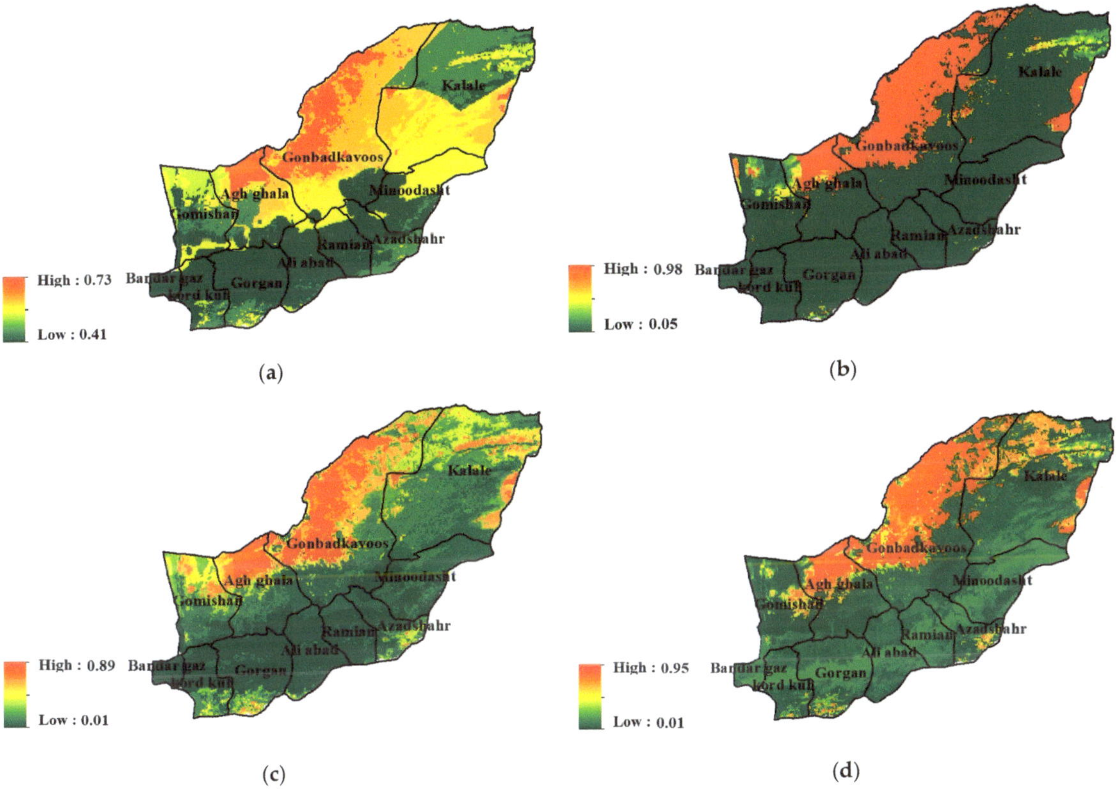

Figure 3. Desertification in northeastern Iran using different models in the SDM package. (**a**), (**b**), (**c**), and (**d**) are models GBM, GLM, RF, and SVM in 2023, respectively.

Table 2. Performance evaluation of models based on various indicators.

Methods	AUC	TSS	KAPPA
GLM	0.79	0.77	0.65
GBM	0.85	0.80	0.72
RF	0.91	0.88	0.90
SVM	0.89	0.86	0.88

The varying performance of the models indicates the presence of uncertainty in their results. Therefore, it is necessary to use ensemble models that leverage the strengths of multiple models. In this study, a weighted average of the four models RF, SVM, GBM, and GLM was used, with the RF model receiving the highest weight and the GLM model the lowest in the ensemble model (Figure 4). The results of the ensemble model also indicate severe desertification in the northern regions of the study area, though the intensity of desertification is shown to be lower compared to the individual models.

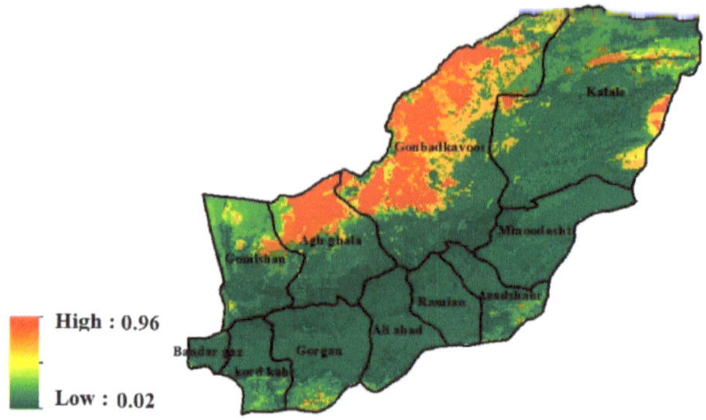

Figure 4. Ensemble model of desertification assessment.

3.3. Importance of Variables in Modeling

Identifying the factors influencing desertification is crucial for managing and preventing this phenomenon. A thorough analysis of these factors allows researchers and environmental managers to model desertification-prone areas more accurately. This, in turn, supports the development of effective strategies to combat desertification. Accordingly, the significance of the indicators used in desertification assessment was determined based on the ensemble model. The process of combining the models was based on the weighted averaging method, where the weight of each model was determined according to its prediction accuracy. Then, the importance of variables was calculated through the weighted combination of the base model results. This approach allowed us to more accurately identify the variables influencing desertification. The variable importance plots presented in the article demonstrate the impact of the key variables that were influential across all models. The results showed that, in order of importance, land-use change, groundwater depth, precipitation, WEHI, NDVI, NDSI, and TGSI are the most influential factors in the region's desertification (Figure 5). Based on these findings, utilizing scenarios such as land-use change and precipitation can be effective in identifying areas prone to desertification and formulating strategies for its control and prevention

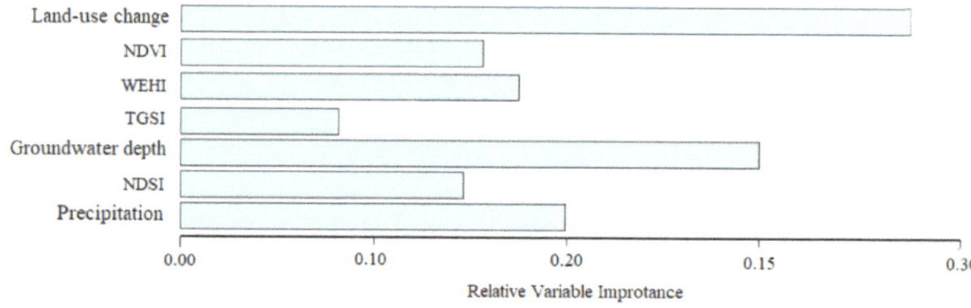

Figure 5. Importance of variables in the study area in 2023.

3.4. Prediction of Future Desertification

To predict desertification in 2040, assuming other parameters remain constant, the average changes in precipitation during the period of 2031–2051 and land use were simulated. Results from the CNRM-CM6 model indicate an increase in precipitation across Golestan Province in the future period. The findings showed that under the SSP1-2.6, SSP2-4.5, and SSP5-8.5 scenarios, annual precipitation at the synoptic station in Gorgan is expected to increase by 78.5, 58.11, and 109.78 mm, respectively, during the 2031–2050 period. At the Gonbad station, precipitation is predicted to increase by 59.99, 56.20, and 78.12 mm under these scenarios. Similarly, at the Kalaleh and Maraveh Tappeh stations, precipitation is expected to rise, with increases of 55.26, 50.55, and 57.99 mm at Kalaleh, and 33.55, 28.42, and 41.30 mm at Maraveh Tappeh, assuming the realization of the climate change scenarios (Figures 6 and 7).

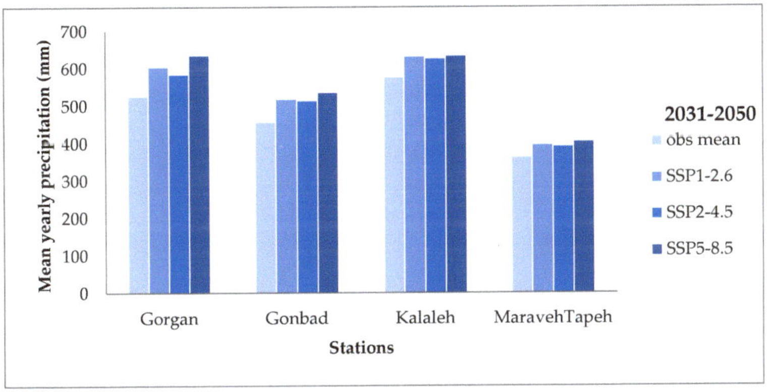

Figure 6. Precipitation prediction in regional stations using different scenarios for the period of 2031–2050.

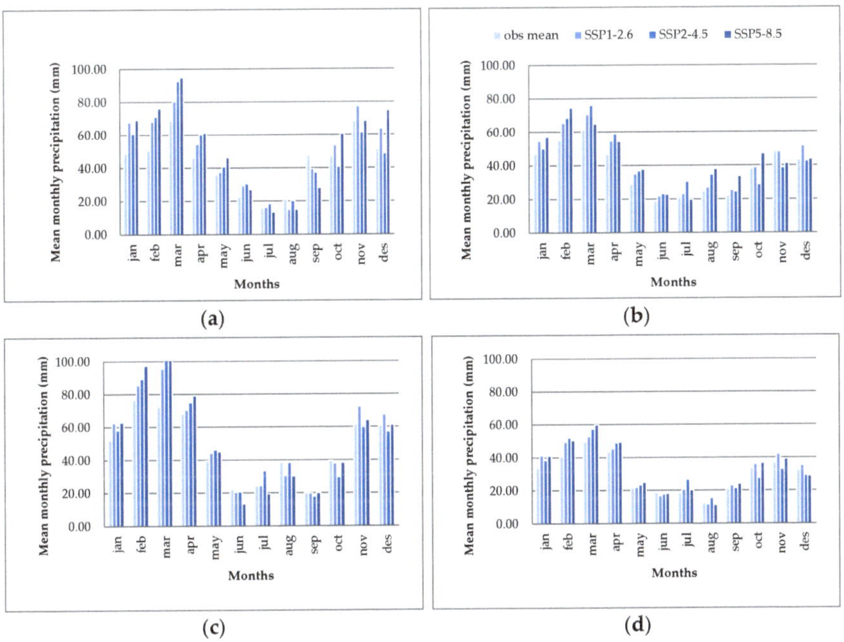

Figure 7. Monthly precipitation prediction using different scenarios for the period of 2031–2050 in (**a**) Gorgan station, (**b**) Gonbad station, (**c**) Kalaleh station and (**d**) Maraveh Tappeh station.

The land-use prediction in 2040, using the Markov Chain model and comparing it with current land use (2023), indicates a reduction in forest areas by 7093 hectares (0.35 percent), rangeland areas by 109,182 hectares (5.38 percent), and wetland areas by 840 hectares (0.04 percent), along with an increase in agricultural land by 4800 hectares (0.24 percent), barren land by 104,998 hectares (5.18 percent), water bodies by 1000 hectares (0.05 percent), and residential areas by 6317 hectares (0.31 percent) (Table 3 and Figure 8). Assuming constant indices for soil surface texture, soil salinity, groundwater depth, and vegetation cover density, and also considering the projected average precipitation for the 2031–2050 period and land use in 2040, desertification was projected using an ensemble model for 2040 (Figure 8). The 2040 desertification map indicates an expansion of this phenomenon in the northern regions of the area. The results showed that desertification in 2040 is expected to increase by approximately 2% (40,562 hectares), with the advance occurring in the Northwest regions adjacent to the Gorgan wetland and the Sangi Tappeh desert (Table 4 and Figure 9).

Table 3. Investigating the trend of land-use class changes.

Class	Land Use	2023 (Hectares)	2040 (Hectares)	Percent Changes
1	Forest	338,308	331,215	−0.35
2	Pasture	708,316	599,134	−5.38
3	Agricultural	804,025	808,825	0.24
4	Barren lands	124,644	229,642	5.18
5	Water	8201	9201	0.05
6	Wetland	8718	7878	−0.04
7	Residential	35,901	42,218	0.31
Total		2,028,113	2,028,113	

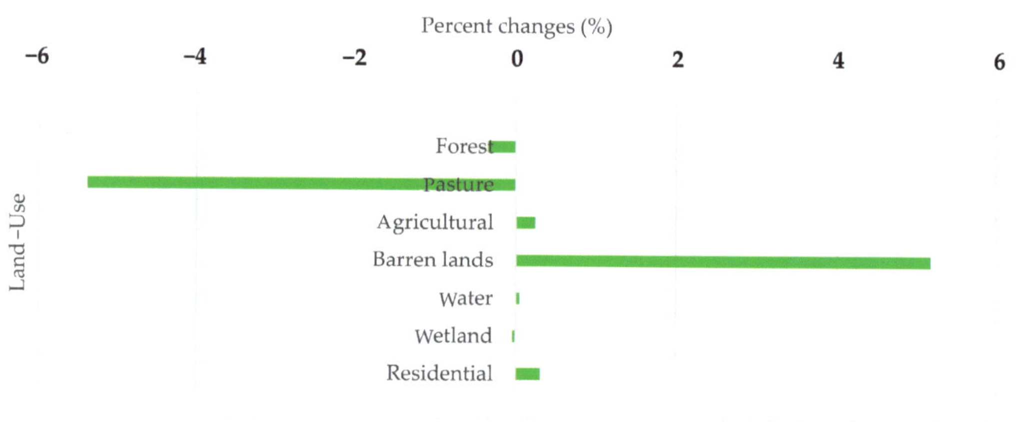

Figure 8. Predicted land-use change between 2023 and 2040.

Table 4. The trend of desertification class changes.

Class	Desertification	2023 (Hectares)	2040 (Hectares)	Percent Changes
1	Low	1,345,809	1,204,281	−6.97
2	Moderate	321,785.44	402,430.4	+3.97
3	Severe	165,359.77	185,680.8	+1
4	Very Severe	195,158.80	235,720.8	+2
Total		2,028,113	2,028,113	

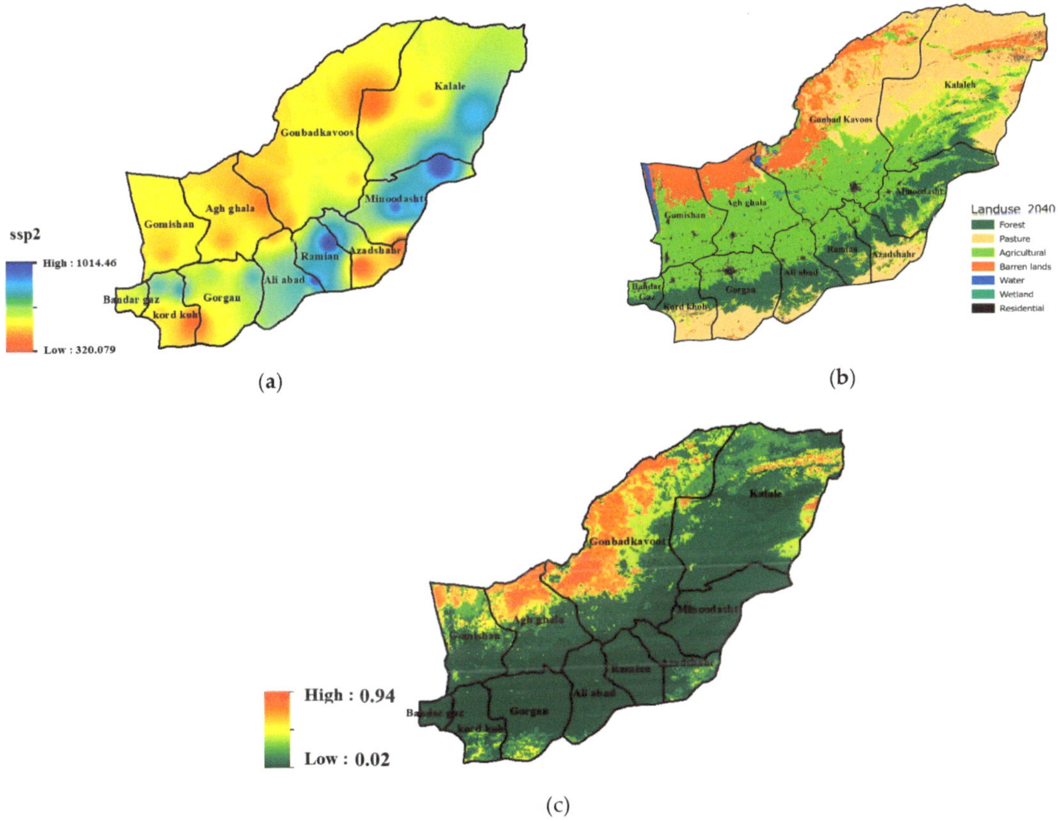

Figure 9. (**a**) Rainfall scenario in 2040, (**b**) Land-use scenario in 2040, and (**c**) Desertification in 2040.

4. Discussion

The novel combination of six distance-based indices with the SDM package in R provides a unique perspective on desertification modeling. This approach enables accurate projections of future desertification trends. Results indicate that the Random Forest (RF) and Support Vector Machine (SVM) models outperformed the Gradient Boosting Machine (GBM) and Generalized Linear Model (GLM) in terms of predictive accuracy. The superior performance of the RF model can be attributed to its ensemble nature, which involves combining multiple decision trees trained on different data samples to enhance diversity and mitigate overfitting.

Random Forest (RF) and Support Vector Machine (SVM) were employed in this study due to their proven effectiveness in handling complex classification problems. RF, with its ensemble nature, reduces variance and improves predictive accuracy by combining multiple decision trees. SVM, on the other hand, excels in high-dimensional space and can effectively separate data classes. While both models offer advantages, they also have limitations. RF can be computationally expensive for large datasets, and SVM can be sensitive to hyper-parameter tuning. To mitigate these limitations and enhance the overall robustness of the model, an ensemble approach was adopted, combining the strengths of both models [25–28].

The Random Forest model outperformed other models in the analyses conducted in this study for several reasons. By combining a large number of decision trees and random sampling, this model reduces data fluctuations and prevents overfitting. Consequently, this method utilizes uncertainty metrics for each tree and ultimately selects the prediction based on majority voting, resulting in higher prediction accuracy. Other studies have also demonstrated the effectiveness of this model. For example, recent research has shown that the Random Forest model, due to its noise-resistant structure, performs exceptionally well, particularly in complex problems and noisy datasets. The findings of this study align with these previous studies, demonstrating that, compared to models such as GLM and BRT, which are more sensitive to data fluctuations, the RF model exhibits greater stability in accuracy [14,29,30].

The ensemble approach, which combines the strengths of Random Forest and Support Vector Machine, proved to be particularly effective in modeling desertification. By mitigating the limitations of individual models, such as overfitting in RF or sensitivity to hyper parameters in SVM, the ensemble model provided a more robust and reliable prediction. The reduced uncertainty in the ensemble predictions allowed for more confident projections of future desertification trends. The results of this study underscore the benefits of ensemble modeling in complex environmental modeling tasks. By combining multiple diverse models, the ensemble approach not only improves predictive accuracy but also enhances the interpretability of results. The ensemble model's ability to identify the most influential factors contributing to desertification (e.g., climate change, land use) provides valuable insights for developing targeted mitigation strategies. Furthermore, the ensemble framework can be adapted to other environmental problems, making it a versatile tool for addressing complex challenges [13,14].

The ensemble model results indicate a higher intensity of desertification in the northern regions of the study area. The analysis reveals that changes in desertification in Golestan Province are influenced by a complex interplay of environmental and anthropogenic factors. Key drivers include land-use change, alterations in groundwater depth, and precipitation patterns. Land-use change, particularly the conversion of agricultural land and pastures to rainfed and fallow lands, significantly accelerates desertification. These changes often result in the loss of natural vegetation, which acts as a barrier against soil erosion. Reduced vegetation cover also leads to increased evaporation and decreased infiltration, disrupting soil moisture balance. Another critical factor is the decline in groundwater levels. Excessive groundwater extraction and climate change can induce soil dryness and reduced agricultural productivity. Additionally, changes in precipitation, a significant climatic factor, play a crucial role. Decreased precipitation in the arid and semi-arid regions of Golestan reduces soil moisture and hampers the regeneration of natural vegetation, exacerbating desertification. Conversely, sudden and intense rainfall events can cause soil erosion and degrade soil structure and biota. Overall, our findings indicate that desertification in Golestan Province is a result of complex interactions between human and natural factors, necessitating appropriate environmental policies and comprehensive conservation measures for effective management and control.

Climate and land-use change simulations reveal complex trends in the study area, with potentially significant environmental consequences. Climate modeling results indicate a general increase in precipitation in the Golestan province. This increase is observed across all synoptic stations but varies spatially, with higher increases observed from east to west. These variations may be attributed to factors such as station location, topography, and changes in large-scale atmospheric circulation patterns. While increased precipitation could have positive impacts like enhanced soil moisture, improved vegetation cover, and reduced drought risk, the increased intensity and frequency of extreme rainfall events may lead to flooding, soil erosion, and land degradation.

Land-use projections from 2023 to 2040 indicate substantial changes that directly influence desertification processes. The significant expansion of barren lands and the decline in pastureland can have detrimental effects on the region, including increased barren lands, resulting from reduced vegetation cover, soil erosion, decreased soil infiltration, and ultimately intensified desertification. Changes in precipitation and land use interact to influence desertification processes. While increased precipitation can improve vegetation cover and reduce soil erosion, it can also lead to flooding and soil degradation if not coupled with proper water and soil management.

The 2040 desertification map indicates an eastward expansion of desert areas from the northeastern regions. The findings suggest that without effective desertification control measures, a 2% increase (40,562 hectares) in desertified areas could occur by 2040. Unsustainable agricultural practices, coupled with land-use changes, are likely to exacerbate desertification in the future. To mitigate these impacts, strategies such as protecting environmentally sensitive areas, conserving natural resources, promoting sustainable agriculture, and optimizing land use are recommended.

While this study provides valuable insights into the spatial and temporal patterns of desertification in Northeast Iran, it is important to acknowledge certain limitations. The accuracy of the results is dependent on the quality and resolution of the input data, particularly remote sensing data. Additionally, the complexity of desertification processes and the influence of various socio-economic factors may not be fully captured by the models employed in this study. Future research could explore more advanced modeling techniques, such as machine learning algorithms that incorporate deep learning, to further improve the accuracy and robustness of the predictions. Furthermore, integrating socio-economic factors into the analysis can provide a more comprehensive understanding of the drivers of desertification and inform targeted interventions.

5. Conclusions

This study highlights the significant role of land-use change and climate variability in driving desertification in Northeast Iran. By integrating advanced machine learning techniques, such as Random Forest and Support Vector Machine, with remote sensing data, this research has enabled accurate mapping and prediction of future desertification trends. The ensemble modeling approach further enhances the robustness and reliability of the results.

The findings indicate that the northern regions of the study area are particularly vulnerable to desertification. The projected increase in desertification by 2040, driven by climate change and land-use change, underscores the urgent need for effective mitigation and adaptation strategies. To address this pressing issue, it is imperative to implement sustainable land management practices, such as afforestation, agroforestry, and soil conservation measures. Additionally, climate change mitigation strategies, including reducing greenhouse gas emissions and promoting renewable energy sources, are crucial to limit the adverse impacts of climate change on desertification.

While this study provides valuable insights, it is important to acknowledge the limitations inherent in the research, such as the reliance on data quality and model assumptions. Future research could explore more advanced modeling techniques, incorporate additional socio-economic factors, and refine the evaluation metrics to further enhance the understanding of desertification dynamics. By addressing the challenges posed by desertification, it is possible to protect ecosystems, safeguard livelihoods, and ensure the long-term sustainability of the region.

Author Contributions: Conceptualization, W.Y.; methodology, W.Y.; software, W.Y.; validation, W.Y.; formal analysis, W.Y.; investigation, W.Y.; resources, W.Y.; data curation, A.B.; writing—original draft preparation, W.Y. and A.B.; writing—review and editing, W.Y., Q.H., J.L., P.H., D.Z. and A.B.; visualization, W.Y.; supervision, W.Y.; project administration; funding acquisition, W.Y., Q.H. and A.B. All authors have read and agreed to the published version of the manuscript.

Funding: This work was supported by the National Natural Science Foundation of China (Grant No. 42277478 and U21A20109).

Data Availability Statement: The data presented in this study are available upon request from the corresponding author.

Conflicts of Interest: The authors declare no conflict of interest.

References

1. Joshi, P.P.; Wynne, R.H.; Thomas, V.A. Cloud Detection Algorithm Using SVM with SWIR2 and Tasseled Cap Applied to Landsat 8. *Int. J. Appl. Earth Obs. Geoinf.* **2019**, *82*, 101898. [CrossRef]
2. Li, J.; Ma, X.; Zhang, C. Predicting the Spatiotemporal Variation in Soil Wind Erosion across Central Asia in Response to Climate Change in the 21st Century. *Sci. Total Environ.* **2020**, *709*, 136060. [CrossRef]
3. Park, S.; Im, J.; Jang, E.; Rhee, J. Drought Assessment and Monitoring through Blending of Multi-Sensor Indices Using Machine Learning Approaches for Different Climate Regions. *Agric. For. Meteorol.* **2016**, *216*, 157–169. [CrossRef]
4. Cha, G.W.; Moon, H.J.; Kim, Y.-C. Comparison of Random Forest and Gradient Boosting Machine Models for Predicting Demolition Waste Based on Small Datasets and Categorical Variables. *Int. J. Environ. Res. Public Health* **2021**, *18*, 8530. [CrossRef]
5. Chan, H.L.; Pan, M. Fishing Trip Cost Modeling Using Generalized Linear Model and Machine Learning Methods—A Case Study with Longline Fisheries in the Pacific and an Application in Regulatory Impact Analysis. *PLoS ONE* **2021**, *16*, e0257027. [CrossRef]
6. Belgiu, M.; Drăgu, L. Random Forest in Remote Sensing: A Review of Applications and Future Directions. *ISPRS J. Photogramm. Remote Sens.* **2016**, *114*, 24–31. [CrossRef]
7. Chan, J.C.W.; Paelinckx, D. Evaluation of Random Forest and Adaboost Tree-Based Ensemble Classification and Spectral Band Selection for Ecotope Mapping Using Airborne Hyperspectral Imagery. *Remote Sens. Environ.* **2008**, *112*, 2999–3011. [CrossRef]
8. Bofana, J.; Zhang, M.; Nabil, M.; Wu, B.; Tian, F.; Liu, W.; Zeng, H.; Zhang, N.; Nangombe, S.S.; Cipriano, S.A.; et al. Comparison of Different Cropland Classification Methods under Diversified Agroecological Conditions in the Zambezi River Basin. *Remote Sens.* **2020**, *12*, 2096. [CrossRef]
9. Yesilnacar, E.; Topal, T. Landslide Susceptibility Mapping: A Comparison of Logistic Regression and Neural Networks Methods in a Medium Scale Study, Hendek Region (Turkey). *Eng. Geol.* **2005**, *79*, 251–266. [CrossRef]
10. Devkota, K.C.; Regmi, A.D.; Pourghasemi, H.R.; Yoshida, K.; Pradhan, B.; Ryu, I.C.; Dhital, M.R.; Althuwaynee, O.F. Landslide Susceptibility Mapping Using Certainty Factor, Index of Entropy and Logistic Regression Models in GIS and Their Comparison at Mugling-Narayanghat Road Section in Nepal Himalaya. *Nat. Hazards* **2013**, *65*, 135–165. [CrossRef]
11. Allouche, O.; Tsoar, A.; Kadmon, R. Assessing the Accuracy of Species Distribution Models: Prevalence, Kappa and the True Skill Statistic (TSS). *J. Appl. Ecol.* **2006**, *43*, 1223–1232. [CrossRef]
12. Naimi, B.; Araújo, M.B. Sdm: A Reproducible and Extensible R Platform for Species Distribution Modelling. *Ecography* **2016**, *39* 368–375. [CrossRef]
13. Ebrahimi, E.; Araújo, M.B.; Naimi, B. Flood Susceptibility Mapping to Improve Models of Species Distributions. *Ecol. Indic.* **2023** *157*, 111250. [CrossRef]
14. Boali, A.; Asgari, H.R.; Behbahani, A.M.; Salmanmahiny, A.; Naimi, B. Remotely Sensed Desertification Modeling Using Ensemble of Machine Learning Algorithms. *Remote Sens. Appl. Soc. Environ.* **2024**, *34*, 101149. [CrossRef]
15. Soleimani-Sardo, M.; Silakhori, E. Land-Use Change Detection in 2040 with CA-Markov Chain Model, Case Study: Jiroft Basin *Watershed Eng. Manag.* **2022**, *14*, 299–311. [CrossRef]
16. Silakhori, E.; Maleki, S.; Pourghasemi, H.R. Spatiotemporal Variations in Land Use of Mahvelat Plain in Iran Using Google Earth Engine from 2011 to 2030. In *Advanced Tools for Studying Soil Erosion Processes*; Elsevier: Amsterdam, The Netherlands, 2024 pp. 413–433.
17. Silakhori, E.; Dahmardeh Ghaleno, M.R.; Meshram, S.G.; Alvandi, E. To Assess the Impacts of Climate Change on Runoff in Golestan Province, Iran. *Nat. Hazards* **2022**, *112*, 281–300. [CrossRef]
18. Xiao, J.; Shen, Y.; Tateishi, R.; Bayaer, W. Development of Topsoil Grain Size Index for Monitoring Desertification in Arid Land Using Remote Sensing. *Int. J. Remote Sens.* **2006**, *27*, 2411–2422. [CrossRef]
19. Meng, X.; Gao, X.; Li, S.; Li, S.; Lei, J. Monitoring Desertification in Mongolia Based on Landsat Images and Google Earth Engine from 1990 to 2020. *Ecol. Indic.* **2021**, *129*, 107908. [CrossRef]
20. Asfaw, E.; Suryabhagavan, K.V.; Argaw, M. Soil Salinity Modeling and Mapping Using Remote Sensing and GIS: The Case of Wonji Sugar Cane Irrigation Farm, Ethiopia. *J. Saudi Soc. Agric. Sci.* **2018**, *17*, 250–258. [CrossRef]
21. Paredes, M.; Bertoldo, S.; Carosso, L.; Lucianaz, C.; Marchetta, E.; Allegretti, M.; Savi, P. Propagation Measurements for a LoRa Network in an Urban Environment. *J. Electromagn. Waves Appl.* **2019**, *33*, 2022–2036. [CrossRef]
22. Moiwo, J.P.; Lu, W.; Tao, F. GRACE, GLDAS and Measured Groundwater Data Products Show Water Storage Loss in Western Jilin, China. *Water Sci. Technol.* **2012**, *65*, 1606–1614. [CrossRef] [PubMed]
23. Yang, X.; Leys, J. Mapping Wind Erosion Hazard in Australia Using MODIS-Derived Ground Cover, Soil Moisture and Climate Data. *IOP Conf. Ser. Earth Environ. Sci.* **2014**, *17*, 012275.
24. Jiang, M.; Lin, Y. Desertification in the South Junggar Basin, 2000–2009: Part I. Spatial Analysis and Indicator Retrieval. *Adv. Space Res.* **2018**, *62*, 1–15. [CrossRef]
25. Wijitkosum, S. The Impact of Land Use and Spatial Changes on Desertification Risk in Degraded Areas in Thailand Sustainable Environment Research The Impact of Land Use and Spatial Changes on Deserti Fi Cation Risk in Degraded Areas in Thailand. *Sustain. Environ. Res.* **2016**, *26*, 84–92. [CrossRef]
26. Zhang, Z.; Hörmann, G.; Huang, J.; Fohrer, N. A Random Forest-Based CA-Markov Model to Examine the Dynamics of Land Use/Cover Change Aided with Remote Sensing and GIS. *Remote Sens.* **2023**, *15*, 2028. [CrossRef]
27. Breiman, L. Random Forests. *Mach. Learn.* **2001**, *45*, 5–32. [CrossRef]

28. Beale, C.M.; Lennon, J.J. Incorporating Uncertainty in Predictive Species Distribution Modelling. *Philos. Trans. R. Soc. B Biol. Sci.* **2012**, *367*, 247–258.
29. Araújo, M.B.; New, M. Ensemble Forecasting of Species Distributions. *Trends Ecol. Evol.* **2007**, *22*, 42–47. [CrossRef]
30. Wahla, S.S.; Kazmi, J.H.; Sharifi, A.; Shirazi, S.A.; Tariq, A.; Joyell Smith, H. Assessing spatio-temporal mapping and monitoring of climatic variability using SPEI and RF machine learning models. *Geocarto Int.* **2022**, *37*, 14963–14982. [CrossRef]

land

MDPI

Article

Deep-Learning-Driven Insights into Nitrogen Leaching for Sustainable Land Use and Agricultural Practices

Caixia Hu, Jie Li *, Yaxu Pang, Lan Luo, Fang Liu, Wenhao Wu, Yan Xu, Houyu Li, Bingcang Tan and Guilong Zhang

Agro-Environmental Protection Institute, Ministry of Agriculture and Rural Affairs, Tianjin 300191, China; 15736785625@163.com (C.H.); pangyaxu@caas.cn (Y.P.); luolan970606@163.com (L.L.); liufang@caas.cn (F.L.); wuwenhao@caas.cn (W.W.); xuyan@caas.cn (Y.X.); lihouyu@caas.cn (H.L.); tanbingchang@caas.cn (B.T.); zhangguilong@caas.cn (G.Z.)

* Correspondence: lijie@caas.cn

Abstract: Nitrate leaching from soil presents a significant threat to soil health, as it can result in nutrient loss, soil acidification, and structural damage. It is crucial to quantify the spatial heterogeneity of nitrate leaching and its drivers. A total of 509 observational data points regarding nitrate leaching in northern China were collected, capturing the spatial and temporal variations across crops such as winter wheat, maize, and greenhouse vegetables. A machine learning (ML) model for predicting nitrate leaching was then developed, with the random forest (RF) model outperforming the support vector machine (SVM), extreme gradient boosting (XGBoost), and convolutional neural network (CNN) models, achieving an R^2 of 0.75. However, the performance improved significantly after integrating the four models with Bayesian optimization (all models had $R^2 > 0.56$), which realized quantitative prediction capabilities for nitrate leaching loss concentrations. Moreover, the XGBoost model exhibited the highest fitting accuracy and the smallest error in estimating nitrate leaching losses, with an R^2 value of 0.79 and an average absolute error (*MAE*) of 3.87 kg/ha. Analyses of the feature importance and SHAP values in the optimal XGBoost model identified soil organic matter, chemical nitrogen fertilizer input, and water input (including rainfall and irrigation) as the main indicators of nitrate leaching loss. The ML-based modeling method developed overcomes the difficulty of the determination of the functional relationship between nitrate loss intensity and its influencing factors, providing a data-driven solution for estimating nitrate–nitrogen loss in farmlands in North China and strengthening sustainable agricultural practices.

Keywords: nitrate; leaching; machine learning; North China

1. Introduction

With the increasing global demand for food, the intensification of agricultural practices has resulted in an over-reliance on nitrogen (N) fertilizers [1,2]. Although it is essential for crop productivity, the widespread application of N fertilizers has also become a significant source of non-point source pollution, posing risks to both agricultural ecosystems and human health [3,4]. Nitrate leaching, the process according to which nitrates are transported from the soil to groundwater, is an important pathway for agricultural non-point source pollution based on slow hydrological transport mechanisms. Therefore, quantifying nitrate accumulation and understanding its driving factors are crucial for enhancing agricultural systems' efficiency and environmental sustainability.

Chemical fertilizers and animal manure are viewed as the main sources of nitrate; however, various factors, including soil types [5], land use patterns [6], and site-specific hydraulic properties [7], significantly influence nitrate concentration and transformation. For instance, regions with high baseline nitrogen levels or those engaged in high-intensity agricultural activities are more likely to experience increased nitrogen leaching [8]. Additionally, chemical processes such as nitrification and denitrification complicate the identification of nitrate sources, as they alter nitrate levels and distributions [9]. These dynamic, interlinked processes contribute to the complexity of nitrate concentration patterns across agricultural landscapes. Although a number of numerical models have been established for the quantitative evaluation of nitrogen leaching, their limitations have become conspicuously evident [10–12]. DSSAT-RZWQM and SWAT model calibration processes are complex and time-consuming, with a limited ability to simulate extreme climate events [13,14]. The highly mechanical nature of the HYDRUS model, its poor representation of spatial heterogeneity, and the need to couple it with crop models limit its widespread use [15]. Thus, there remains a critical need for robust methods to assess nitrate sources and sinks spatially and temporally across diverse agricultural systems.

To address these challenges, recent advances in ML offer promising approaches to predicting and managing N leaching [16,17]. ML algorithms, with their capacity to process complex datasets and identify latent patterns, have proven to be effective in making predictions [18–20]. Techniques such as random forest (RF), XGBoost, and light gradient boosting (LGB) have been applied to exploring the nonlinear relationships between environmental variables and solute transport indicators [21]. For example, Dhaliwal et al. [22] used a decision tree and an RF algorithm to elucidate the factors influencing soil N_2O emissions. Xue et al. [23] employed five ML algorithms to develop predictive models for the N and phosphorus concentrations in Wenzhou, China. Similarly, Ransom et al. [24] applied XGBoost to predicting the nitrate concentrations in groundwater from 3099 wells. Although these studies have contributed valuable insights, there remains a lack of a standardized approach to the model selection for nitrogen leaching prediction, with most of the research concentrating on groundwater quality and solute transport rather than nitrogen leaching loss rates specifically. The integration of nitrogen leaching data with key variables such as climate variability, soil type, land use, and fertilizer practices could significantly enhance the performance and applicability of nitrate leaching models [21,25,26]. Applying ML to analyzing the intrinsic relationships and mechanisms of key factors could improve the prediction accuracy for N leaching and deepen our understanding of its dynamic role, thus providing a solid theoretical and technical basis for the formulation of agricultural N management strategies.

The North China Plain is a crucial grain production area in China, and it is characterized by fluvo-aquic soil, which has rapid organic carbon turnover and limited carbon sequestration capacities [27]. The monsoon climate results in rainfall being primarily concentrated in the summer, and insufficient precipitation has made over-irrigation the norm, subsequently triggering nitrate leaching. Furthermore, under the commonly practiced two-year, three-crop planting system, farmers apply an average of 500 kg/ha, significantly exceeding the recommended amount and increasing the risks of N leaching and non-point source pollution [28]. The objectives of this study are as follows: (1) quantify the spatial variability in nitrate leaching and identify the key factors at the regional scale; (2) evaluate the predictive accuracy of different ML models for nitrate leaching and determine the optimal model; and (3) propose corresponding control strategies and suggestions to provide scientific support for promoting the prevention and control of agricultural non-point source pollution in North China. These findings are intended to contribute valuable scientific insights for sustainable land management and environmental protection in the region.

2. Materials and Methods

2.1. Study Area

North China (113°04′–119°53′ E, 36°01′–42°37′ N) covers the plains area of Beijing Municipality, Tianjin Municipality, Hebei Province, most of Shanxi Province, and the central part of the Inner Mongolia Autonomous Region. The climate is dominated by a temperate monsoon climate, with the mean annual temperatures ranging from 8 °C to 13 °C. Annual precipitation varies between 400 mm and 1000 mm, although the Inner Mongolia Autonomous Region experiences less than 400 mm of precipitation. The primary crops cultivated in North China include winter wheat, maize, and facility vegetables.

2.2. Data Management

2.2.1. Data Acquisition and Predictor Parameters

In this study, data on nitrate leaching intensity from farmland in North China were obtained from the Web of Science (SCI) and China National Knowledge Infrastructure (CNKI) databases using the keywords TS = ((“nitrogen leaching” OR “nitrate nitrogen”) AND (“facility vegetables” OR “farmland”)). To ensure the relevance and quality of the data, the literature sources excluded reviews, laboratory simulations, and watershed simulations, focusing instead on field-based studies. With respect to the data shown in the tables, they were obtained directly from the text in the original literature, and graphical data were extracted from graphs published in the literature using the GetData Graph Digitizer tool. A comprehensive literature search was conducted for missing data points using multiple source databases and journals, and missing data points were supplemented through cross-referencing and carrying out comparisons to ensure the completeness of the data and reduce bias. The missing values in this study were mainly related to soil clay, sand, and nitrate–nitrogen loss data. For the missing data on soil clay content and bulk density, relevant data with the same latitude and longitude as those corresponding to the experimental area were examined to supplement them by referring to spatial distribution data on the soil texture in China. For data from the research that only provided information on the total nitrogen loss, nitrate–nitrogen loss was converted into 95 percent of the total nitrogen loss. Moreover, for missing meteorological data, we referred to the China Meteorological Data Network.

Ultimately, 509 datasets on nitrate leaching intensity (kg/ha/year) from farmland across North China were compiled (Table 1). The key parameters collected included soil physicochemical properties, climate data (such as mean annual temperature and precipitation), N fertilizer application rates, and irrigation practices (Table 2).

Table 1. Type of crop and number of datasets collected on nitrate–nitrogen leaching loss intensity.

Crop Cultivation Type	Number of Data Groups
Winter wheat	39
Summer maize	30
Spring maize	67
Greenhouse vegetables	373

Table 2. Summary of data obtained for training the machine-learning-based nitrate–nitrogen leaching loss prediction model.

No.	Data	Variables	Unit
1	The properties of the surface soil (0–20 cm)	Soil organic matter	g/kg
		Total N	g/kg
		Soil pH	
		Bulk density	g/cm^3
		Soil texture	
		Soil clay	%
2	Climate	Average annual rainfall	mm
		Average annual temperature	°C
3	Nitrogen input	Chemical N fertilizer input	kg/ha
		Organic N input	kg/ha
4	Irrigation methods	Drip irrigation	mm
		Flood irrigation	mm

2.2.2. Feature Extraction

Feature extraction involves selecting the most relevant and informative subset of features from the original feature set for model training in order to improve the model's performance, reduce over-fitting, and reduce computational costs [29]. We adopt a filtering feature selection method that implements filtering based on the statistical associations between features and target variables without relying on specific machine learning models. In this study, Pearson's correlation coefficient was calculated to select features that were significantly correlated with the target variable in order to preliminarily determine the subset of features and lay the foundation for the subsequent model construction and analysis.

2.2.3. Data Cleaning

Data cleaning can identify and correct data errors, anomalies, and inconsistencies; improve data quality; remove noise and interference; and help models accurately capture internal patterns. Outliers are removed from the original data; missing values are imputed using the mean, median, and mode of the adjacent data in order to optimize the model's accuracy and performance.

2.2.4. Data Transformation

Data transformation is a crucial step during the preprocessing stage which involves transforming raw data into a format suitable for machine learning algorithms to process, including data standardization/normalization, feature encoding, and dimensionality reduction [30].

Data standardization: In order to remove the influence of dimensionality and improve the model's performance, the Z-score normalization method was used in this study. Based on the mean and standard deviation of the features, normalization was performed in order to map the data to the [0,1] or [−1,1] interval [31].

Feature encoding: For classification variables such as soil texture and irrigation method in this study, we used one-hot encoding to process them, converting them into encoding forms of 0 and 1 according to their categories, which facilitated accurate calculation of the sample spacing or similarity based on distance measurement algorithms.

2.3. *Model Framework*

In this study, we developed four nonlinear regression models, including an SVM, RF, XGBoost, and a CNN for the prediction of nitrogen leaching.

2.3.1. The Support Vector Machine (SVM)

An SVM is a supervised learning algorithm that can be used for classification and regression problems [32]. It separates different classes of data by finding the optimal hyperplane and introduces a kernel function that allows the SVM to handle nonlinear problems efficiently. The SVM allows for a certain degree of error by introducing an ε-insensitive loss function, which helps the model to focus on the data points that are the most difficult to predict. The complexity of the model is controlled by regularization terms to prevent over-fitting. In the SVM, only a few data points (support vectors) determine the performance of the model, which makes the model sparse and helps improve the computational efficiency.

2.3.2. Random Forest Regression (RF)

The random forest (RF) algorithm indeed addresses some limitations of the classification and regression tree (CART) method by employing an ensemble learning approach [33]. The RF creates multiple CART models, known as trees, to form an ensemble. This ensemble approach helps reduce over-fitting and variance, resulting in more stable and accurate predictions. Each tree in the RF is trained on a bootstrap sample, which is a random sample with replacement from the original dataset. The samples that are not selected in the bootstrap sampling process form the Out-of-Bag (OOB) dataset. These samples are used to estimate the model's performance without the need for a separate validation set. By averaging the predictions of multiple trees, the RF reduces the variance in the model, quantifying the impact of each predictor variable on the model's predictive performance.

2.3.3. Extreme Gradient Boosting (XGBoost)

XGBoost is a gradient boosting algorithm that iteratively trains a series of weak learners, adjusting the errors of the previous round of models at each iteration to progressively improve the performance of the overall model [33]. The algorithm introduces a regularization term to avoid over-fitting and is highly optimized for computational efficiency. Various optimization techniques, such as block-based gradient computation, cache optimization, and multi-threaded processing, are used to make the algorithm computationally efficient when dealing with large-scale datasets. XGBoost uses a depth-first strategy during tree growth and considers gains when splitting each node, which helps generate cleaner models.

2.3.4. The Convolutional Neural Network (CNN)

Convolutional neural networks (CNNs) can be used for data modeling, especially when dealing with data with distinct spatial structures [34]. The convolutional layers in the CNN capture the local features of the input data through filters (convolution kernels). These filters typically only cover a small portion of the input data, allowing the network to detect local patterns and features. Because each neuron is only connected to a local region of the input data, CNNs have far fewer connections than fully connected networks, rendering these networks more efficient and reducing the risk of over-fitting. A CNN is composed of multiple convolutional and pooling layers stacked together, with each layer capturing features at different scales to form a hierarchical feature representation. Moreover, it can be adjusted according to the needs of different tasks, such as changing the number of layers, the size and quantity of the filters, and adding regularization terms.

2.3.5. Model Training

Using the random sampling method, the collected data samples are divided into 80:20 (80 training and 20 testing samples) in order to construct a regional total N leaching intensity prediction model based on each machine learning algorithm. Ten-fold cross-validation was used five times to avoid the model over-fitting during the fitting process.

2.3.6. Model Validation

The ten-fold cross-validation method is employed five times in the model construction process to evaluate the parameters [35]. The original dataset is partitioned into 10 equally sized and mutually exclusive subsets, termed folds. Subsequently, 5 independent ten-fold cross-validation procedures are carried out. Prior to each iteration of validation, the folds are randomly re-partitioned. During a single validation round, 9 of the folds are successively selected for training purposes, while the remaining 1 fold is designated for testing, thereby generating 10 distinct sets of training–test combinations. The model training and evaluation phases yield 10 performance indicator values, and the average of these values is computed to obtain the final performance indicator for that particular validation. This process is repeated 5 times, yielding 5 sets of final performance indicators, which are then subjected to a comprehensive analysis. Such an approach enables a comprehensive assessment of the model's performance stability, facilitates the screening of optimal models, and aids performance estimation.

2.4. *Hyperparameter Tuning*

Bayesian hyperparameter optimization speeds up the optimization process by building a probabilistic model that predicts which combinations of hyperparameters are likely to result in a better performance, thus reducing the number of hyperparameter combinations that need to be evaluated. The key steps include initialization, modeling, acquisition functions, and iterative optimization [36].

2.5. Evaluation of Model Performance

The predictive accuracy of each model was evaluated using the R^2, *RMSE*, and *MAE*. R^2 measures the proportion of the variance in the dependent variable that is predictable from the independent variables [37]. The following equations are used to evaluate the performance of the model:

$$R^2 = 1 - \frac{\sum_{i=1}^{n}(y_i - \hat{y}_i)^2}{\sum_{i=1}^{n}(y_i - \overline{y})^2} \quad (1)$$

$$RMSE = \sqrt{1/n\sum_{i=1}^{n}(Y_i - X_i)^2} \quad (2)$$

$$MAE = 1/n\sum_{i=1}^{n}\lceil Y_i - X_i \rceil \quad (3)$$

where x_i and y_i denote the measured and predicted values. The quality of the data points is n. The value of R^2 ranges from 0 to 1; the closer the value is to 1, the better the model fit. The *RMSE* provides the average error size in the same unit as the original data, and the *MAE* measures the average absolute difference between the predicted and actual values, and the smaller its value, the better the model's estimation performance.

3. Results

3.1. Analysis of the Correlation of the Independent Variables

Quantile–quantile (QQ) plots were employed to assess the normality of both the raw data and processed data for each variable (Figure 1). Normalizing the data improved their distribution, filtered or aggregated the raw data, and enhanced their suitability for statistical analysis and modeling. In Figure 1B,D,F,H–L, the processed data align more closely with the diagonal, indicating enhanced normality post-processing. A slight deviation from the normal distribution showed that non-parametric tests or additional transformations would be necessary. In contrast, Figure 1A,C,E,G showed significant deviations from the diagonal in both the raw and processed data, indicating these variables may require further transformation or potential exclusion. The distributions of soil type and irrigation method also deviated from normality, and this was likely due to their categorical nature or the coding methods.

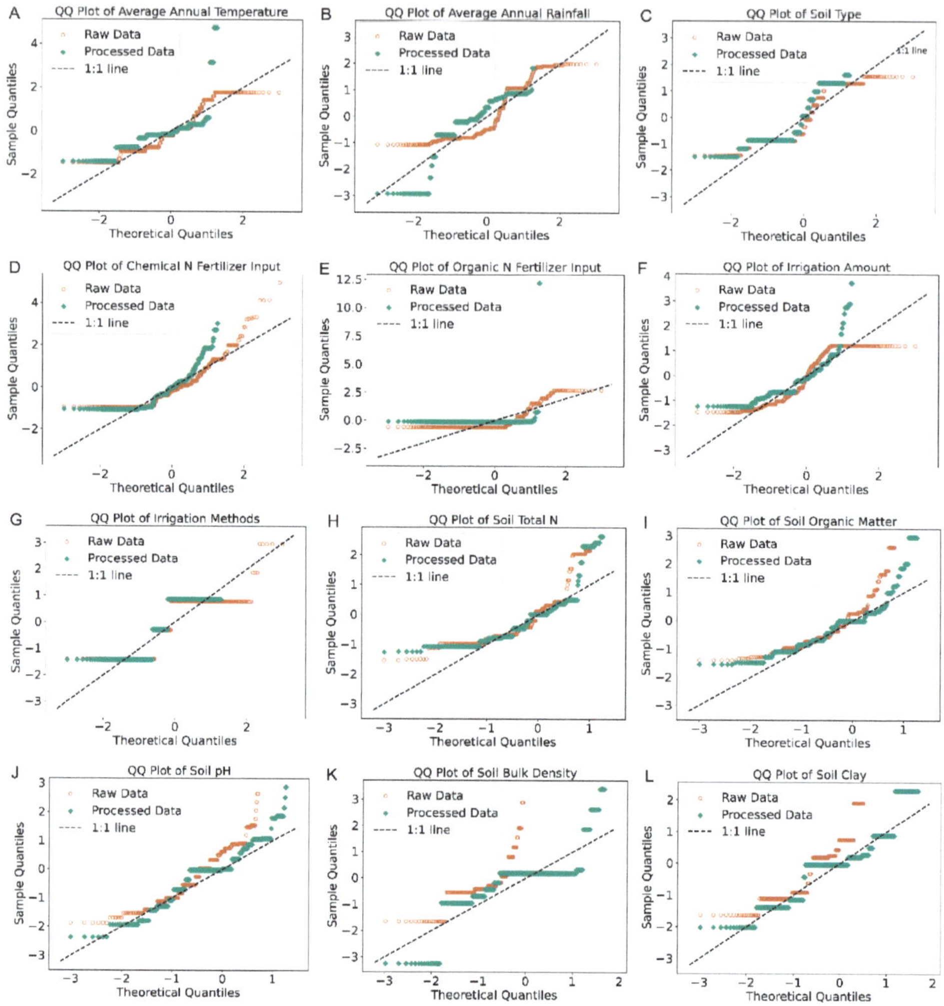

Figure 1. Raw data and processed data QQ plots of average annual temperature (**A**), average annual rainfall (**B**), soil type (**C**), chemical N fertilizer input (**D**), organic N fertilizer input (**E**), irrigation amount (**F**), irrigation methods (**G**), soil total N (**H**), soil organic matter (**I**), soil pH (**J**), soil bulk density (**K**) and soil clay (**L**).

The Pearson's correlation coefficient was used to measure the linear correlation between continuous variables (Figure 2). This analysis allows for screening predictor variables based on their influence on and contribution to the target variable. High correlations between certain features may justify the removal of one to reduce redundancy. In the Pearson's correlation analysis (Figure 2), the nitrate N loss rate showed positive correlations with chemical N fertilizer input, soil organic matter, and total N, and negative correlations were observed relative to soil pH, bulk density, clay content, and irrigation methods. The correlations between nitrate N loss and the other variables were weak. Positive correlations were also observed between average annual rainfall and temperature, as well as between irrigation method and irrigation amount; notably, the QQ plots for rainfall and irrigation amount align closely with a normal distribution. Based on the QQ plots and the correlation analysis, eight key features—average annual rainfall, chemical N fertilizer input, irrigation amount, soil total N, organic matter, pH, bulk density, and clay content—were selected for further analysis.

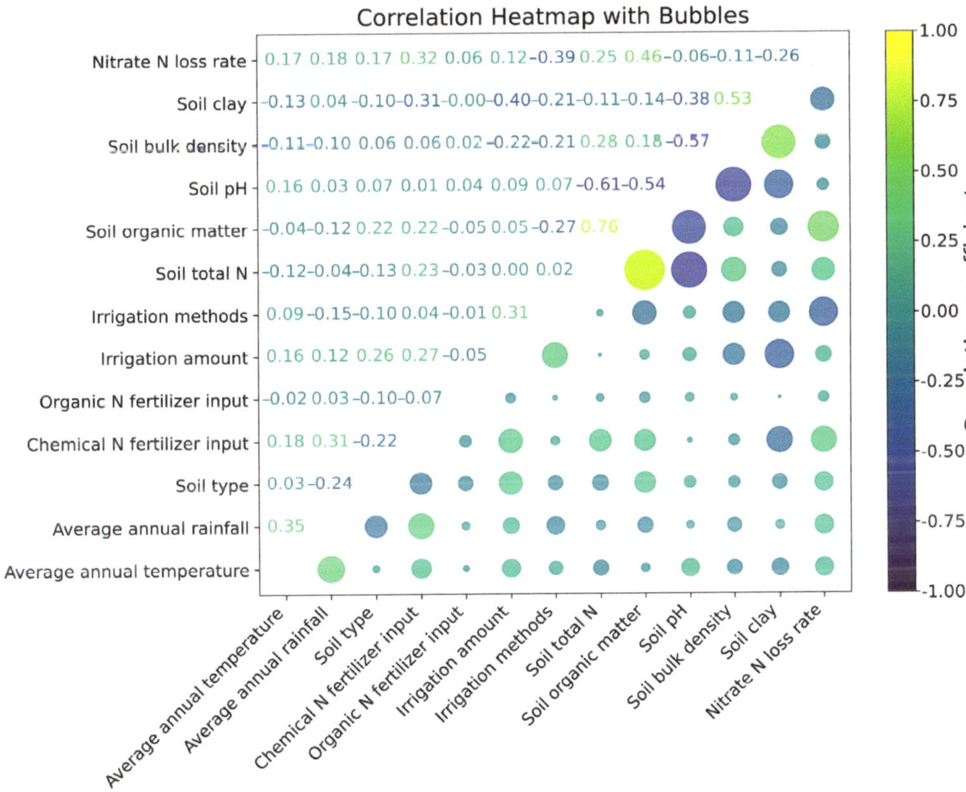

Figure 2. Pearson's correlation matrix of independent variables.

3.2. Training and Testing of the Model

The models for nitrate N loss rate prediction were tested across four machine learning algorithms (the SVM, RF, XGBoost, and the CNN) (Figure 3). According to the models' results, XGBoost achieved the highest performance on the training set, with an *MAE* value of 0.97, an *RMSE* of 5.26, and a high R^2 of 0.94, closely followed by RF, which attained an *MAE* of 2.42, an *RMSE* of 6.58, and an R^2 of 0.90. For the test sets, RF emerged as the best performing algorithm among the four, achieving the lowest *RMSE* of 10.01 and the highest R^2 of 0.75. In contrast, the performance of the SVM and CNN models exhibited a lower prediction accuracy for both the training and test sets.

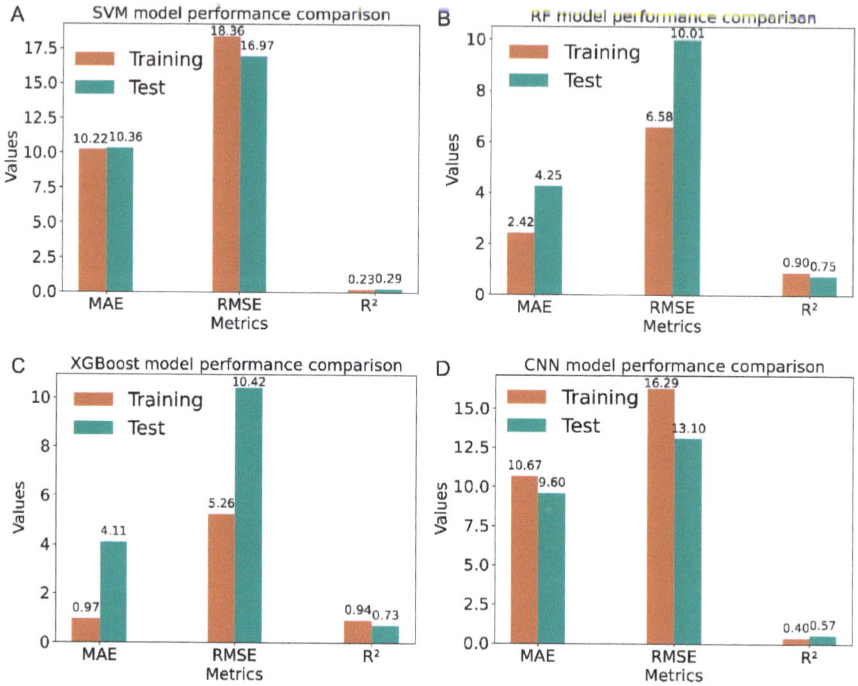

Figure 3. Comparison of R^2, *RMSE*, and *MAE* using the SVM (**A**), RF (**B**), XGBoost (**C**), and CNN (**D**) models for nitrate–nitrogen loss rate prediction on training and test datasets. Abbreviations: SVM, support vector machine; RF, random forest; XGBoost, extreme gradient boosting; CNN, convolutional neural network. Moving average error (*MAE*), root mean square error (*RMSE*), and coefficient of determination (R^2).

Overall, RF and XGBoost demonstrated the strongest linear relationship between the predicted and actual values and had the fewest outliers (Figure S1). While the SVM model performed slightly worse than RF, especially in high-value regions, the CNN model performed well in low-value regions but exhibited reduced accuracy at higher values. These results highlighted the importance of selecting a machine learning model suited to specific dataset characteristics and prediction tasks. Further research could explore the factors driving the performance differences among the models and examine potential improvements through hyperparameter tuning.

3.3. Model Optimization

Figure 4 demonstrates the fitting statistics for each model after Bayesian optimization with 10-fold cross-validation, which was evaluated using the *MAE*, *RMSE*, and R^2 metrics. The optimization improved the performance of all four models. Among them, XGBoost achieved the highest predictive accuracy (R^2 = 0.79) and the lowest *MAE* (3.87) across the test sets, followed closely by the RF and SVM, with R^2 values of 0.75 and 0.76, respectively. The CNN model displayed the lowest R^2 (0.56) and highest *MAE* (6.50) and *RMSE* (9.99), indicating the weakest performance among the models. These results suggested that tree-based models such as XGBoost were particularly well suited to this regression task due to their capability to capture complex nonlinear relationships in the data.

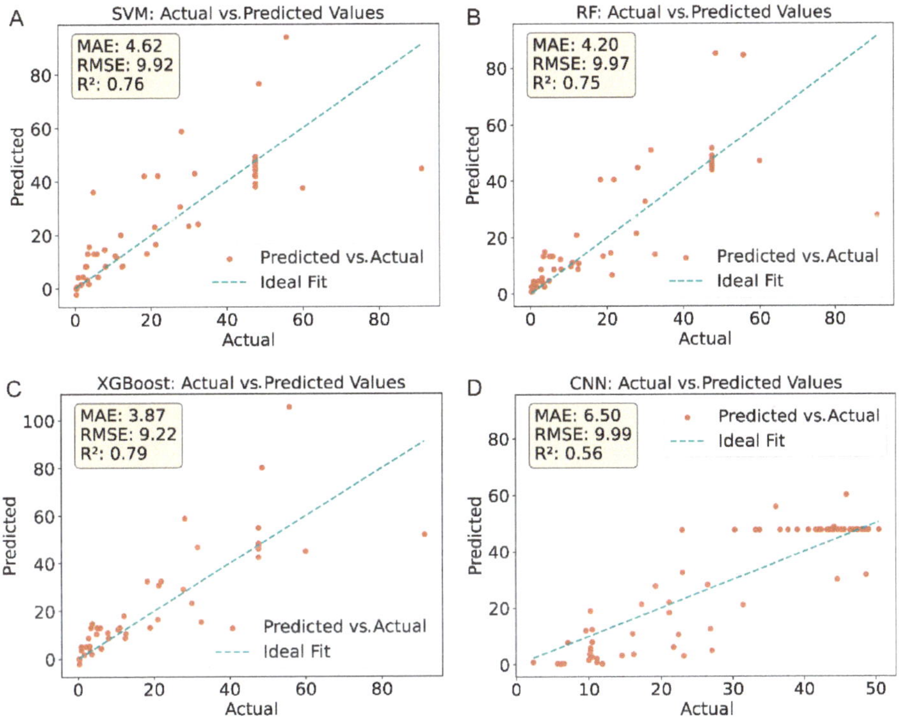

Figure 4. Result of Bayesian-optimized hyperparameters in SVM (**A**), RF (**B**), XGBoost (**C**), and CNN (**D**) models for nitrate–nitrogen loss rate prediction. Abbreviations: SVM, support vector machine; RF, random forest; XGBoost, extreme gradient boosting; CNN, convolutional neural network. Moving average error (*MAE*), root mean square error (*RMSE*), and coefficient of determination (R^2).

Residual plots (Figure S2) further assessed the models' accuracy by comparing their residuals (differences between predicted and actual values). For the SVM and CNN, the residuals were dispersed across the predicted range, with a tendency toward overestimating at low predicted values and underestimating at high predicted values. In contrast, RF and XGBoost exhibited more concentrated residuals near the zero line in the low-value regions, indicating higher accuracy in these areas. However, in high predicted value regions, the residuals began to deviate from zero, suggesting a slight bias in these higher ranges.

3.4. Variable Importance Analysis

Feature importance, which measures the influence of each feature on a model's predictive accuracy, was evaluated based on internal metrics (Figure 5A). The analysis in this study indicated that soil organic matter was the most influential factor affecting nitrate N levels, followed by the chemical N fertilizer input and average annual rainfall. Irrigation amount also played a significant role, while soil pH, bulk density, total N, and clay content had lesser impacts. These findings highlight the critical roles of soil organic matter, N management, and water balance in controlling nitrate leaching and the necessity of tailored agricultural practices to reduce environmental impacts.

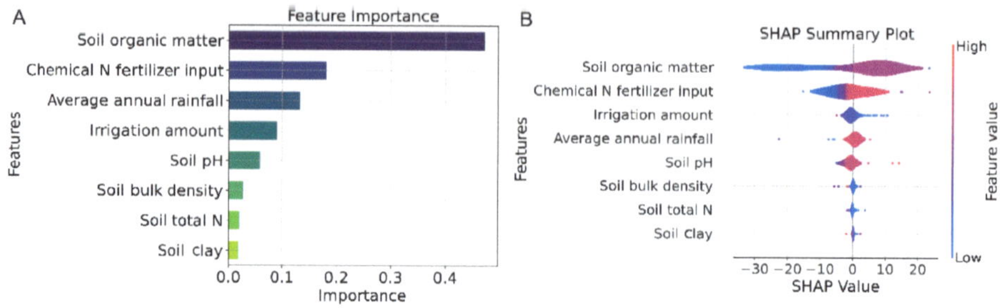

Figure 5. Ranking of the importance of input features (**A**) and the SHAP value for a particular variable (**B**).

SHAP (Shapley Additive Explanations) values provide further insight by assigning scores to each feature, reflecting its contribution to model predictions (Figure 5B). Positive SHAP values suggest that an increase in features increases the model outputs, while negative values indicate a decrease. The high SHAP values for soil organic matter and chemical N fertilizer input underlined their substantial impact on nitrate concentrations. Moderate to high SHAP values for irrigation practices showed that irrigation volume significantly influenced nitrate levels, while moderate values for annual rainfall indicated that rainfall affects nitrate leaching and transport in the soil. Additionally, soil pH, bulk density, clay content, and total N exhibited lower SHAP values, signifying their minimal impact on the model predictions. These analyses implied the importance of integrating these factors into nitrate management strategies in order to optimize the N levels in agricultural soils.

Table 3 shows the predictive performance of the model with various feature combinations. It is observed that as the number of features increases, both the *MAE* and *RMSE* values exhibit a decreasing trend, whereas R^2 exhibits a corresponding increase. Specifically, when all of the features are incorporated, the *MAE* decreases to 3.87, the *RMSE* decreases to 9.22, and R^2 increases to 0.79. These results suggest that the addition of features can significantly enhance the predictive performance of the model.

Table 3. Predictive performance as a function of each predictor's contribution.

Features	*MAE*	*RMSE*	R^2
Average annual rainfall	6.35	13.39	0.56
Average annual rainfall and chemical N fertilizer input	5.78	12.26	0.63
Average annual rainfall, chemical N fertilizer input, and irrigation amount	5.81	12.07	0.64
Average annual rainfall, chemical N fertilizer input, irrigation amount, and soil total N	4.47	10.74	0.71
Average annual rainfall, chemical N fertilizer input, irrigation amount, soil total N, and soil organic matter	4.30	10.85	0.71
Average annual rainfall, chemical N fertilizer input, irrigation amount, soil total N, soil organic matter, and soil pH	4.15	10.41	0.73
Average annual rainfall, chemical N fertilizer input, irrigation amount, soil total N, soil organic matter, soil pH, and soil bulk density	4.03	9.79	0.76
Average annual rainfall, chemical N fertilizer input, irrigation amount, soil total N, soil organic matter, soil pH, soil bulk density, and soil clay	3.87	9.22	0.79

4. Discussion

4.1. Model Performance Evaluation and Comparison

Pearson's correlation analysis aids in model construction by identifying and assessing the linear relationships among variables, thereby guiding the feature selection and model optimization [38]. Normalization is a feature transformation technique that scales the data to a fixed range (e.g., [0,1] or [−1,1]), improving the data distribution and visualization and the model's learning capacity [29]. QQ plots are employed to verify whether the data adhere to theoretical distributions, such as a normal distribution, ensuring the validity of the data transformations. In this study, based on Pearson's correlation and QQ plot analyses, eight variables—average annual rainfall, chemical N fertilizer input, irrigation, soil total N, organic matter, pH, bulk density, and clay—demonstrated a normalized data distribution (Figures 1 and 2) and strong correlations with nitrate loss, rendering them suitable input variables.

Traditional N leaching monitoring methods, such as using ceramic suction cup samplers [39] and lysimeters [40], often introduce boundary effects (disturbing the hydraulic gradient) and soil disturbances [41]. These issues complicate the evaluation of the N concentrations at specific depths or times and may potentially impact the N mass balance [42]. In contrast, ML approaches, including RF, SVMs, XGBoost, and CNNs, offer significant advantages for N leaching monitoring by handling complex nonlinear relationships, learning from large datasets, and providing accurate predictions [43]. However, ML models also have certain limitations, such as reliance on field research data, potential biases, decision opacity, and challenges in generalization [44]. Despite these limitations, they can learn potential patterns through feature selection, dimensionality reduction, and noise tolerance, thereby providing accurate predictions. Moreover, continuous optimization and parameter adjustment can further enhance the model performance. Therefore, selecting an appropriate ML model requires comprehensive consideration of the specific application scenarios, data characteristics, and task objectives [45,46].

In this study, we used eight key variables, including average annual rainfall, chemical N fertilizer input, and irrigation amount, to predict the nitrate leaching losses in North China. Four ML models were employed for this prediction: an SVM, RF, XGBoost, and a CNN (Figure 3). Among these models, the RF demonstrated the strongest explanatory and fitting capabilities for the target variable, followed closely by XGBoost. In addition, the R^2 of the other models (SVM and CNN) was slightly lower than these two values. This difference is primarily attributed to the inherent characteristics of the data and the algorithms of the models. The data exhibited complex nonlinear relationships and diverse distributions, and there were interaction effects among the key variables (Table 3). For instance, different combinations of annual average rainfall and chemical N fertilizer inputs can influence the nitrate leaching losses, and irrigation practices can alter the transport pathways of nitrates [41,45,47]. The RF and XGBoost, being ensemble models, can effectively capture nonlinearities and interaction effects by constructing multiple decision trees and training them based on various samples and feature subsets. These models are capable of efficiently segmenting and learning from high-dimensional and highly variable data, thus demonstrating strong interpretation and fitting capabilities [33]. However, the SVM's performance may be slightly weaker due to its reliance on finding the optimal hyperplane for data partitioning [32]. Inappropriate kernel functions or parameters can impede the accurate prediction of nitrate leaching losses when processing data with eight key variables. The CNN performs well with spatially structured data such as images, but nitrate leaching loss data lack spatial locality and a hierarchical structure. Moreover, the limited data available in this study were insufficient for effective feature learning. Additionally, the numerous hyperparameters involved in the CNN's multi-layer convolution and pooling operations, if improperly set, can negatively impact feature learning and model performance [39,40].

Although the RF exhibited a strong linear relationship between the predicted and actual values, it produced several outliers at high nitrate concentration levels (Figure S1); this observation aligns with previous findings by Knoll et al. [16]. When constructing RF models, random selection of the features and sampling techniques are typically employed to mitigate the variance in the model. In regions with a high concentration of data, due to the relatively sparse nature of the data in such areas, certain decision trees might over-fit the local characteristics, thereby resulting in the emergence of outliers during the integration process. In contrast, XGBoost demonstrated a stronger linear correlation with the actual values and fewer outliers, indicating its better performance at high concentration levels. The difference in the R^2 values between the training (0.94) and test (0.73) datasets for

XGBoost suggests a tendency for ensemble methods, including both RF and XGBoost, to over-fit the data. This over-fitting phenomenon likely inflates the model's performance metrics and points to challenges in generalizing to unseen data [48]. This discrepancy can be attributed to XGBoost's adaptation of a gradient boosting strategy. However, this approach places excessive emphasis on the errors within the training data. Consequently, the model's performance may be overly tuned and optimized to the training set, thereby sacrificing a certain degree of generalization ability. Despite this, the XGBoost model achieved the lowest *MAE* of 4.11 kg/ha, indicating minimal prediction errors and underscoring its superior performance relative to that of the other models. Further hyperparameter tuning is recommended in order to enhance the model performance and robustness.

Feature engineering, cross-validation, and Bayesian optimization are critical techniques for enhancing the accuracy and generalization of ML models [36,49]. Bayesian optimization is especially effective due to its ability to address data variability and improve model generalization [50]. By continuously adjusting the model parameters, it reduced model over-fitting and increased the model's prediction accuracy for new data. Li et al. [51] developed a model for predicting the ammonium nitrogen concentrations in groundwater in Northeast China, significantly improving the model performance with hyperparameter tuning and achieving an *MAE* of 0.09.

Following Bayesian optimization, the four models demonstrated improved training outcomes. In contrast, the CNN model underperformed, with R^2, *RMSE*, and *MAE* values of 0.56, 9.99, and 6.50, respectively (Figure 4). The remaining three models exhibited strong linear relationships between their predicted and measured values, with the test set R^2 values exceeding 0.75. This suggests that the optimized models provide reliable predictions for nitrate leaching loss rates. Feng et al. [52] predicted the nitrate leaching in North China's aeration zone, achieving a Willmott index of 0.71 and an *RMSE* of 42.28 for all soil layers via a regional water and nitrogen transport model. Similarly, Zang et al. [53] developed a quasi-3D model for the nitrate dynamics in the Beijing Plain, achieving an R^2 of 0.73, while Liu et al. [54] observed that the RF model outperformed traditional SMLR for the nitrogen use efficiency in Northeastern farmland, reaching an R^2 of 0.89 versus 0.67 for SMLR. Notably, the XGBoost model delivered the best results, achieving an R^2 of 0.79, an *MAE* of 3.87, and a more concentrated residual distribution in the test set (Figure S2) and indicating strong predictive performance for nitrate leaching loss. The superior stability of the XGBoost model can be attributed to optimized loss functions and regularization terms, which enhance its fitting performance. To summarize, these results implied machine learning's advantages over traditional models in predicting N leaching.

4.2. Importance Analysis

To assess the key variables influencing nitrate leaching in the study area, we utilized feature importance and SHAP values as the criteria. The results indicate that soil organic matter, chemical N fertilizer input, and water inputs (rainfall and irrigation amount) are the most significant factors affecting nitrate leaching (Figure 5). This finding highlights the critical role of soil properties, agricultural practices, and climatic factors in determining nitrate leaching intensity.

The climate in North China is relatively humid, mainly consisting of moist soil, which results in the rapid turnover of soil organic carbon (SOC) and a limited carbon sequestration capacity [55]. The high concentration of soluble carbon and N increases the risk of N loss through leaching. Long-term cultivation accelerates the decomposition of surface SOC [56], changes the physical and chemical properties of the soil, and indirectly affects N adsorption and migration. For instance, it promoted the formation of soil aggregates, enhanced water retention, and affected clay content and microbial carbon accumulation [57]. Nitrate accumulation is notably higher in coarse-grained soil profiles compared to medium-coarse-grained profiles, with the moisture content correlating positively with clay content and impacting solute transport. The flow rate in coarse-grained profiles was approximately 7.5 times higher than that in fine-grained profiles [58–60]. Increased soil organic matter typically might support a more diverse and abundant microbial community, thereby affecting N mineralization, nitrification, and denitrification processes [61]. Research has shown that organic inputs, including green fertilizers [62], a combination application of organic–inorganic fertilizers [63], and organic amendments [64], can increase the soil organic matter content and enhance the biotic and abiotic fixation of ammonium and nitrate N in soil, reducing N leaching by 15% without affecting the crop yield.

Water input, including rainfall and irrigation, significantly influences nitrate leaching by controlling soil drainage [65]. Studies have shown that over-irrigation is a primary driver of nitrate leaching in North China [66]. In 1999 and 2000, drainage losses reached 84 and 121.3 mm, with nitrate leaching losses of 28.6 and 81.8 kg/ha as soil nitrate was transported to deeper layers during the growing season [67]. Li et al. [27] used NLEAP-GIS to simulate the nitrate leaching in the Jing–Jin–Ji region, identifying rainfall and irrigation as the primary factors affecting leaching variations. Similarly, Wang et al. [68] showed that nitrate leaching is higher in rainy years than dry or normal years in North China based on HYDRUS-1D simulations with weather data from 1986 to 2006. Moreover, farmers apply an average of 500 kg/ha of N annually—far exceeding the recommended rates of 240 and 326 kg/ha for maize and wheat, respectively—resulting in substantial N loss [69,70]. Zhang et al. [71] confirmed the positive correlation between nitrate leaching and N application levels in the Beijing–Tianjin–Tangshan region, which is consistent with our findings. Yang et al. [65] found that furrow fertilization resulted in a greater increase in nitrate leaching than sprinkler irrigation. However, due to the data limitations, our model did not account for specific factors such as irrigation methods.

Other factors, such as soil pH, bulk density, and soil total N, also impact nitrate leaching. Lower soil pH values may enhance the conversion of ammonium N (NH_4^+) into nitrate nitrogen (NO_3^-), while lower soil bulk density values increase the water permeability, increasing the risk of nitrate leaching [72]. Additionally, a higher soil total N content is correlated with increased nitrate leaching, as microbial processes such as N mineralization respond dynamically to N levels [73]. Therefore, optimizing fertilizer management, adjusting soil pH, and enhancing the soil's physical properties are essential strategies for reducing nitrate leaching and mitigating soil pollution risks.

The leaching of agricultural nitrate varies by region and is influenced by multiple factors. In temperate, rainfed, cereal crop regions, extreme precipitation and high rainfall intensity are the primary drivers of nitrate leaching, while soil texture and N fertilizer management play lesser roles [74,75]. In intensive agricultural areas, the primary nitrate sources include chemical fertilizers, soil N, and manure or sewage [76]. North China is characterized by concentrated rainfall, a winter wheat–maize double-cropping system, and excessive N fertilizer use. Here, precipitation, soil organic matter, and fertilizer input collectively influence nitrate leaching (Table 3). Excessive applications of fertilizer risk nutrient loss, particularly following heavy rains in the rainy season (June to August), while flood irrigation exacerbates leaching during dry seasons. Soil organic matter promotes macropore formation, enhancing preferential flow and accelerating N transport, thereby increasing leaching losses [77]. Meanwhile, farmers often apply up to 500 kg/ha of N fertilizer, resulting in substantial N loss. These findings underscore the need for region-specific strategies to manage nitrate leaching. Key recommendations include the following:

(1) Improving soil properties and water retention capacity: implement conservation tillage, reduced tillage, or no-till practices to enhance soil organic matter and reduce nitrogen loss.
(2) Optimizing nitrogen fertilizer management: reduce nitrogen fertilizer use, apply controlled-release fertilizers and nitrification inhibitors, and adopt optimized fertilization methods to curb nitrate leaching.
(3) Controlling irrigation amount: employ drip irrigation, microsprinklers, or subsurface irrigation to enhance water and fertilizer efficiency.

Comprehensive considerations of these strategies can optimize agricultural practices, enhance soil health, and mitigate environmental impacts.

5. Conclusions

This study employed four machine learning models—RF, an SVM, XGBoost, and a CNN—to analyze the spatial variability in the nitrate leaching loss from agricultural fields in North China. Among these, the Bayesian-optimized XGBoost model achieved the highest predictive accuracy ($R^2 = 0.79$) and the lowest error ($MAE = 3.87$ kg/ha), outperforming RF, the SVM, and the CNN. Analyses of the variable importance identified soil organic matter, chemical nitrogen fertilizer input, and water inputs (rainfall and irrigation) as the primary drivers of nitrate leaching. Effective strategies to reduce nitrate leaching include improving the soil's properties and water retention capacity, optimizing N fertilizer management, and controlling the amount of irrigation. These results support the conclusion that improving the soil's properties and water retention, optimizing nitrogen fertilizer management, and controlling irrigation can effectively reduce nitrate leaching. Additionally, this data-driven approach provides valuable guidance for developing targeted agricultural management practices to control nitrogen non-point source pollution in North China's farmlands.

Supplementary Materials: The following supporting information can be downloaded at: https://www.mdpi.com/article/10.3390/land14010069/s1, Figure S1: Result in SVM (A), RT (B), XGBoost (C), and CNN (D) models for nitrate nitrogen loss rate prediction; Figure S2: Residual plot in SVM (A), RF (B), XGBoost (C), and CNN (D) models for nitrate nitrogen loss rate prediction.

Author Contributions: C.H.: writing—original draft; visualization; methodology; and formal analysis. J.L.: investigation; supervision; funding acquisition; and validation. Y.P.: data curation. L.L.: visualization. F.L.: investigation. W.W.: data curation. Y.X.: writing—review and editing and supervision. H.L.: software. B.T.: data curation. G.Z.: resources; project administration; and conceptualization. All authors have read and agreed to the published version of the manuscript.

Funding: This study was funded by the National Key R&D Program of China (grant number 2023YFD1701805).

Data Availability Statement: The data in this article are available upon request.

Acknowledgments: The authors extend their appreciation to the Chinese Academy of Agricultural Sciences Innovation Program (Agro-Environmental Protection Institute).

Conflicts of Interest: The authors declare no conflicts of interest. The funder had no role in the design of this study; in the collection, analysis, or interpretation of the data; in the writing of the manuscript; or in the decision to publish the results.

References

1. Hou, L.; Ma, C.; Liu, T. Revealing Greenhouse Gas Emission and Nitrogen Fertilizer Destination: A Case Study in Chengdu Plain Cultivation Industry. *Sustainability* **2024**, *16*, 6073. [CrossRef]
2. Zhao, H.; Li, X.; Jiang, Y. Response of Nitrogen Losses to Excessive Nitrogen Fertilizer Application in Intensive Greenhouse Vegetable Production. *Sustainability* **2019**, *11*, 1513. [CrossRef]
3. Aryal, B.; Gurung, R.; Camargo, A.F.; Fongaro, G.; Treichel, H.; Mainali, B.; Angove, M.J.; Ngo, H.H.; Guo, W.; Puadel, S.R. Nitrous oxide emission in altered nitrogen cycle and implications for climate change. *Environ. Pollut.* **2022**, *314*, 120272. [CrossRef] [PubMed]
4. Bai, Z.; Huang, J.; Zhu, L.; Cao, X.; Zhu, C.; Zhong, C.; Jin, Q.; Zhang, J. Effects of N application strategies on N leaching loss in paddy soil and N use characteristics in different super hybrid rice cultivars. *Paddy Water Environ.* **2020**, *18*, 27–41. [CrossRef]
5. Coucheney, E.; Eckersten, H.; Hoffmann, H.; Jansson, P.; Gaiser, T.; Ewert, F.; Lewan, E. Key functional soil types explain data aggregation effects on simulated yield, soil carbon, drainage and nitrogen leaching at a regional scale. *Geoderma* **2018**, *318*, 167–181. [CrossRef]
6. Wang, Z.H.; Wan, L.; Xu, C.M. Spectrums of Nitrogen and Phosphorus Leaching Loss from Different Land-Use Types. *Soil Air Water* **2018**, *46*, 1700538. [CrossRef]
7. Ersahin, S.; Rüstü Karaman, M. Estimating potential nitrate leaching in nitrogen fertilized and irrigated tomato using the computer model NLEAP. *Agric. Water Manag.* **2001**, *51*, 1–12. [CrossRef]
8. Lei, Z.; Li, Q.; Song, X.; Wang, W.; Zhang, Z.; Peng, C.; Tian, L. Biochar mitigates dissolved organic carbon loss but does not affect dissolved organic nitrogen leaching loss caused by nitrogen deposition in Moso bamboo plantations. *Glob. Ecol. Conserv.* **2018**, *16*, e00494. [CrossRef]

9. Xue, D.; Botte, J.; De Baets, B.; Accoe, F.; Nestler, A.; Taylor, P.; Van Cleemput, O.; Berglund, M.; Boeckx, P. Present limitations and future prospects of stable isotope methods for nitrate source identification in surface- and groundwater. *Water Res.* **2009**, *43*, 1159–1170. [CrossRef]
10. Li, S.; Bhattarai, R.; Cooke, R.A.; Verma, S.; Huang, X.; Markus, M.; Christianson, L. Relative performance of different data mining techniques for nitrate concentration and load estimation in different type of watersheds. *Environ. Pollut.* **2020**, *263*, 114618. [CrossRef]
11. Almasri, M.N.; Kaluarachchi, J.J. Implications of on-ground nitrogen loading and soil transformations on groundwater quality management. *J. Am. Water Resour. Assoc.* **2004**, *1*, 165–186. [CrossRef]
12. Van der Laan, M.; Annandale, J.G.; Bristow, K.L.; Stirzaker, R.J.; Preez, C.C.D.; Thorburn, P.J. Modelling nitrogen leaching: Are we getting the right answer for the right reason? *Agric. Water Manag.* **2014**, *133*, 74–80. [CrossRef]
13. Cameira, M.R.; Sousa, P.L.; Farahani, H.J.; Ahuja, L.R.; Pereira, L.S. Evaluation of the RZWQM for the Simulation of Water and Nitrate Movement in Level-Basin, Fertigated Maize. *J. Agric. Eng. Res.* **1998**, *69*, 331–341. [CrossRef]
14. Gassman, P.W.; Sadeghi, A.M.; Srinivasan, R. Applications of the SWAT Model Special Section: Overview and Insights. *J. Environ. Qual.* **2014**, *43*, 1–8. [CrossRef]
15. Hu, K.; Li, B.; Chen, D.; Zhang, Y.; Edis, R. Simulation of nitrate leaching under irrigated maize on sandy soil in desert oasis in Inner Mongolia, China. *Agric. Water Manag.* **2008**, *95*, 1180–1188. [CrossRef]
16. Knoll, L.; Breuer, L.; Bach, M. Large scale prediction of groundwater nitrate concentrations from spatial data using machine learning. *Sci. Total Environ.* **2019**, *668*, 1317–1327. [CrossRef]
17. Mahlknecht, J.; Torres-Martinez, J.A.; Kumar, M.; Mora, A.; Kaown, D.; Loge, F.J. Nitrate prediction in groundwater of data scarce regions: The futuristic fresh-water management outlook. *Sci. Total Environ.* **2023**, *905*, 166863. [CrossRef]
18. Feng, B.; Ma, J.; Liu, Y.; Wang, L.; Zhang, X.; Zhang, Y.; Zhao, J.; He, W.; Chen, Y.; Weng, L. Application of machine learning approaches to predict ammonium nitrogen transport in different soil types and evaluate the contribution of control factors. *Ecotoxicol. Environ. Saf.* **2024**, *284*, 116867. [CrossRef]
19. Awais, M.; Aslam, B.; Maqsoom, A.; Khalil, U.; Ullah, F.; Azam, S.; Imran, M. Assessing Nitrate Contamination Risks in Groundwater: A Machine Learning Approach. *Appl. Sci.* **2021**, *11*, 10034. [CrossRef]
20. Mohammadi, B. A review on the applications of machine learning for runoff modeling. *Sustain. Water Resour. Manag.* **2021**, *7*, 98. [CrossRef]
21. Jacob, C.; Masilamani, U.S. Performance evaluation of different machine learning algorithms for prediction of nitrate in groundwater in Thiruvannamalai District. *Clean Soil Air Water* **2024**, 2400060. [CrossRef]
22. Dhaliwal, J.K.; Panday, D.; Robertson, G.P.; Saha, D. Machine learning reveals dynamic controls of soil nitrous oxide emissions from diverse long-term cropping systems. *J. Environ. Qual.* **2024**. [CrossRef] [PubMed]
23. Xue, J.; Yuan, C.; Ji, X.; Zhang, M. Predictive modeling of nitrogen and phosphorus concentrations in rivers using a machine learning framework: A case study in an urban-rural transitional area in Wenzhou China. *Sci. Total Environ.* **2024**, *910*, 168521. [CrossRef]
24. Ransom, K.M.; Nolan, B.T.; Stackelberg, P.E.; Belitz, K.; Fram, M.S. Machine learning predictions of nitrate in groundwater used for drinking supply in the conterminous United States. *Sci. Total Environ.* **2022**, *807*, 151065. [CrossRef]
25. Rakotovololona, L.; Beaudoin, N.; Ronceux, A.; Venet, E.; Mary, B. Driving factors of nitrate leaching in arable organic cropping systems in Northern France. *Agric. Ecosyst. Environ.* **2019**, *272*, 38–51. [CrossRef]
26. Li, J.; Hu, W.; Chau, H.W.; Beare, M.; Cichota, R.; Teixeira, E.; Moore, T.; Di, H.; Cameron, K.; Guo, J.; et al. Response of nitrate leaching to no-tillage is dependent on soil, climate, and management factors: A global meta-analysis. *Glob. Chang. Biol.* **2023**, *29*, 2172–2187. [CrossRef]
27. Li, Z.T.; Wen, X.M.; Xu, C.S.; Li, X.X.; Hu, B.Q. Regional simulation of nitrate leaching potential from winter wheat-summer maize rotation croplands on the North China Plain using the NLEAP-GIS model. *Agric. Ecosyst. Environ.* **2020**, *294*, 106861. [CrossRef]
28. Zhu, Z.L.; Chen, D.L. Nitrogen fertilizer use in China—Contributions to food production, impacts on the environment and best management strategies. *Nutr. Cycl. Agroecosyst.* **2002**, *63*, 117–127. [CrossRef]
29. Salau, A.O.; Jain, S. *Feature Extraction: A Survey of the Types, Techniques, Applications*; IEEE: Piscataway, NJ, USA, 2019; Volume 1, pp. 158–164.
30. Maharana, K.; Mondal, S.; Nemade, B. A review: Data pre-processing and data augmentation techniques. *Glob. Transit. Proc.* **2022**, *3*, 91–99. [CrossRef]
31. Shalabi, L.A.; Shaaban, Z.; Kasasbeh, B. Data Mining: A Preprocessing Engine. *J. Comput. Sci.* **2006**, *2*, 735–739. [CrossRef]
32. Wang, X.; Huang, F.; Cheng, Y. Computational performance optimization of support vector machine based on support vectors. *Neurocomputing* **2016**, *211*, 66–71. [CrossRef]
33. Wang, Y.; Su, J.; Zhao, X. Penalized semiparametric Cox regression model on XGBoost and random survival forests. *Commun. Stat. Simul. Comput.* **2023**, *52*, 3095–3103. [CrossRef]
34. Alzubaidi, L.; Zhang, J.; Humaidi, A.J.; Al-Dujaili, A.; Duan, Y.; Al-Shamma, O.; Santamaría, J.; Fadhel, M.A.; Al-Amidie, M.; Farhan, L. Review of deep learning: Concepts, CNN architectures, challenges, applications, future directions. *J. Big Data* **2021**, *8*, 53. [CrossRef] [PubMed]
35. Ramezan, A.C.; Warner, A.T.; Maxwell, E.A. Evaluation of Sampling and Cross-Validation Tuning Strategies for Regional-Scale Machine Learning Classification. *Remote Sens.* **2019**, *11*, 185. [CrossRef]

36. Yin, J.; Li, N. Ensemble learning models with a Bayesian optimization algorithm for mineral prospectivity mapping. *Ore Geol. Rev.* **2022**, *145*, 104916. [CrossRef]
37. Walpole, R.E.; Myers, R.H.; Myers, S.L.; Ye, K. *Probability and Statistics for Engineers and Scientists*; Macmillan: New York, NY, USA, 1993.
38. Schober, P.; Boer, C.; Schwarte, L.A. Correlation Coefficients: Appropriate Use and Interpretation. *Anesth. Analg.* **2018**, *5*, 1763–1768. [CrossRef]
39. Poss, R.; Noble, A.D.; Dunin, F.X.; Reyenga, W. Evaluation of ceramic cup samplers to measure nitrate leaching in the field. *Eur. J. Soil Sci.* **1995**, *46*, 667–674. [CrossRef]
40. Hirschi, M.; Michel, D.; Lehner, I.; Seneviratne, S.I. A site-level comparison of lysimeter and eddy covariance flux measurements of evapotranspiration. *Hydrol. Earth Syst. Sci.* **2017**, *21*, 1809–1825. [CrossRef]
41. Abdou, H.M.; Flury, M. Simulation of water flow and solute transport in free-drainage lysimeters and field soils with heterogeneous structures. *Eur. J. Soil Sci.* **2004**, *55*, 229–241. [CrossRef]
42. Wang, Q.; Cameron, K.; Buchan, G.; Zhao, L.; Zhang, E.H.; Smith, N.; Carrick, S. Comparison of lysimeters and porous ceramic cups for measuring nitrate leaching in different soil types. *N. Z. J. Agric. Res.* **2012**, *55*, 333–345. [CrossRef]
43. Chlingaryan, A.; Sukkarieh, S.; Whelan, B. Machine learning approaches for crop yield prediction and nitrogen status estimation in precision agriculture: A review. *Comput. Electron. Agric.* **2018**, *151*, 61–69. [CrossRef]
44. Adombi, A.V.D.P.; Chesnaux, R.; Boucher, M. Review: Theory-guided machine learning applied to hydrogeology—State of the art, opportunities and future challenges. *Hydrogeol. J.* **2021**, *29*, 2671–2683. [CrossRef]
45. Azad, N.; Behmanesh, J.; Rezaverdinejad, V.; Khodaverdiloo, H.; Thompson, S.E.; Mallants, D.; Ramos, T.B.; He, H. CNN deep learning performance in estimating nitrate uptake by maize and root zone losses under surface drip irrigation. *J. Hydrol.* **2023**, *625*, 130148. [CrossRef]
46. Karimi, Y.; Prasher, S.O.; Patel, R.M.; Kim, S.H. Application of support vector machine technology for weed and nitrogen stress detection in corn. *Comput. Electron. Agric.* **2006**, *51*, 99–109. [CrossRef]
47. Parizad, A.; Hatziadoniu, C. Deep Learning Algorithms and Parallel Distributed Computing Techniques for High-Resolution Load Forecasting Applying Hyperparameter Optimization. *IEEE Syst. J.* **2022**, *16*, 3758–3769. [CrossRef]
48. Ram, H.; Grinfeld, G.; Liberman, N. Anticipated variability increases generalization of predictive learning. *NPJ Sci. Learn.* **2024**, *9*, 55. [CrossRef]
49. Thanh, N.N.; Chotpantarat, S.; Ngu, N.H.; Thunyawatcharakul, P.; Kaewdum, N. Integrating machine learning models with cross-validation and bootstrapping for evaluating groundwater quality in Kanchanaburi province, Thailand. *Environ. Res.* **2024**, *252*, 118952. [CrossRef]
50. Ouyang, L.; Park, C.; Ma, Y.; Ma, Y.; Wang, M. Bayesian hierarchical modelling for process optimisation. *Int. J. Prod. Res.* **2021**, *59*, 4649–4669. [CrossRef]
51. Li, W.; Ye, X.; Du, X. Imputation of Ammonium Nitrogen Concentration in Groundwater Based on a Machine Learning Method. *Water* **2022**, *14*, 1595. [CrossRef]
52. Feng, W.; Wang, S.; Tan, K.; Ma, L.; Hu, C. Simulation of spatial and temporal variation of nitrate leaching in the vadose zone of alluvial regions on a large regional scale. *Sci. Total Environ.* **2024**, *916*, 170114. [CrossRef]
53. Zang, Y.; Hou, X.; Li, Z.; Li, P.; Sun, Y.; Yu, B.; Li, M. Quantify the effects of groundwater level recovery on groundwater nitrate dynamics through a quasi-3D integrated model for the vadose zone-groundwater coupled system. *Water Res.* **2022**, *226*, 119213. [CrossRef] [PubMed]
54. Liu, Y.; Heuvelink, G.B.M.; Bai, Z.; He, P.; Jiang, R.; Huang, S.; Xu, X. Statistical analysis of nitrogen use efficiency in Northeast China using multiple linear regression and Random Forest. *J. Integr. Agric.* **2022**, *21*, 3637–3657. [CrossRef]
55. Ju, X.; Liu, X.; Zhang, F.; Roelcke, M. Nitrogen fertilization, soil nitrate accumulation, and policy recommendations in several agricultural regions of China. *Ambio* **2004**, *33*, 300–305. [CrossRef] [PubMed]
56. Wang, H.; Wang, S.; Zhang, Y.; Wang, X.; Wang, R.; Li, J. Tillage system change affects soil organic carbon storage and benefits land restoration on loess soil in North China. *Land Degrad. Dev.* **2018**, *29*, 2880–2887. [CrossRef]
57. Liao, L.; Green, C.T.; Bekins, B.A.; Böhlke, J.K. Factors controlling nitrate fluxes in groundwater in agricultural areas. *Water Resour. Res.* **2012**, *48*. [CrossRef]
58. Min, L.; Shen, Y.; Pei, H. Estimating groundwater recharge using deep vadose zone data under typical irrigated cropland in the piedmont region of the North China Plain. *J. Hydrol.* **2015**, *527*, 305–315. [CrossRef]
59. Liu, M.; Min, L.; Wu, L.; Pei, H.; Shen, Y. Evaluating nitrate transport and accumulation in the deep vadose zone of the intensive agricultural region, North China Plain. *Sci. Total Environ.* **2022**, *825*, 153894. [CrossRef]
60. Min, L.; Shen, Y.; Pei, H.; Wang, P. Water movement and solute transport in deep vadose zone under four irrigated agricultural land-use types in the North China Plain. *J. Hydrol.* **2018**, *559*, 510–522. [CrossRef]
61. Coonan, E.C.; Kirkby, C.A.; Kirkegaard, J.A.; Amidy, M.R.; Strong, C.L.; Richardson, A.E. Microorganisms and nutrient stoichiometry as mediators of soil organic matter dynamics. *Nutr. Cycl. Agroecosyst.* **2020**, *117*, 273–298. [CrossRef]
62. Hu, Z.; Zhao, Q.; Zhang, X.; Ning, X.; Liang, H.; Cao, W. Winter Green Manure Decreases Subsoil Nitrate Accumulation and Increases N Use Efficiencies of Maize Production in North China Plain. *Plants* **2023**, *12*, 311. [CrossRef]
63. Zhang, Z.; Zhang, X.; Mahamood, M.; Zhang, S.; Huang, S.; Liang, W. Effect of long-term combined application of organic and inorganic fertilizers on soil nematode communities within aggregates. *Sci. Rep.* **2016**, *6*, 31118. [CrossRef] [PubMed]

64. Wei, Z.; Hoffland, E.; Zhuang, M.; Hellegers, P.; Cui, Z. Organic inputs to reduce nitrogen export via leaching and runoff: A global meta-analysis. *Environ. Pollut.* **2021**, *291*, 118176. [CrossRef] [PubMed]
65. Yang, W.; Jiao, Y.; Yang, M.; Wen, H.; Gu, P.; Yang, J.; Liu, L.; Yu, J. Minimizing Soil Nitrogen Leaching by Changing Furrow Irrigation into Sprinkler Fertigation in Potato Fields in the Northwestern China Plain. *Water* **2020**, *12*, 2229. [CrossRef]
66. Lv, H.; Zhou, W.; Dong, J.; He, S.; Chen, F.; Bi, M.; Wang, Q.; Li, J.; Liang, B. Irrigation amount dominates soil mineral nitrogen leaching in plastic shed vegetable production systems. *Agric. Ecosys. Environ.* **2021**, *317*, 107474. [CrossRef]
67. Zhu, A.; Zhang, J.; Zhao, B.; Cheng, Z.; Li, L. Water balance and nitrate leaching losses under intensive crop production with Ochric Aquic Cambosols in North China Plain. *Environ. Int.* **2005**, *31*, 904–912. [CrossRef]
68. Wang, H.; Ju, X.; Wei, Y.; Li, B.; Zhao, L.; Hu, K. Simulation of bromide and nitrate leaching under heavy rainfall and high-intensity irrigation rates in North China Plain. *Agric. Water Manag.* **2010**, *97*, 1646–1654. [CrossRef]
69. Ju, X.T.; Kou, C.L.; Zhang, F.S.; Christie, P. Nitrogen balance and groundwater nitrate contamination; comparison among three intensive cropping systems on the North China Plain. *Environ. Pollut.* **2006**, *143*, 117–125. [CrossRef]
70. Liu, X.; Ju, X.; Zhang, F.; Pan, J.; Christie, P. Nitrogen dynamics and budgets in a winter wheat–maize cropping system in the North China Plain. *Field Crops Res.* **2003**, *83*, 111–124. [CrossRef]
71. Zhang, W.L.; Tian, Z.X.; Zhang, N.; Li, X.Q. Nitrate pollution of groundwater in northern China. *Agric. Ecosys. Environ.* **1996**, *59*, 223–231. [CrossRef]
72. Wen, S.; Cui, N.; Wang, Y.; Gong, D.; Xing, L.; Wu, Z.; Zhang, Y.; Wang, Z. Determining effect of fertilization on reactive nitrogen losses through nitrate leaching and key influencing factors in Chinese agricultural systems. *Agric. Water Manag.* **2024**, *303*, 109055. [CrossRef]
73. Verma, P.; Sagar, R. Effect of nitrogen (N) deposition on soil-N processes: A holistic approach. *Sci. Rep.* **2020**, *10*, 10470. [CrossRef] [PubMed]
74. Zhu, X.; Miao, P.; Zhu, H.; Li, W.; Liang, X.; Wang, L.; Chen, Z.; Zhou, J. Extreme precipitation accelerates nitrate leaching in the intensive agricultural region with thick unsaturated zones. *Sci. Total Environ.* **2024**, *918*, 170789. [CrossRef] [PubMed]
75. Tamagno, S.; Eagle, A.J.; McLellan, E.L.; van Kessel, C.; Linquist, B.A.; Ladha, J.K.; Lundy, M.E.; Pittelkow, C.M. Predicting nitrate leaching loss in temperate rainfed cereal crops: Relative importance of management and environmental drivers. *Environ. Res. Lett.* **2022**, *17*, 64043. [CrossRef]
76. Kou, X.; Ding, J.; Li, Y.; Li, Q.; Mao, L.; Xu, C.; Zheng, Q.; Zhuang, S. Tracing nitrate sources in the groundwater of an intensive agricultural region. *Agric. Water Manag.* **2021**, *250*, 106826. [CrossRef]
77. Min, L.; Shen, Y.; Pei, H.; Jing, B. Characterising deep vadose zone water movement and solute transport under typical irrigated cropland in the North China Plain. *Hydrol. Process.* **2017**, *31*, 1498–1509. [CrossRef]

land

Article

Study on Annual Signals of Greenland Ice Sheet Mass and Associated Influencing Factors Based on GRACE/GRACE-FO Data

Kaifeng Ma, Jing Han, Zhen Li, Junzhen Meng, Qingfeng Hu *, Peipei He and Changxu Yao

College of Surveying and Geo-Informatics, North China University of Water Resources and Electric Power, Zhengzhou 450046, China; makaifeng@ncwu.edu.cn (K.M.); 201926134@stu.ncwu.edu.cn (J.H.); lizhen@ncwu.edu.cn (Z.L.); mengjunzhen@ncwu.edu.cn (J.M.); hepei@ncwu.edu.cn (P.H.); 201801326@stu.ncwu.edu.cn (C.Y.)

* Correspondence: huqingfeng@ncwu.edu.cn

Abstract: As global temperatures rise, the Greenland ice sheet (GrIS) is undergoing accelerating mass loss, with significant implications for sea level rise and climate systems. Using GRACE and GRACE Follow-On (GRACE-FO) RL06 data from April 2002 to May 2023, alongside MARv3.14 regional climate model outputs (ice melting, runoff, rainfall, snowfall, and land surface temperature (LST)), we investigated the drivers of GrIS mass changes. Continuous wavelet transform analysis revealed significant annual signals in all variables except snowfall, with wavelet decomposition showing the largest annual amplitudes for ice melting (58.8 Gt/month) and runoff (44.5 Gt/month), surpassing those of GRACE/GRACE-FO (31.1 Gt/month). Cross-correlation analysis identified ice melting, runoff, rainfall, snowfall, and LST as significantly correlated with GrIS mass changes, with ice melting, runoff, and LST emerging as primary drivers, while snowfall and runoff exerted secondary influences. Temporal lags of 3, 4, 4, 7, and 4 months were observed for ice melting, runoff, rainfall, snowfall, and LST, respectively. These findings highlight the complex interplay of climatic and hydrological processes driving GrIS mass loss.

Keywords: GrIS; GRACE/GRACE-FO; mass change; driving factors; lagging

1. Introduction

As global temperatures continue to rise, the melting of polar ice sheets has become a significant contributor to the rapid rise in global sea levels observed since the beginning of the 21st century [1]. Among these ice sheets, the Greenland ice sheet (GrIS), the largest in the Northern Hemisphere, holds the potential to raise sea levels by approximately 7.4 m if it were to melt completely [2]. According to previous studies, GrIS mass loss is by far the most significant contributor to sea level rise in recent years [3,4]. Consequently, monitoring GrIS mass change is critical for understanding not only global sea level rise but also the global water cycle and the broader dynamics of climate change [5].

Traditional measurement methods encounter difficulty when directly monitoring the mass change of the GrIS and providing accurate quantitative analysis [6]. Advancements in space-based geodetic observation technologies have revolutionized our ability to monitor ice sheet dynamics. Specifically, missions such as the Ice, Cloud, and Land Elevation Satellite (ICESat), Cryosphere Satellite (CryoSat), and the Gravity Recovery and Climate Experiment (GRACE) satellites have provided increasingly accurate measurements of mass changes in polar regions [7–9]. Launched in March 2002, the first GRACE satellite mission concluded successfully in October 2017. Its successor, GRACE Follow-On (GRACE-FO), was launched in May 2018 to continue monitoring global gravity field variations and provides high-precision data for monitoring ice sheet mass [10]. Over nearly two decades, the combined GRACE and GRACE-FO time series have generated an invaluable dataset for studying the GrIS and its evolving mass changes [11].

Numerous studies have employed GRACE data to investigate GrIS mass change over different time scales. For example, Barletta et al. [12] found that the mass change of Greenland and Antarctic ice sheets and its influence on sea level are closely related to GIA. After removing the influence of GIA, the mass change rates of Greenland and Antarctic ice sheets are -101 Gt·a^{-1} and -171 Gt·a^{-1}, respectively. Ran et al. [13] explored the seasonal variations in GrIS mass, revealing that the peak mass anomaly occurs in July, reaching between 80 and 120 Gt. Zou et al. [14] further analyzed the secular, seasonal, and interannual mass changes of the GrIS, finding that the majority of mass loss occurs during the summer months. Some researchers have also focused on long-term trends using different data sources. Forsberg et al. [7] estimated a mass loss of 265 Gt/yr from 2002 to 2015, contributing to a global mean sea level rise of 0.72 mm per year. The summer of 2012 was marked by a record mass loss in the GrIS following a period of accelerated melting. Mankoff et al. [15], using the regional climate model MARv3.12, observed a significant decline in the surface mass balance of the GrIS, from 450 Gt/yr in 1860 to 260 Gt/yr by 2010, while emissions increased substantially during the same period. More recently, Gao et al. [16] analyzed GrIS mass loss from April 2002 to May 2020, incorporating LST, rainfall, and GRACE/GRACE-FO data, and found that anomalous increases in LST and runoff, coupled with decreased rainfall, exacerbated melting in Greenland during the 2010–2017 period.

Much of the current research on GrIS mass change has focused on linear and secular mass change trends, as well as monthly cumulative mass loss [17–22]. However, GRACE/GRACE-FO data also reveal important annual signals, which are essential for understanding the temporal dynamics of GrIS mass changes and their relationship with atmospheric and oceanic climate conditions. These changes in ice sheet mass are influenced by climatic factors, including temperature shifts and atmospheric circulation patterns [5]. While several studies have examined the seasonal variation in GrIS mass and its drivers, much of this research is concentrated on the winter and summer seasons, with fewer studies addressing the full annual cycle. Most studies on the drivers of GrIS mass change have focused on quantitative assessments of single drivers (e.g., melting, precipitation) [23]. However, the strength and time lag of the effects of different drivers (e.g., ice melting, runoff, rainfall, snowfall, et al.) of the GrIS mass change have not yet been systematically analyzed. Furthermore, most studies have failed to make full use of time–frequency analysis methods (e.g., wavelet transforms) to analyze driving mechanisms on different time scales. The annual variation characteristics of ice sheet mass and the influence of complex climate dynamics cannot be fully revealed.

To address the above shortcomings, this study is based on GRACE/GRACE-FO satellite data, combining multiple methods such as continuous wavelet transform, wavelet decomposition, and cross-correlation analysis. The relationship between anniversary signals of GrIS mass change and their drivers is systematically investigated. By analyzing the strength and lag effects of LST, ice melting, runoff, rainfall, and snowfall on changes in GrIS mass, uncovering the main drivers influencing changes in GrIS mass, filling the research gap in this area. Through this analysis, we seek to explore the contributions of each of these variables to GrIS mass change, offering a more comprehensive understanding of the complex interactions driving ice sheet dynamics.

The study is organized as follows; firstly, GRACE/GRACE-FO satellite data were used to construct a time series of the GrIS mass change, and this was combined with regional climate models to obtain time series of ice melting, runoff, rainfall, and snowfall. Then, GrIS mass change time series were fitted using linear and quadratic fitting methods, and we compared their performance in capturing the rate of change in mass. In addition, continuous wavelet transform (CWT) was used to analyze the cyclic nature of GrIS mass change and drivers. Afterwards, the annual signals of each variable were extracted by wavelet decomposition and analyzed for comparison. Finally, inter-correlation analyses were used to assess the lags of the drivers to changes in ice sheet mass.

2. Materials and Methods

2.1. Data

2.1.1. Study Area

The Greenland region is located between 59° N and 83° N latitude and 11° W and 74° W longitude. It is bordered to the east by the Greenland Sea, to the west by Baffin Bay, to the north by the Arctic Ocean, and to the southeast by the North Atlantic Ocean. This region represents the largest expanse of continental ice outside Antarctica. Approximately 80% of Greenland lies within the Arctic Circle and is covered in ice. The Greenland region accounts for 10% of the global ice sheet and contributes between 7% and 8% of the world's freshwater resources [24]. As one of the largest freshwater reservoirs in the world, the continental ice sheet covers an area of 1,833,900 square kilometers, with its northern and southern extremities approximately 2200 km apart. The entire ice sheet is shaped like an inverted bowl and rests upon bedrock. On average, the thickness of the ice sheet measures 1600 m, reaching a maximum thickness of 3367 m [25]. The average elevation of the peripheral region ranges from 1000 to 2000 m, encompassing a total volume of ice and snow amounting to 2.9 million cubic kilometers [26].

2.1.2. GRACE/GRACE-FO RL06 Data

Currently, the Jet Propulsion Laboratory (JPL), the Center for Space Research (CSR), and the Helmholtz Centre Potsdam-German Research Centre for Geosciences (GFZ) provide GRACE/GRACE-FO gravity satellite data [27]. This study primarily utilizes the latest GRACE/GRACE-FO RL06 data released by CSR, with spherical harmonic coefficients truncated to the 60th order. The dataset covers the period from April 2002 to June 2017 for GRACE and from June 2018 to May 2023 for GRACE-FO, totaling 254 months of data. Notably, there are some data gaps, particularly a near one-year gap between GRACE and GRACE-FO.

It is important to note that the reference frame used in the calculation of GRACE and GRACE-FO gravity satellite data is centered on the Earth's centroid. In this reference frame, the first-order spherical harmonic coefficients (S_{11}, C_{11}, C_{10}), which correspond to surface mass migration and load deformations, are all zero. Therefore, the first-order coefficients (S_{11}, C_{11}, C_{10}) proposed by Swenson et al. [28] are applied to correct for the Earth's centroid variation during data preprocessing. Due to the influence of satellite orbit, ocean, atmospheric models, and other factors [29], the C_{20} term calculated from satellite data is often inaccurate. Thus, the C_{20} term derived from satellite laser ranging (SLR) is used to replace the C_{20} term in the GRACE/GRACE-FO spherical harmonic coefficients [30].

Furthermore, due to errors from satellite orbit, ocean and atmosphere models, and spherical harmonic terms in the global gravity field, significant north–south stripe errors occur when using time-varying gravity field data to invert surface mass. To mitigate these errors, a combined approach using CHENP4M6 deconvolution filtering and Gaussian filtering with a 250 km radius was employed [31,32]. GRACE/GRACE-FO data are also sensitive to glacial isostatic adjustment (GIA) signals, which must be subtracted when studying surface mass changes. Accordingly, a GIA model was applied to correct for the post-ice rebound effect in the time-varying GRACE/GRACE-FO signals [33,34]. To ensure a complete dataset for analyzing ice sheet mass changes, singular spectrum analysis (SSA) was used to interpolate missing data [35], resulting in 254 months of continuous data.

2.1.3. Surface Mass Balance (SMB) Data

Mass balance methods are widely used to study changes in ice sheet mass. These methods determine mass changes by calculating both surface and basal mass inputs (e.g., rainfall, snowfall) and outputs (e.g., ice melting, runoff, evaporation, ice discharge). Snowfall, rainfall, ice melting, and runoff are referred to as the ice sheet surface mass balance (SMB). The SMB data used in this study are derived from the regional climate model (Modèle Atmosphérique Régional, MARv3.14), developed by the Climatology Laboratory at the University of Liège's Faculty of Geography [36]. MARv3.14 data, which can be accessed at http://ftp.climato.be/fettweis/MARv3.14/Greenland/ (accessed on 1 January 2025), provide monthly 1 × 1 km SMB data for ice sheet and tundra regions. The model has a spatial resolution of 1 km in Greenland, and the time span used in this study ranges from 2002 to 2023. The SMB data from the MAR model primarily include ice melting, runoff, rainfall, snowfall, evaporation, and surface water. Evaporation and surface water exhibit less variability compared to other variables and are therefore excluded from this analysis [37].

2.1.4. Land Surface Temperature (LST) Data

To explore the relationship between GrIS mass change and LST variations, this study uses temperature data from the Global Historical Climatology Network version 2 (GHCN) and the Climate Anomaly Monitoring System (CAMS), published by the National Oceanic and Atmospheric Administration (NOAA) [38]. This dataset differs from several existing surface air temperature datasets in two key aspects: (1) it utilizes a combination of two extensive individual datasets of station observations sourced from the Global Historical Climatology Network version 2 and the Climate Anomaly Monitoring System (GHCN + CAMS). This allows for regular updates in near real-time, supported by a substantial number of stations; and (2) it employs unique interpolation methods, including the anomaly interpolation approach that incorporates spatially and temporally varying temperature lapse rates derived from observation-based reanalysis for topographic adjustment. The LST data, which can be accessed at https://psl.noaa.gov/data/gridded/data.ghcncams.html (accessed on 1 January 2025), have a spatial resolution of 0.5° × 0.5° and cover the period from April 2002 to May 2023.

2.2. Methodology of Analysis

Gaussian filtering is a filtering technique that utilizes a spatial smoothing kernel function. It is primarily employed to mitigate the higher-order errors associated with spherical harmonic coefficients. Define W as a function that depends only on the angle of α between two points (θ,λ) and (θ',λ'), as follows $W(\theta,\lambda,\theta\prime,\lambda\prime) = W(\alpha)$, among $\cos\alpha = sin\theta sin\theta' + cos\theta cos\theta\prime\cos(\lambda-\lambda')$. The formula is:

$$\overline{\Delta\rho}(\theta,\lambda) = \frac{2a\rho_{ave}\pi}{3}\sum_{l=0}^{\infty}\sum_{m=0}^{l}\frac{2l+1}{1+k_l}W_l\widetilde{P_{lm}}(sin\theta)[\Delta C_{lm}\cos(m\lambda)+\Delta S_{lm}\sin(m\lambda) \tag{1}$$

Among them:

$$W_l = \int_0^{\pi} W(\alpha)P_l(cos\alpha)sin\alpha\, d\alpha \tag{2}$$

And $P_l = \widetilde{P_{lm=0}} / \sqrt{2l+1}$ is a Legendre polynomial.

The aforementioned smoothing method was initially proposed by Jekeli [39] to address the issue of poor precision in the high-order terms of GRACE spherical harmonic coefficients, with the aim of enhancing estimates of changes in the Earth's gravitational field. Wahr et al. [40] were the first to explore the potential application of this method for processing GRACE spherical harmonic coefficient data, yielding favorable results. Jekeli [39] introduced a normalized Gaussian mean kernel function, which ensures that the global integral of W is equal to 1, and it is expressed as follows:

$$W(\alpha) = \frac{b}{2\pi} \frac{exp[-b(1 - cos\,\alpha)]}{1 - e^{-2b}} \tag{3}$$

Among them:

$$b = \frac{ln2}{1 - cos(r/\alpha)} \tag{4}$$

where r represents the weight at $\alpha = 0$, which corresponds to the distance from the Earth's surface when reduced by half; this is referred to as the Gaussian filter radius and is employed recursive relations to compute coefficient W_l:

$$\begin{cases} W_0 = \frac{1}{2\pi} \\ W_1 = \frac{1}{2\pi}\left[\frac{1+e^{-2b}}{1-e^{-2b}} - \frac{1}{b}\right] \\ W_{l+1} = -\frac{2l+1}{b} W_l + W_{l-1} \end{cases} \tag{5}$$

As demonstrated in Equations (3) and (4), the Gaussian smoothing weight coefficient is influenced by both the filtering radius and its order. Consequently, when selecting the filter radius, it is essential to not only minimize banding errors but also to retain as much of the genuine signals from the higher-order terms as possible.

The spatial correlation error in the spherical harmonic coefficients of GRACE significantly impacts the inversion results, particularly contributing to the north–south banding error. To address this issue of spatial correlation in GRACE errors, Swenson and Wahr proposed a method to mitigate the north–south banding error by reducing the correlation between spherical harmonic potential coefficients [41], employing a polynomial decorrelation filtering technique. The fundamental principle of this decorrelation filtering method is as follows: first, a specific number m is determined; subsequently, odd-order and even-order coefficients are fitted with polynomials in succession. The resulting fitting values are treated as errors and are then subtracted from the original spherical harmonic coefficients. The specific derivation process is outlined as follows: when fitting ΔC_{lm} of a fixed order, if the sliding window is set to 5, the requirements for $\Delta C_{l-4,m}$, $\Delta C_{l-2,m}$, $\Delta C_{l,m}$, and $\Delta C_{l+4,m}$ are determined. Subsequently, a cubic polynomial fitting is applied.

$$\Delta C_m(l) = a_0 + a_1 l + a_2 l^2 + a_3 l^3 \tag{6}$$

Since the fitting value is considered an error, the spherical harmonic coefficient after decorrelation filtering is denoted as $\Delta\hat{C}_{lm} = \Delta C_{lm} - \Delta C_m(l)$. We apply the same procedure for ΔS_{lm}, which will yield $\Delta\hat{S}_{lm} = \Delta S_{lm} - \Delta S_m(l)$.

Chambers [42] and Chen et al. [43] have enhanced the previously described sliding decorrelation filtering method. In this paper, we propose a novel non-sliding decorrelation filtering method (PnMm). The fundamental concept of this approach is as follows: the low-order term coefficients (specifically, the first m order coefficients) remain unchanged, while a polynomial fit is conducted on all odd-order terms (and even-order terms) that are equal to or exceed the degree of the m-th order coefficients. Subsequently, these fitted values are subtracted from the original spherical harmonic coefficients, with these fitting values being treated as errors.

Although Gaussian filtering is effective in suppressing high-frequency errors, and decorrelation filtering can eliminate correlation errors, each method has its inherent limitations when applied independently. To achieve a more effective reduction of banding and correlation errors, the strengths of both filtering techniques can be fully utilized by integrating Gaussian filtering with decorrelation filtering. In this combined approach, decorrelation filtering is initially employed to remove coefficient correlations, followed by Gaussian smoothing to mitigate high-frequency noise.

CWT can be used to infer the periodic characteristics of the signal by observing the distribution of the wavelet coefficients, so this paper uses CWT to study the periodic change in ice sheet mass. The CWT of the time series X_m (m = 1, 2,..., N) is defined as [44,45].

$$W_m^x(S) = \sqrt{\delta t/s}\sum_{m'=1}^{N} x_{m'}\psi_{\omega_0}[(m'-m)\frac{\delta t}{s}] \quad (7)$$

where $W_m^x(S)$ wavelet coefficients, δt and ψ_{ω_0} are time scales and the wavelet generating function, s is the wavelet scales, and m' and $\sqrt{\frac{\delta t}{s}}$ are inverse time and normalization factors, respectively. The main feature of the CWT is its ability to provide both time and frequency information about the signal. The basic idea is to select a suitable wavelet function based on the signal characteristics of the time series; the wavelet is then used as a bandpass filter, and the characteristic period of the filter is linear with the wavelet scale. It is worth noting that the Morlet wavelet is used in this paper. It is expressed as follows:

$$\psi_{\omega_0}(\eta) = Ke^{i\omega_0\eta}e^{-\frac{1}{2}\eta^2} \quad (8)$$

where K ($K = \pi^{-1/4}$) denotes the normalization parameter, and η ($\eta = s \cdot t$) and ω_0 denote time and frequency, respectively. When $\omega_0 = 6$, it provides a good balance between the local resolution of time and frequency.

To further analyze the annual variability in GrIS mass, this study employs the wavelet decomposition method to extract the annual signals of various variables. Wavelet decomposition allows a signal to be separated into time series with different frequency components by stretching and transforming the wavelet function. This method decomposes the original signal (S) into low-frequency approximations (A) and high-frequency details (D). The approximation sequence captures the trend characteristics of the original signal, while the detailed sequence represents its finer variations. After the first-level decomposition, the signal is expressed as S = A1 + D1. The second-level decomposition further breaks down the approximate component A1 from the first-level decomposition into A2 and D2, resulting in S = A2 + D2 + D1, where A1 = A2 + D2. This process continues for N levels, with the original signal decomposed as S = AN + DN + ... + D2 + D1, where D1, D2,..., DN represent the detail components obtained at each level of decomposition, and AN represents the final approximation at level N [46]. For this study, the time series are decomposed into scales corresponding to the following frequencies: D1 (2–4 months), D2 (4–8 months), D3 (8–16 months), and D4 (16–32 months). The approximation A4 represents the long-term trend component. To effectively capture the annual signals, we apply a four-level decomposition, with the corresponding frequencies summarized in Table 1 [47].

Table 1. Frequencies corresponding to the scales of the Meyer wavelet.

Level	n	Period (Months)	Range (Months)
D1	1	2	2–4
D2	2	4	4–8
D3	3	8	8–16
D4	4	16	16–32
A4	5	32	>32

3. Results

3.1. GRACE Data Processing Results

When using GRACE/GRACE-FO data to derive changes in Earth's surface mass, these changes are typically represented as equivalent water height [10]. Taking the February 2004 CSR RL06 data as an example, this paper describes the processing of spherical harmonic coefficients from the GRACE/GRACE-FO gravity field using Gaussian filtering, decorrelation filtering, and combined filtering methods. Figure 1a shows the inversion results of global surface mass changes using the raw, unfiltered GRACE/GRACE-FO data. Figure 1b–d display the inversion results using spherical harmonic coefficients with Gaussian filter radii of 200 km, 250 km, and 300 km, respectively.

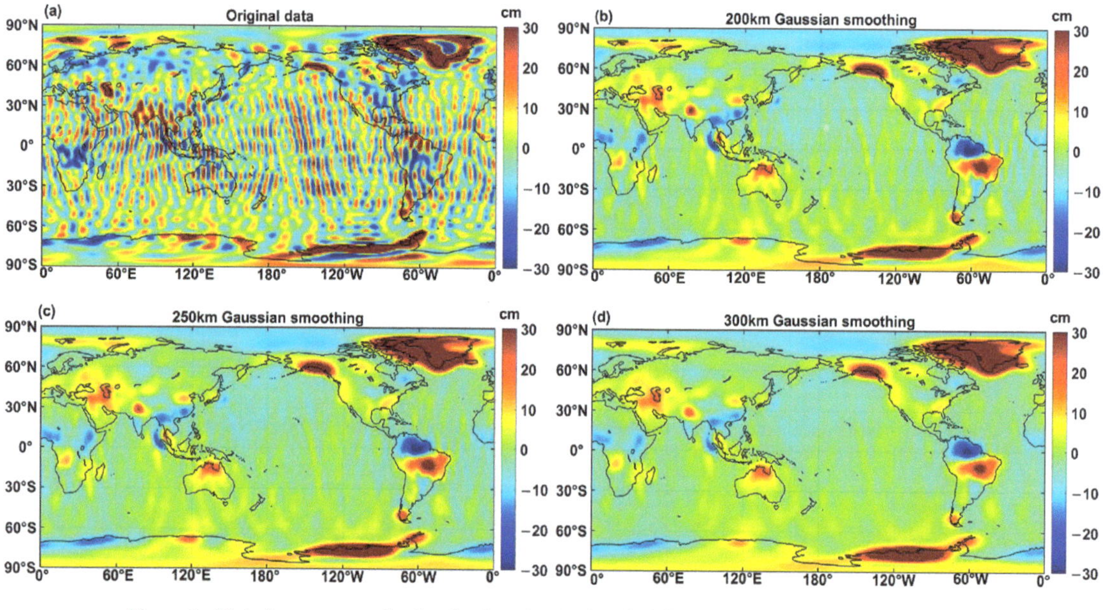

Figure 1. Global mass anomaly distribution obtained with different Gaussian smoothing radii in February 2004. (**a**) Unfiltered; (**b**) 200 km; (**c**) 250 km; (**d**) 300 km.

The results in Figure 1 indicate that as the Gaussian filter radius increases, the suppression of north–south stripe errors and high-frequency noise becomes more pronounced. At a filter radius of 250 km, the geophysical signal is clear, but some residual stripe errors and high-frequency noise remain in mid-latitude regions. A 300 km, the Gaussian filter radius effectively eliminates stripe errors and high-frequency noise, resulting in a clearer geophysical signal, although the signal strength at high latitudes (Greenland and Antarctica) weakens as the filter radius increases. In conclusion, Gaussian filtering is highly effective in suppressing north–south stripe errors (mainly high-frequency noise), and its suppression effect strengthens with increasing filter radius. However, this comes at the cost of slightly weakening the signal itself. Therefore, a 250 km Gaussian filter radius offers a balance, effectively suppressing stripe errors while retaining a valid signal.

Figure 2 presents the global surface mass changes derived from the GRACE/GRACE-FO raw data and processed using the decorrelation filtering method. The results show a significant reduction in north–south stripe errors compared to the unfiltered raw data. However, due to higher-order term errors in the spherical harmonic coefficients, some residual north–south stripe errors persist near the equator. Combining the results from Figures 1 and 2 reveals that the north–south stripes generated by different filtering methods exhibit varying error distributions and densities, which contribute to different signal leakage patterns. Specifically, signal leakage in high-latitude regions (e.g., Antarctica and Greenland) is notably weaker in Figure 2a,b compared to Figure 2c, whereas north–south stripe errors near the equator are more pronounced in Figure 2a,b than in Figure 2c. These findings indicate that decorrelation filtering alone is insufficient to fully suppress stripe errors.

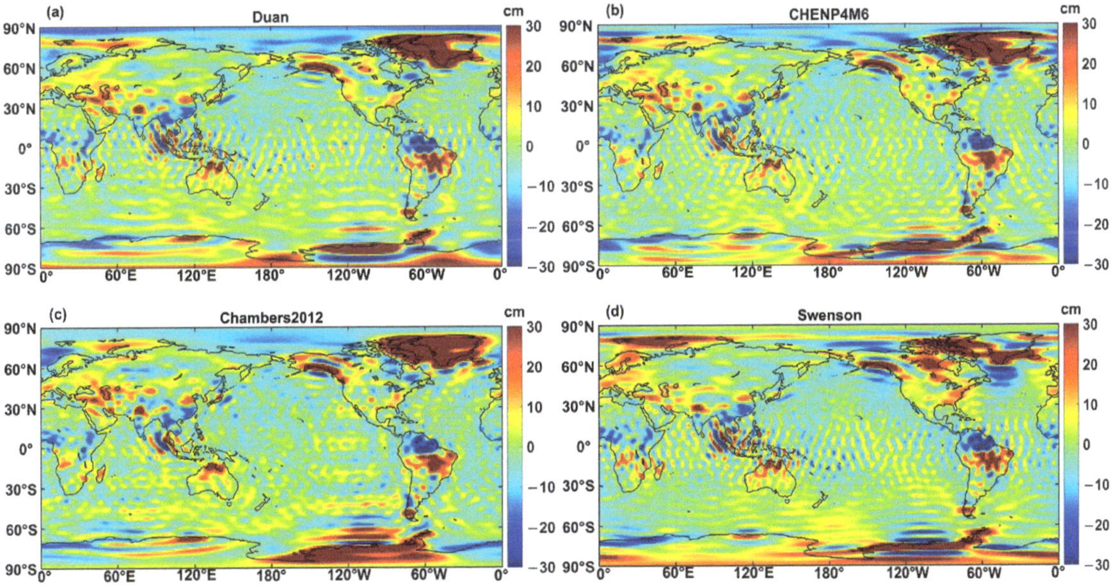

Figure 2. Global mass anomaly distributions obtained with different decorrelation filters in February 2004. (**a**) Duan; (**b**) CHENP4M6; (**c**) Chambers2012; (**d**) Swenson.

The combined Gaussian and decorrelation filtering results demonstrate that high-frequency errors and systematic correlation errors in the spherical harmonic coefficients of the GRACE/GRACE-FO gravity field are challenging to suppress simultaneously with a single filtering method. However, when spatial smoothing and decorrelation filtering are used in combination, a more effective suppression is achieved. To this end, several filter combination modes were tested by integrating Gaussian filtering with a 250 km radius and different decorrelation filtering methods, including Duan, CHENP4M6, Chambers2012, and Swenson. The results, shown in Figure 3, reveal that Figure 3a contains only minor north–south stripe errors, while signal leakage is more pronounced in Figure 3c,d. In contrast, Figure 3b exhibits no obvious stripe errors or signal leakage. Consequently, for subsequent GrIS mass change analyses, a combination model of 250 km Gaussian filtering and CHENP4M6 decorrelation filtering was selected.

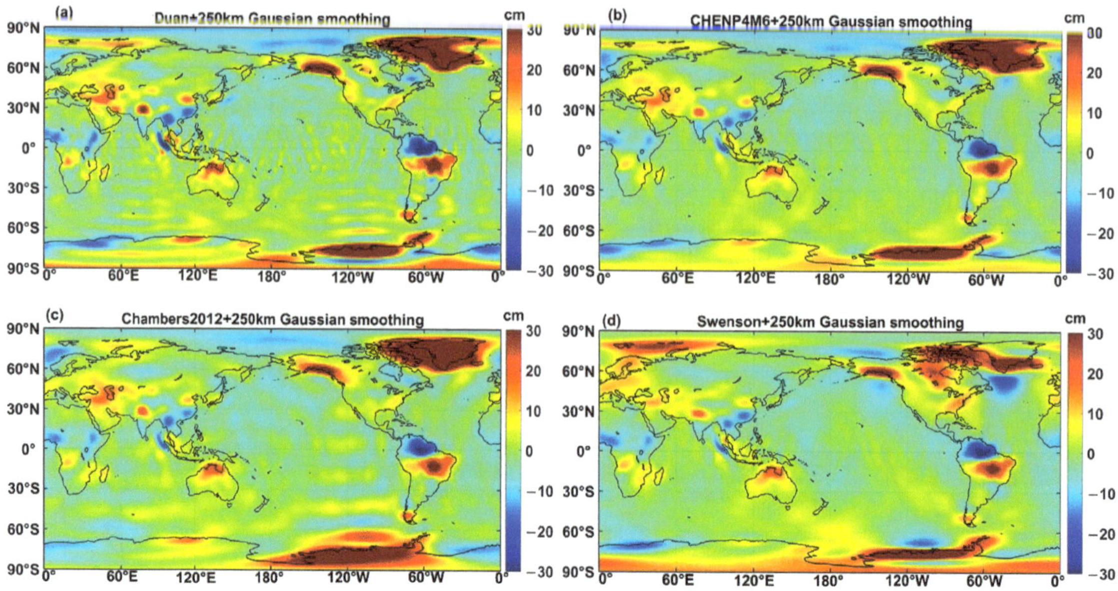

Figure 3. Global mass anomaly distribution obtained by Gaussian smoothing with a radius of 250 km by combining different decorrelation filtering methods in February 2004. (**a**) Duan; (**b**) CHENP4M6; (**c**) Chambers2012; (**d**) Swenson.

3.2. Trend Analysis of Ice Sheet Mass Changes

Based on the research of both domestic and international scholars on GRACE/GRACE-FO, the time series of GrIS mass change can be obtained by applying the combination filtering process and selecting the appropriate boundary for the study area, followed by GIA correction. Assuming no errors, the time series derived from GRACE/GRACE-FO ice sheet mass changes can be classified into secular, seasonal, and interannual signals according to their spectral characteristics. Secular signals primarily reflect long-term mass trends or periodic fluctuations lasting more than 10 years. Seasonal signals include annual and semi-annual cycles, which are driven by factors such as the Earth's orbit. Interannual signals fall between seasonal and secular time scales and primarily represent fluctuations in ice sheet mass from year to year.

To further analyze trends in the GrIS mass time series, linear and quadratic curve fitting methods based on least squares estimation were applied to the GrIS mass change data from 2002 to 2023, as shown in Figure 4. The analysis reveals both positive and negative mass changes, where positive values indicate an increase in GrIS mass, and negative values indicate a decrease. Overall, the GrIS has been in mass deficit throughout the study period, with the rate of loss following a pattern of slow to fast changes, and then slow again, before accelerating once more. The linear rate of mass change for the GrIS is -131.4 ± 2 Gt/yr, while the quadratic fit yields a rate of -152.7 ± 8 Gt/yr, where ± 2 and ± 8 represent fitting error estimates. Notably, GrIS mass loss sharply increased in 2012, stabilized in the two years following, and continued to decrease thereafter. However, results may vary due to differences in data sources, time periods, and data processing methods.

The effectiveness of the linear and quadratic curve fits can be assessed based on the coefficients obtained. The coefficient of determination for the quadratic fit is 0.987, while the linear fit yields 0.885, indicating that the quadratic fit provides a better representation of the data.

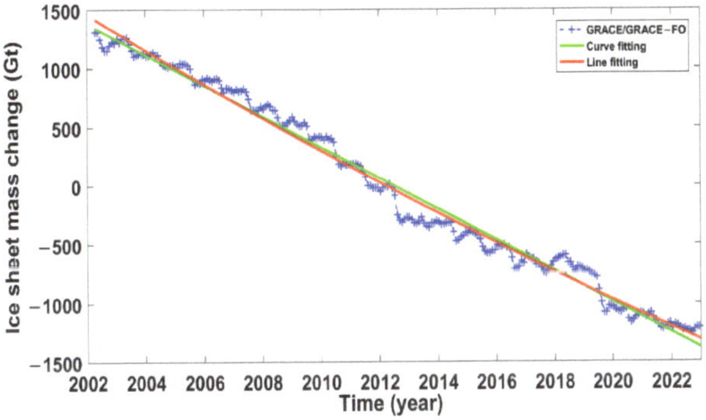

Figure 4. GRACE/GRACE-FO-derived ice sheet mass change time series from April 2002 to May 2023.

To further investigate the drivers of GrIS mass change, Figure 5 displays the time series of ice melting, runoff, rainfall, snowfall, LST, and ice sheet mass change after removing secular trends. This adjustment allows for a clearer reflection of short-term fluctuations and anomalous changes in ice sheet mass. The figure reveals that ice melting and runoff exhibit similar trends. Ice melting and GRACE/GRACE-FO show opposite trends, while runoff and GRACE/GRACE-FO share similar trends. Snowfall and rainfall have less influence on ice sheet mass changes compared to ice melting, runoff, and GRACE/GRACE-FO. Specifically, GrIS mass changes are primarily driven by ice melting and runoff, with snowfall and rainfall exerting secondary effects.

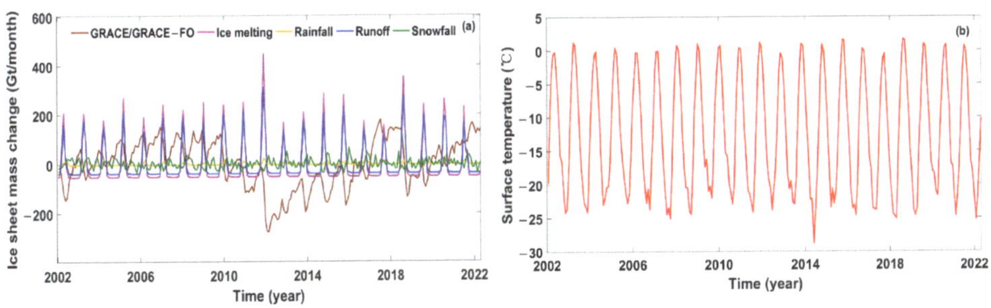

Figure 5. Time series of various variables. (**a**) GRACE/GRACE-FO, ice melting, runoff, rainfall, and snowfall; (**b**) LST.

Additionally, LST trends in the opposite direction to ice sheet mass, with an increase in LST corresponding to a decrease in ice sheet mass. This suggests that rising temperatures contribute to accelerated melting of the ice sheet. Notably, the GrIS mass exhibits an accelerated decline between 2010 and 2012, with the most significant decrease occurring in 2012.

3.3. Periodic Signal Analysis of GrIS Mass Change and Its Drivers

To further explore the relationship between GrIS mass change and both SMB and LST from 2002 to 2023, the CWT method was applied to identify cyclical signals in each variable. The results are presented in Figure 6, which displays the CWT energy spectra for each time series. In the figure, a thick black solid line or thicker indicates regions where the red noise test passed the 95% confidence level. The area below the thin black solid line represents the wavelet influence cones, where edge effects may significantly impact the data [44].

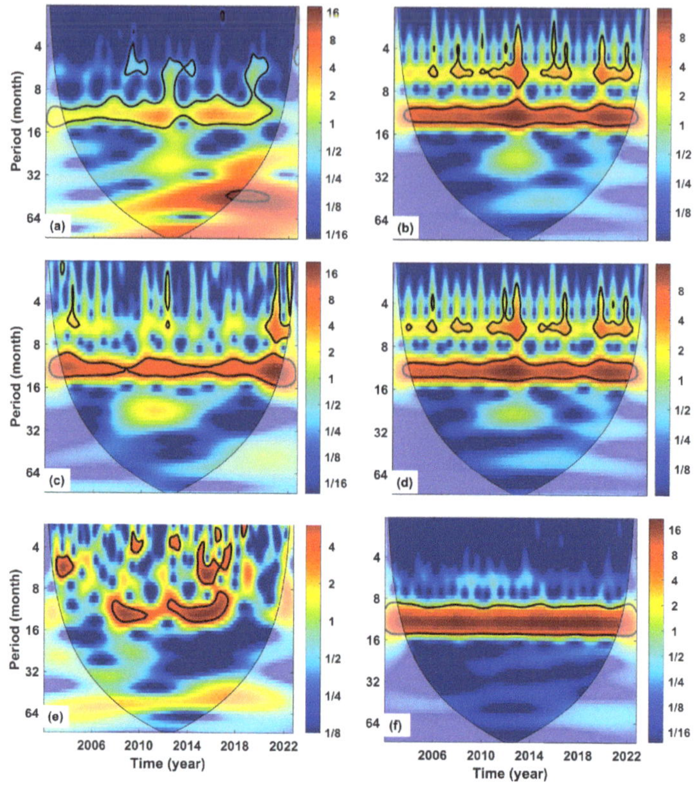

Figure 6. CWT of variables from April 2002 to May 2023. (**a**) GrIS mass change; (**b**) ice melting; (**c**) rainfall; (**d**) runoff; (**e**) snowfall; (**f**) LST.

Figure 6 presents the results of the continuous wavelet transform (CWT) analysis for GrIS mass, ice melting, rainfall, runoff, snowfall, and LST time series. The wavelet power spectra for LST and GrIS mass changes show a high degree of similarity, with the signal peaks predominantly occurring in the 10- to 14-month period range. A distinct annual signal is observed across the entire time domain, and the corresponding spectral energy passes the 95% confidence test, confirming significant annual signals in both LST and GrIS mass changes. These findings align with those of Bian et al. [31], who also identified strong annual signals in LST and GrIS mass variations using the CWT method.

Additionally, ice melting, rainfall, and runoff show distinct annual signals throughout the observation period, while snowfall exhibits less pronounced annual variability over several years. GrIS mass changes, ice melting, runoff, rainfall, and snowfall also display periodic signals in the 4- to 6-month range; however, these signals have weaker energy spectra and are less significant. In summary, the results highlight that the annual signal is the predominant feature across the entire time domain.

3.4. Comparison of Annual Signals for GrIS Mass Change and Control Variables

According to Section 3.3, the annual signals of each variable dominate the entire time domain; to investigate the correlation of GrIS mass changes with the annual signals of various climate variables, wavelet decomposition method is used to decompose the original time series of each variable. As can be seen from Table 1, the D3 band contains annual signals, and Figure 7 shows the time series of each variable annual signal in the D3 band.

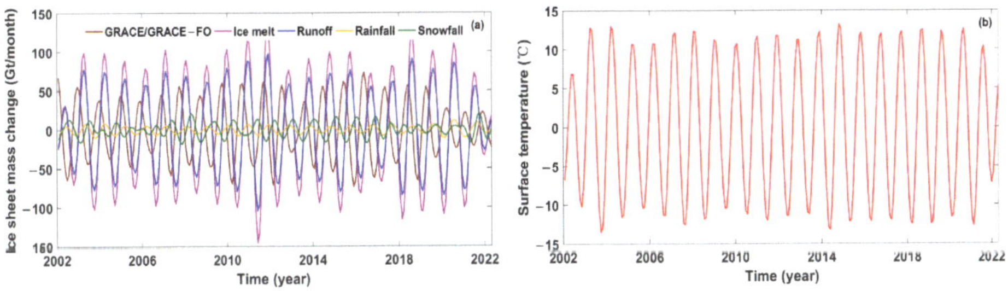

Figure 7. Annual changes in several variables: (**a**) GrIS mass change, ice melting, rainfall, runoff, and snowfall; (**b**) LST.

Figure 7 illustrates substantial annual amplitude variations in GRACE/GRACE-FO, ice melting, and runoff compared to smaller variations in rainfall and snowfall. As shown in Table 2, the annual amplitudes of ice melting (58.8 Gt/month) and runoff (44.5 Gt/month) are significantly larger, being 1.9 and 1.4 times that of GRACE/GRACE-FO (31.1 Gt/month), respectively. Additionally, ice melting and runoff exhibit similar annual phase changes, and their amplitudes surpass those of rainfall (4.1 Gt/month) and snowfall (7.4 Gt/month). These findings highlight that runoff and ice melting are the primary drivers of GrIS mass changes, with snowfall and rainfall having secondary effects.

Table 2. The mean annual amplitudes between GRACE/GRACE-FO-derived ice sheet masses change, ice melting, rainfall, runoff, and snowfall.

Variables	GRACE/GRACE-FO	Ice Melting	Runoff	Rainfall	Snowfall
Annual mean amplitude (Gt/month)	31.1	58.8	44.5	4.1	7.4

Further analysis of Figure 7a,b reveals that the annual phase of LST (land surface temperature) is opposite to that of GRACE/GRACE-FO, indicating a clear link between rising temperatures and increased ice mass loss. Notably, during the 2022–2023 period, Greenland experienced sustained temperature increases, culminating in a peak in May 2023, when GrIS mass loss also reached its maximum. These observations underscore the critical role of temperature in accelerating GrIS mass loss.

Correlation analysis was used to further explore the correlation between the annual signals of the variables and GrIS mass changes. For the time series $x_1(t)$ and $x_2(t)$, the cross-correlation function is provided by [48]:

$$\rho(T) = \frac{\sigma_{12}(t)}{\sqrt{\sigma_{11}\sigma_{22}}} \tag{9}$$

where $\rho(T)$ is between -1 and 1, the absolute magnitude responds to the degree of correlation between two time series, T is the delay of the time series $x_1(t)$ relative to the time series $x_2(t)$, $\sigma_{12}(t)$ is the mutual covariance of time series $x_1(t)$ and $x_2(t)$, and σ_{11} and σ_{22} are the autocovariances of time series $x_1(t)$ and $x_2(t)$, respectively. By studying the $\rho(T)$ distribution of the time series, it is possible to analyze the correlation of the time series in the time domain. Generally, the correlation coefficient ρ_{max} with the largest absolute value from $\rho(T)$ is used to represent the degree of correlation between the two sets of time series; ρ_{max} with the corresponding delay T then reflects the delay in time of the two timeseries [49].

Figure 8a–e illustrate the normalized cross-correlation between GrIS mass changes and ice melting, rainfall, runoff, snowfall, and LST in the D3 frequency band. Table 3 summarizes the corresponding lags and correlation coefficients obtained from the cross-correlation analysis. The results indicate that ice melting, rainfall, runoff, snowfall, and LST are significantly correlated with GrIS mass changes, with maximum correlation coefficients of 0.863, 0.790, 0.852, 0.712, and 0.899, respectively. These peak correlations correspond to lag times of 3, 4, 4, 7, and 4 months for ice melting, rainfall, runoff, snowfall, and LST, respectively. Furthermore, the analysis confirms that ice melting, runoff, and LST exert the strongest influence on GrIS mass changes, while snowfall and rainfall play secondary roles.

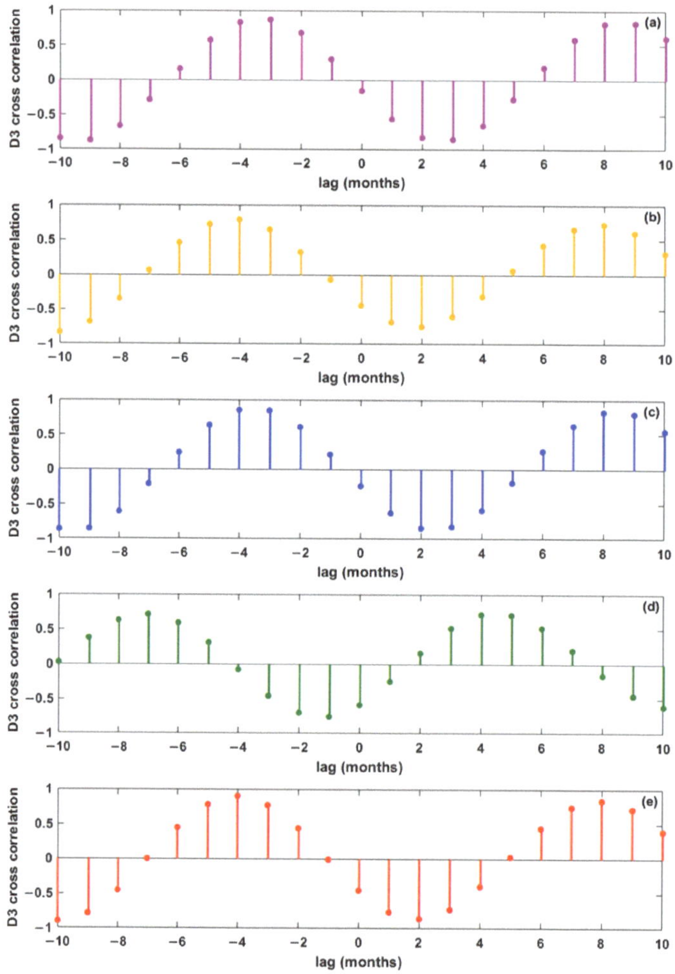

Figure 8. Cross-correlation between GRACE/GRACE-FO-derived ice mass change and ice melting, rainfall, runoff, snowfall, and LST with different lags. (**a**) Ice melting; (**b**) rainfall; (**c**) runoff; (**d**) snowfall; (**e**) LST.

According to the results of cross-correlation analysis, the influence of driving factors on the lag in mass change of the ice sheet is further examined. Melting typically coincides with rising temperatures; during warmer seasons, the surface of the ice sheet begins to melt. Water then penetrates through the surface or flows along cracks within the ice sheet, and it takes time for this process to manifest significant changes in mass. Some of this meltwater may percolate or refreeze within the ice sheet or underground, contributing to a hysteretic effect on mass change. Runoff refers to the outflow of meltwater from the ice sheet into oceanic or groundwater systems. The flow of meltwater from the surface can be delayed; particularly during periods when meltwater accumulates, there exists a lag between water reaching the edge of the ice sheet and runoff that ultimately enters oceanic bodies. Due to complex drainage systems within the ice sheet, complete outflow can take months. Portions of this meltwater may even remain stored in groundwater systems for extended periods, further exacerbating delays in mass changes associated with melting processes. The impact of rainfall on ice sheet mass is primarily manifested through an acceleration in snowmelt and a direct increase in water volume. While rainfall hastens melting processes, its effects on ice sheet mass are not immediately observable but become apparent over time. Rainfall rapidly melts the snow layer atop the ice sheet, while groundwater infiltration and outflow along the edges further contribute to this dynamic. Collectively, these processes result in a lag in changes to ice sheet mass. In contrast, snowfall typically requires an extended period before its effects are reflected in alterations to ice sheet mass. As snow accumulates, it forms multiple layers that influence the surface characteristics of the ice sheet. The accumulation of snow undergoes seasonal deposition and compaction processes that ultimately translate into changes in ice sheet mass. Land surface temperature directly influences both melting of the ice sheet and accumulation of snow. As temperatures rise, melting begins at the surface of the ice sheet; however, melted ice and snow can temporarily refreeze, forming either an additional layer of ice or a water body. This phenomenon may take several months before resulting in significant changes to overall ice sheet mass.

Table 3. Correlation of GRACE/GRACE-FO-derived ice masses change with ice melting, rainfall, runoff, snowfall, and LST.

Lags \ Variables	Ice Melting	Rainfall	Runoff	Snowfall	LST
−10	−0.841	−0.788	−0.851	0.023	−0.888
−9	−0.861	−0.685	−0.851	0.372	−0.788
−8	−0.668	−0.355	−0.612	0.627	−0.459
−7	−0.292	0.065	−0.211	0.712	−0.005
−6	0.160	0.459	0.244	0.594	0.450
−5	0.568	0.724	0.632	0.302	0.782
−4	0.825	0.790	0.852	−0.084	0.899
−3	0.863	0.644	0.844	−0.456	0.771
−2	0.672	0.329	0.610	−0.705	0.435
−1	0.299	−0.070	0.213	−0.711	−0.018
0	−0.155	−0.446	−0.242	−0.591	−0.462
1	−0.565	−0.688	−0.627	−0.248	−0.768
2	−0.821	−0.746	−0.842	0.160	−0.866
3	−0.854	−0.608	−0.828	0.513	−0.732
4	−0.654	−0.312	−0.589	0.711	−0.406
5	−0.275	0.062	−0.191	0.707	0.024
6	0.177	0.417	0.256	0.511	0.443
7	0.576	0.659	0.629	0.189	0.739
8	0.810	0.725	0.825	−0.165	0.835
9	0.819	0.597	0.793	−0.457	0.709
10	0.603	0.310	0.544	−0.620	0.396

4. Discussion

Previous studies on GrIS mass changes have predominantly focused on short-term time series and seasonal trends, with most investigations emphasizing winter and summer variations. In contrast, periodic and annual signals have received comparatively less attention. For instance, Shamshiri et al. [50] employed continuous wavelet transform and cross-wavelet transform to investigate the common frequencies and relative phases between the GrIS mass change time series and the IST time series. Their analysis revealed a strong shared frequency between these variables. Furthermore, by fitting the GrIS mass time series, it was determined that a quadratic fit provided superior results, which aligns with the conclusions drawn in this study.

Bevis et al. [9] indicate that the negative phase of the North Atlantic Oscillation (NAO) is linked to an increase in surface melting of ice sheets and a corresponding acceleration in mass loss. Additionally, this phase is associated with heightened flow rates and thinning of the ice within the sheet, which are extending further inland. Their results indicated that the topographic characteristics of southwest Greenland make it more sensitive to atmospheric forcing, predicting this region will become a major contributor to sea level rise within 20 years. Ramillien et al. [51] conducted an analysis that revealed significant interannual fluctuations in ice sheet mass change throughout the study period, which are closely associated with climatic factors. It was emphasized that subsequent studies should incorporate additional climate, temperature, and precipitation data to achieve a more comprehensive understanding of the various influences on changes in ice sheet mass.

This study utilized the latest GRACE/GRACE-FO RL06 data (2002–2023) provided by CSR, complemented by the MARv3.14 regional climate model and GHCN + CAMS global land surface temperature (LST) data, to investigate GrIS mass changes and their periodic variations. Firstly, an in-depth analysis of the time series of GrIS mass changes was carried out. The results of the study demonstrate that the GrIS shows significant mass loss trend over the past decades; however, the rate of change exhibits phase characteristics. In particular, in 2012, the GrIS experienced a rapid melting event. Subsequently, the rate of loss of ice mass slowed between 2013 and 2018. However, in 2019, the Greenland ice sheet once again experienced a rapid melting phenomenon. The above results show that GrIS mass changes exhibit significant volatility across time, which was consistent with the results of Sasgen et al. [52].

To further explore the annual signal of mass change in the GrIS and its driving factors, periodic signals in GrIS mass change, ice melting, runoff, rainfall, snowfall, and LST were analyzed using continuous wavelet transform (CWT), and annual signals were extracted via wavelet decomposition. The results show that ice melting and runoff are the main factors affecting the mass change of the GrIS. For example, a sharp increase in ice melting and runoff in 2012 led to a severe loss of mass of the GrIS that year, and a decrease in precipitation and an increase in ice melting and runoff in 2019 combined to cause a severe loss of mass of the GrIS that year. In addition, in 2013, ice melting and reduced runoff allowed the GrIS to experience a slight mass accumulation during the year. LST contributes to the mass loss of the GrIS to some extent: the main manifestation is that the increase in temperature will lead to an increase in runoff and ice melting, which will affect the mass change of the GrIS. Ruan [37] analyzed the relationship between the mass change of the GrIS and the variation in MAR model variables in summer and winter: it was found that rainfall and snowfall are not the main factors affecting the mass loss of the GrIS; the amount of summer melt is the main factor affecting the mass loss of the GrIS. The conclusions are basically the same as those obtained in this study, but this study combined temperature data; thus, the driving factors affecting the mass change of the GrIS are analyzed more comprehensively. Furthermore, the refreezing of melted ice and snow significantly impacts the mass change of the GrIS. During the summer months, the influence of refreezing on surface moisture is particularly pronounced. The GrIS experiences melting of ice and snow, resulting in the formation of surface water (meltwater). If this meltwater is not drained or absorbed into the ice sheet promptly, it will refreeze during fall and winter as temperatures decline. The layer formed by this refreezing process affects the rate of mass change within the ice sheet. The melting process absorbs energy from solar radiation, while refreezing releases heat. Although refreezing temporarily mitigates meltwater loss, it also alters the thermodynamic properties of the ice sheet. In the short term, refreezing may slow down mass loss; however, over a longer period, this newly formed layer could accelerate changes in mass within the ice sheet. If surface water undergoes repeated cycles of melting and refreezing throughout an extended seasonal cycle, it may lead to continuous adjustments in the structure of the ice sheet—ultimately exacerbating its overall mass change.

Moreover, the analysis demonstrated that various factors affect GrIS mass changes differently and that lag effects are present, with ice melting, runoff, rainfall, snowfall, and LST influencing ice mass change at varying temporal offsets. By analyzing the lag of each driving factor relative to the mass change of the GrIS, we can better help people to improve the understanding of the mass change of the GrIS. Based on previous studies, the annual signal and driving factors of mass change of the GrIS are studied in detail from the perspective of climate and temperature factors. Although some preliminary conclusions were drawn in this study, the limitations of the data, as noted by Florent Cambier et al. [53], suggest that further investigation is warranted regarding the primary patterns of ice sheet mass change and their relationship with climate indices (e.g., North Atlantic Oscillation (NAO), Greenland Blockage Index (GBI), Atlantic Multi-Year Intergenerational Oscillation (AMO)) as well as meteorological parameters (e.g., temperature, precipitation, and albedo) in future research. Such efforts could further elucidate the complex interactions between climatic, hydrological, and geophysical factors affecting GrIS mass dynamics.

5. Conclusions

As global climate change intensifies, the loss of mass of the GrIS has become one of the main drivers of global sea level rise. Assessing and analyzing the impact of climatic factors on changes in the mass of the GrIS has become a central issue in current research. In this context, the present study, based on GRACE/GRACE-FO satellite data and characterization of mass changes in the monitoring of the GrIS, focuses on analyzing the drivers of changes in the mass of the GrIS and its annual changes. The key findings of this study are as follows: firstly, the time series data for land surface temperature (LST), ice melting, runoff, rainfall, snowfall, and GrIS mass changes exhibit significant cyclical variations. Further analysis utilizing continuous wavelet transform (CWT) revealed that GrIS mass changes, ice melting, runoff, rainfall, and LST all display prominent annual signals; however, the annual signal for snowfall is relatively weak and negligible. To further investigate the annual variations in these variables, wavelet decomposition was employed to extract their respective annual signals. The results indicate that the annual amplitudes of ice melting and runoff demonstrate substantial variability that significantly exceeds those of rainfall and snowfall. Moreover, LST exhibited an annual phase that is inversely related to GrIS mass changes, suggesting a strong correlation between rising temperatures and increased ice mass loss. Finally, cross-correlation analysis indicated that ice melting, runoff, rainfall, snowfall, and LST all exert significant influences on GrIS mass changes with lag times of 3, 4, 4, 7, and 4 months, respectively. Among these factors, ice melting, runoff, and LST were identified as primary drivers of mass loss while snowfall and rainfall had secondary effects.

While this study highlights the interplay of climatic and hydrological processes influencing GrIS mass loss, the dominant role of warmer temperatures, ice melting, and runoff in mass loss was emphasized. These findings provide important insights for further understanding of GrIS mass dynamics. However, reliance on regional climate models and LST data still has certain limitations. Future research should incorporate ice discharge data, observed precipitation, and ice core records from meteorological stations, such as those of the Danish Meteorological Institute (DMI), to refine our understanding of GrIS mass dynamics and improve predictions of its contribution to global sea level rise.

Author Contributions: Conceptualization, K.M. and J.H.; methodology, K.M. and Z.L.; validation, J.H.; formal analysis, Q.H. and Z.L.; investigation, J.M. and P.H.; data curation, C.Y.; writing—original draft preparation, J.H.; writing—review and editing, K.M. and Z.L.; visualization, J.H.; supervision, P.H.; project administration, Q.H.; funding acquisition, J.M. All authors have read and agreed to the published version of the manuscript.

Funding: This research was funded by National Key Research and Development Program of China (No. 2024YFC3212200), the Henan Science Foundation for Distinguished Young Scholars of China (No. 242300421041), the Henan Key Research and Development Program of China (No. 241111321100), the National Natural Science Foundation of China (No. 42277478), and the Henan Province Colleges and Universities Young Backbone Teacher Training program (Grant No. 2023GGJS073).

Data Availability Statement: The data used to support the findings of this study are available from the corresponding author upon request.

Acknowledgments: The authors would like to thank all websites for sharing the data used in the paper and the reviewers for their insightful comments.

Conflicts of Interest: The authors declare no conflicts of interest.

References

1. Shepherd, A.; Ivins, E.; Rignot, E.; Smith, B.; Broeke, M.; Velicogna, I.; Whitehouse, P.; Briggs, K.; Joughin, I.; Krinner, G.; et al. Mass Balance of the Greenland Ice Sheet from 1992 to 2018. *Nature* **2020**, *579*, 233–239.
2. Morlighem, M.; Williams, C.N.; Rignot, E.; An, L.; Arndt, J.E.; Bamber, J.L.; Catania, G.; Chauché, N.; Dowdeswell, J.A.; Dorschel B.; et al. BedMachine v3: Complete Bed Topography and Ocean Bathymetry Mapping of Greenland from Multibeam Echo Sounding Combined with Mass Conservation. *Geophys. Res. Lett.* **2017**, *44*, 11051–11061.
3. Van Tricht, K.; Lhermitte, S.; Lenaerts, J.T.; Gorodetskaya, I.V.; L'Ecuyer, T.S.; Noël, B.; van den Broeke, M.R.; Turner, D.D.; van Lipzig, N.M. Clouds Enhance Greenland Ice Sheet Meltwater Runoff. *Nat. Commun.* **2016**, *7*, 10266. [CrossRef]
4. Huai, B.; van den Broeke, M.R.; Reijmer, C.H. Long-term Surface Energy Balance of the Western Greenland Ice Sheet and the Role of Large-Scale Circulation Variability. *Cryosphere* **2020**, *14*, 4181–4199. [CrossRef]
5. Zou, X.W.; Li, Z.; Yang, D.Y.; Wang, Z.; Zhou, Z.K.; Shan, Q.; Sun, W.J.; Ding, M.H.; Jiang, W.P. Advance of Mass Change and Driving Mechanism Research on Greenland Ice Sheet. *Geomat. Inf. Sci. Wuhan Univ.* **2023**, *48*, 1192–1203.
6. Feng, G.P.; Wang, Q.; Song, Q.T. Greenland Ice Sheet Mass Variations Based on GRACE Satellite Gravity Data. *Haiyang Xuebao* **2018**, *40*, 73–84.
7. Forsberg, R.; Sørensen, L.; Simonsen, S. Greenland and Antarctica Ice Sheet Mass Changes and Effects on Global Sea Level. *Surv. Geophys.* **2017**, *38*, 89–104.
8. Su, X.; Shum, C.K.; Guo, J.; Duan, J.; Howat, I.; Yi, Y. High Resolution Greenland Ice Sheet Inter-annual Mass Variations Combining GRACE Gravimetry and Envisat Altimetry. *Earth Planet. Sci. Lett.* **2015**, *422*, 11–17. [CrossRef]
9. Bevis, M.; Harig, C.; Khan, S.A.; Brown, A.; Simons, F.J.; Willis, M.; Fettweis, X.; van den Broeke, M.R.; Madsen, F.B.; Kendrick, E.; et al. Accelerating Changes in Ice Mass within Greenland, and the Ice Sheet's Sensitivity to Atmospheric Forcing. *Proc. Natl. Acad. Sci. USA* **2019**, *116*, 1934–1939.
10. Byron, D.T.; Srinivas, B.; John, C.R.; Paul, F.T.; Michael, M.W. GRACE Measurements of Mass Variability in the Earth System. *Science* **2004**, *305*, 503–505.
11. Shang, P.; Su, X.; Luo, Z. Characteristics of the Greenland Ice Sheet Mass Variations Revealed by GRACE/GRACE Follow-On Gravimetry. *Remote Sens.* **2022**, *14*, 4442. [CrossRef]
12. Barletta, V.R.; Sabadini, R.; Bordoni, A.V. Isolating the PGR Signal in the GRACE Data: Impact on Mass Balance Estimates in Antarctica and Greenland. *Geophys. J. Int.* **2008**, *172*, 18–30.

13. Ran, J.; Vizcaino, M.; Ditmar, P.; van den Broeke, M.R.; Moon, T.; Steger, C.R.; Enderlin, E.M.; Wouters, B.; Noël, B.; Reijmer, C.H.; et al. Seasonal Mass Variations Show Timing and Magnitude of Meltwater Storage in the Greenland Ice Sheet. *Cryosphere* **2018**, *12*, 2981–2999.
14. Zou, F.; Tenzer, R.; Fok, H.S.; Nichol, J.E. Mass Balance of the Greenland Ice Sheet from GRACE and Surface Mass Balance Modelling. *Water* **2020**, *12*, 1847. [CrossRef]
15. Mankoff, K.D.; Fettweis, X.; Langen, P.L.; Stendel, M.; Fausto, R.S. Greenland Ice Sheet Mass Balance from 1840 through Next Week. *Earth Syst. Sci. Data* **2021**, *13*, 5001–5025.
16. Gao, Y.; Wang, Z.; Li, F.; Chao, N. The Spatio-temporal Feature of Greenland Mass Changes as Determined by GRACE, Swarm, and GRACE-FO. *Chin. J. Geophys.* **2021**, *64*, 2405–2416.
17. Velicogna, I.; Wahr, J. Greenland Mass Balance from GRACE. *Geophys. Res. Lett.* **2005**, *32*, L18505.
18. Chen, J.L.; Wilson, C.R.; Tapley, B.D. Satellite Gravity Measurements Confirm Accelerated Melting of Greenland Ice Sheet. *Science* **2006**, *313*, 1958–1960.
19. Velicogna, I. Increasing Rates of Ice Mass Loss from the Greenland and Antarctic Ice Sheets Revealed by GRACE. *Geophys. Res. Lett.* **2009**, *36*, L19503.
20. Velicogna, I.; Sutterley, T.C.; van den Broeke, M.R. Regional Acceleration in Ice Mass Loss from Greenland and Antarctica Using GRACE Time-variable Gravity Data. *Geophys. Res. Lett.* **2014**, *41*, 8130–8137.
21. Jin, S.; Zou, F. Re-estimation of Glacier Mass Loss in Greenland from GRACE with Correction of Land-ocean Leakage Effects. *Glob. Planet. Change* **2015**, *135*, 170–178. [CrossRef]
22. Groh, A.; Horwath, M.; Horvath, A.; Meister, R.; Shepherd, A. Evaluating GRACE Mass Change Time Series for the Antarctic and Greenland Ice Sheet—Methods and Results. *Geosciences* **2019**, *9*, 415. [CrossRef]
23. Rignot, E.; Mouginot, J.; Scheuchl, B.; Broeke, M.V.; Wessem, M.J.; Morlighem, M. Four Decades of Antarctic Ice Sheet Mass Balance from 1979–2017. *Proc. Natl. Acad. Sci. USA.* **2019**, *116*, 1095–1103. [CrossRef] [PubMed]
24. Goldhar, C.; Ford, J.D. Climate Change Vulnerability and Food Security in Qeqertarsuaq, Greenland. In *Community Adaptation and Vulnerability in Arctic Regions*; Springer: Dordrecht, The Netherlands, 2010; pp. 263–283.
25. Thomas, R.; Csatho, B.; Davis, C.; Kim, C.; Krabill, W.; Manizade, S.; Mcconnell, J.; Sonntag, J. Mass Balance of Higher-Elevation Parts of the Greenland Ice Sheet. *J. Geophys. Res. Atmos.* **2001**, *106*, 33707–33716. [CrossRef]
26. Bamber, J.L.; Ekholm, S.; Krabill, W.B. A New, High-resolution Digital Elevation Model of Greenland Fully Validated with Airborne Laser Altimeter Data. *J. Geophys. Res. Sol. Earth* **2001**, *106*, 6733–6745. [CrossRef]
27. Chen, J.; Tapley, B.; Tamisiea, M.E.; Save, H.; Wilson, C.; Bettadpur, S.; Seo, K.W. Error Assessment of GRACE and GRACE Follow-on Mass Change. *J. Geophys. Res.* **2021**, *126*, e2021JB022124. [CrossRef]
28. Swenson, S.; Chambers, D.; Wahr, J. Estimating Geocenter Variations from a Combination of GRACE and Ocean Model Output. *J. Geophys. Res.* **2008**, *113*, B08410. [CrossRef]
29. Sasgen, I.; Vanden, B.M.; Bamber, J.L.; Eric, R.; Louise, S.S.; Bert, W.; Zdenek, M.; Isabella, V.; Sebastian, B.S. Timing and Origin of Recent Regional Ice-Mass Loss in Greenland. *Earth Planet. Sci. Lett.* **2012**, *333*, 293–303. [CrossRef]
30. Chen, J.L.; Rodell, M.; Wilson, C.R.; Famiglietti, J.S. Low Degree Spherical Harmonic Influences on Gravity Recovery and Climate Experiment (GRACE) Water Storage Estimates. *Geophys. Res. Lett.* **2005**, *32*, L14405. [CrossRef]
31. Zou, F.; Tenzer, R.; Fok, H.S.; Nichol, J.E. Recent Climate Change Feedbacks to Greenland Ice Sheet Mass Changes from GRACE. *Remote Sens.* **2020**, *12*, 3250. [CrossRef]
32. Bian, Y.K.; Yue, J.P.; Gao, W.; Li, Z.; Lu, D.K.; Xiang, Y.F.; Chen, J. Analysis of the Spatiotemporal Changes of Ice Sheet Mass and Driving Factors in Greenland. *Remote Sens.* **2019**, *11*, 862. [CrossRef]
33. Geruo, A.; Wahr, J.; Zhong, S. Computations of the Viscoelastic Response of a 3-D Compressible Earth to Surface Loading: An application to Glacial Isostatic Adjustment in Antarctica and Canada. *Geophys. J. Int.* **2013**, *192*, 557–572.
34. Feng, W. GRAMAT: A Comprehensive Matlab Toolbox for Estimating Global Mass Variations from GRACE Satellite Data. *Earth. Sci. Inform.* **2019**, *12*, 389–404. [CrossRef]
35. Yi, S.; Sneeuw, N. Filling the Data Gaps Within GRACE Missions Using Singular Spectrum Analysis. *J. Geophys. Res. Solid Earth* **2021**, *126*, e2020JB021227.
36. Fettweis, X.; Box, J.E.; Agosta, C.; Amory, C.; Kittel, C.; Lang, C.; van As, D.; Machguth, H.; Gallée, H. Reconstructions of the 1900–2015 Greenland Ice Sheet Surface Mass Balance Using the Regional Climate MAR Model. *Cryosphere Discuss.* **2017**, *11*, 1015–1033.
37. Ruan, R. Analysis on the Characteristics of Greenland Ice Sheet Mass Change and a Preliminary Study on Mechanism. *Clim. Change Res. Lett. 2019. 8*, 473–483.
38. Fan, Y.; Dool, H.V.D. A Global Monthly Land Surface Air Temperature Analysis for 1948-present. *J. Geophys. Res. Atmos.* **2008**, *113*, 18.
39. Jekeli, C. *Alternative Methods to Smooth the Earth's Gravity Field*; Scientific Report 327, School of Earth Science, The Ohio State University: Columbus, OH, USA, 1981.

40. Wahr, J.; Molenaar, M.; Bryan, F. Time Variability of the Earth's Gravity Field: Hydrological and Oceanic Effects and Their Possible Detection using GRACE. *J. Geophys. Res. Sol. Earth* **1998**, *103*, 30205–30229.
41. Swenson, S.; Wahr, J. Post-Processing Removal of Correlated Errors in GRACE Data. *Geophys. Res. Lett.* **2006**, *33*, 08402.
42. Chambers, D.P. Evaluation of New GRACE Time-Variable Gravity Data over the Ocean. *Geophys. Res. Lett.* **2006**, *33*, 17603 [CrossRef]
43. Chen, J.L.; Wilson, C.R.; Tapley, B.D.; Grand, S. GRACE Detects Coseismic and Postseismic Deformation from the Sumatra-Andaman Earthquake. *Geophys. Res. Lett.* **2007**, *34*, 173–180.
44. Cooper, G.R.J.; Cowan, D.R. Comparing Time Series Using Wavelet-based Semblance Analysis. *Comput. Geosci.* **2008**, *34*, 95–102
45. Grinsted, A.; Moore, J.C.; Jevrejeva, S. Application of the Cross Wavelet Transform and Wavelet Coherence to Geophysical Time Series. *Nonlinear Proc. Geophys.* **2004**, *11*, 561–566.
46. Li, Z.; Yue, J.; Xiang, Y.; Chen, J.; Bian, Y.; Chen, H. Multiresolution Analysis of the Relationship of Solar Activity, Global Temperatures, and Global Warming. *Adv. Meteorol.* **2018**, *2018*, 2078057.
47. Souza, M.P.; Echer, E.; Rigozo, N.R.; Brum, C.G.; Nordemann, D.J.; Gonzalez, W.D. On the Relationship Between Global, Hemispheric and Latitudinal Averaged Air Surface Temperature (GISS Time Series) and Solar Activity. *J. Atmos. Sol. Terr. Phys.* **2012**, *74*, 87–93.
48. Jenkins, G.M.; Watts, D.G. *Spectral Analysis and Its Applications*; Holden Day: San Francisco, CA, USA, 1968; Volume 36, p. 554.
49. Gao, C.C.; Lu, Y.; Shi, H.L.; Zhang, Z.Z.; Xu, C.Y.; Tan, B. Detection and Analysis of Ice Sheet Mass Changes over 27 Antarctic Drainage Systems from GRACE RL06 Data. *China J. Geophys.* **2019**, *62*, 864–882.
50. Shamshiri, R.; Nahavandchi, H.; Joodaki, G. Seasonal Variation Analysis of Greenland Ice Mass Time-series. *Acta Geod. Geophys.* **2018**, *53*, 1–14.
51. Ramillien, G.; Lombard, A.; Cazenave, A.; Ivins, E.R.; Llubes, M.; Remy, F.; Biancale, R. Interannual Variations of the Mass Balance of the Antarctica and Greenland Ice Sheets from GRACE. *Global Planet. Change* **2006**, *53*, 1980–2008.
52. Sasgen, I.; Wouters, B.; Gardner, A.S.; Michalea, D.K.; Marco, T.; Felix, W.L.; Christoph, D.; Himanshu, S.; Xavier, F. Return to Rapid Ice Loss in Greenland and Record Loss in 2019 Detected by the GRACE-FO Satellites. *Commun. Earth Environ.* **2020**, *1*, 1–8.
53. Cambier, F.; Darrozes, J.; Llubes, M.; Seoane, L.; Ramillien, G. Greenland's Annual and Interannual Mass Variations from GRACE/GRACE-FO Linked with Climatic Indices. *EGUsphere* **2025**, *2025*, 1–23.

land

MDPI

Article

Artificial-Intelligence-Based Investigation on Land Use and Land Cover (LULC) Changes in Response to Population Growth in South Punjab, Pakistan

Tanweer Abbas [1], Muhammad Shoaib [1,*], Raffaele Albano [2,*], Muhammad Azhar Inam Baig [1], Irfan Ali [3], Hafiz Umar Farid [1] and Muhammad Usman Ali [1]

1 Department of Agricultural Engineering, Bahauddin Zakariya University, Multan 60800, Pakistan; engr.tanweeer@gmail.com (T.A.); azharinam@bzu.edu.pk (M.A.I.B.); hufarid@bzu.edu.pk (H.U.F.); engr.usmanali@bzu.edu.pk (M.U.A.)
2 Department of Health Science, University of Basilicata, 85100 Potenza, Italy
3 Pakistan Agricultural Research Council, Islamabad 44000, Pakistan; irfanali@parc.gov.pk
* Correspondence: msho127@aucklanduni.ac.nz (M.S.); raffaele.albano@unibas.it (R.A.)

Abstract: Land use and land cover (LULC) changes are significantly impacting the natural environment. Human activities and population growth are negatively impacting the natural environment. This negative impact directly relates to climate change, sustainable agriculture, inflation, and food security at local and global levels. Remote sensing and GIS tools can provide valuable information about change detection. This study examines the correlation between population growth rate and LULC dynamics in three districts of South Punjab, Pakistan—Multan, Bahawalpur, and Dera Ghazi Khan—over a 30-year period from 2003 to 2033. Landsat 7, Landsat 8, and Sentinel-2 satellite imagery within the Google Earth Engine (GEE) cloud platform was utilized to create 2003, 2013, and 2023 LULC maps via supervised classification with a random forest (RF) classifier, which is a subset of artificial intelligence (AI). This study achieved over 90% overall accuracy and a kappa value of 0.9 for the classified LULC maps. LULC was classified into built-up, vegetation, water, and barren classes in Multan and Bahawalpur, with an additional "rock" class included for Dera Ghazi Khan due to its unique topography. LULC maps (2003, 2013, and 2023) were prepared and validated using Google Earth Engine. Future predictions for 2033 were generated using the MOLUSCE model in QGIS. The results for Multan indicated substantial urban expansion as built-up areas increased from 8.36% in 2003 to 25.56% in 2033, with vegetation and barren areas displaying decreasing trends from 82.96% to 70% and 7.95% to 3.5%, respectively. Moreover, areas containing water fluctuated and ultimately changed from 0.73% in 2003 to 0.9% in 2033. In Bahawalpur, built-up areas grew from 1.33% in 2003 to 5.80% in 2033, while barren areas decreased from 79.13% to 74.31%. Dera Ghazi Khan expressed significant increases in built-up and vegetation areas from 2003 to 2033 as 2.29% to 12.21% and 22.53% to 44.72%, respectively, alongside reductions in barren and rock areas from 32.82% to 10.83% and 41.23% to 31.2%, respectively. Population projections using a compound growth model for each district emphasize the demographic impact on LULC changes. These results and findings focus on the need for policies to manage unplanned urban sprawl and focus on environmentally sustainable practices. This study provides critical awareness to policy makers and urban planners aiming to balance urban growth with environmental sustainability.

Keywords: climate change; sustainable agriculture; food security; inflation; artificial intelligence; remote sensing; GIS

1. Introduction

Land cover tells us how the land is covered by distinct natural features that are present on a piece of land, whereas land use represents the change in those natural features due to human activity. Landscape patterns are affected by physical as well as socioeconomic factors [1]. Significant transformations in land cover are caused by anthropogenic activities [2]. People have long altered land cover to meet their needs, affecting the natural environment. The natural environment is important for sustainable agricultural practices, but it is being negatively impacted by human activities, leading to climate change becoming a significant problem. Continuous land use monitoring is essential to protect the natural environment. To maintain urban ecosystem sustainability, there is a need to manage rapid urbanization [3]. Monitoring the dynamics of LULC change over time is essential for sustainable development and natural resource management [4]. Satellite images help us to understand landscape dynamics at specific locations and times [5]. The population of Pakistan is 241.49 million, with the highest average annual growth rate of 2.55% compared to neighboring countries' average annual growth rates such as those of India (0.81%), Bangladesh (1.1%), Iran (0.7%), and Afghanistan (2.5%). This population expansion is also producing unplanned urban sprawl, industrialization, developmental activities, and road constructions, which are not only shifting farm land to built-up regions but also increasing the land surface temperature (LST) owing to deforestation. The increasing population demands more food to fulfill the food requirements of the people, and food security problems are emerging [6]. Climate change and decreases in the agricultural area are causing LST to increase, exacerbating the problem of decreasing crop production [7]. These factors are putting pressure on food productivity, leading to inflation. The growth rate of the population is directly related to land use and land cover changes. A United Nations study conducted on future urban growth expressed that, in the future, there will be more urban growth in Asian megacities. Therefore, it is very important to have actual and up-to-date land use and land cover maps of this region. Classified maps will provide us with valuable information on the landscape dynamics and changing patterns in different classes. Change detection maps play an important role, providing valuable statistics [8]. Therefore, this study intended to develop an up-to-date map of the study area to predict future LULC and to know the impact of human activities on the environment. Despite technological advancements, researchers still have difficulties in accurately producing maps for the upcoming year based on population increases. However, artificial intelligence (AI) and machine learning (ML) algorithms are prominent techniques for land use and land cover prediction. Accurate LULC maps are being prepared via supervised classification [9,10]. To produce LULC maps, a combination of different models with cellular automata (CA) simulation models is being used by researchers. Artificial neural networks (ANNs) and multi-layer perceptrons (MLPs), which are AI algorithms, are widely being used for prediction of LULC [5,11–15]. Many researchers have also used the Modules for Land Use Change Scenario Estimation (MOLUSCE) plug-in, an extension of QGIS, which provides tools for LULC modeling and prediction [16,17]. TerrSet software also provides a land change module for prediction, which is also being used by researchers [18,19]. In this study, an attempt was made to prepare up-to-date LULC maps of three districts of South Punjab with different geographical features. The present study prepared LULC maps for the first time using Google Earth Engine (GEE) with a random forest (RF) classifier, which is a machine learning (ML) algorithm and is subset of artificial intelligence (AI). Furthermore, predictions were made in MOLUSCE for three entire districts, i.e., Multan, Bahawalpur, and Dera Ghazi Khan. A comparative analysis of four land cover classes, built-up, vegetation, water, and barren, was also performed for each district within the study area. Moreover, the impact of population growth on these classes was analyzed.

2. Materials and Methods

2.1. Study Area

The study area comprises three prominent districts of South Punjab: Multan, Bahawalpur, and Dera Ghazi Khan (Figure 1). The region has an arid climate with hot summers and mild winters. Multan, known as the City of Saints and the oldest city of the region, lies on the eastern bank of the Punjab River, expanding over an area of 3647.45 km^2, and has a population of 5.36 million people, according to the latest 7th Digital Census held in 2023. It is home to historical shrines and fertile soil, which is considered alluvial and mainly comes with the deposition of River Chenab sediments. It is located between the longitude of 71°00′54″ E to 72°58′43″ E and a latitude of 29°27′21″ N to 30°45′30″ N. Bahawalpur District consists of an area of 22,353.5 km^2 and is the largest district in the province of Punjab. It lies between 70°54′ to 72°50′ E longitude and 27°48′ to 29°50′ N latitude. It is located south of Multan. A large portion, which is present on the southern side, is within the Cholistan Desert. Its northern side has good potential for agriculture and is also home to most of the population, which was 4.28 million people in 2023. Dera Ghazi Khan is situated at 29.5° N to 31.2° N latitude and 69.3° E to 70.9° E longitude, with an area of 16,061.5 km^2 and a population of 3.39 million people. The topography of the district consists of fertile plains of the Indus River, the range of Suleiman Mountain, and part of the Thal Desert. Annual rainfall is about 200–250 mm, mainly in monsoon. The eastern part of the district is located near the bank of the Indus River. The study area consists of fertile soil which hosts different crops such as wheat, cotton, sugarcane, and rice. An increase in population, unplanned urban sprawl, and conversion of agricultural land into housing is affecting climate change. Furthermore, food security and inflation are more challenging issues based on research data. Up-to-date LULC maps of the study areas will help policymakers and planners manage their natural environment effectively and support sustainable agriculture.

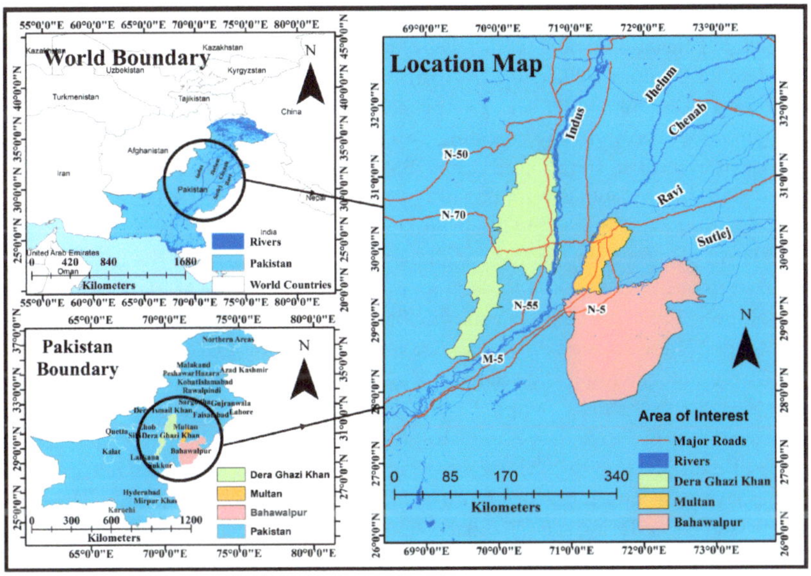

Figure 1. Map of the study area.

2.2. Data Requirement

For this research, essential data sets include shape files, satellite data, population data, and other relevant factors to comprehensively analyze LULC changes. These resources are critical for accurately identifying changing patterns and transformations.

2.2.1. Shape Files

DIVA-GIS data were obtained from the website https://diva-gis.org (accessed on 5 July 2023) to extract shape files of the study areas with specific boundary lines.

2.2.2. Satellite Data

Satellite images for the last 20 years were collected to detect the changes in the study area. The satellite images were processed in GEE using coding. The images were clipped to the desired study areas for a specific period for better visualization and understanding. The Landsat 7 satellite image was used to prepare the LULC map for 2003, the Landsat 8 image for 2013 and the Sentinel-2 satellite image for 2023. Detailed information about the satellite data is given below in Table 1.

Table 1. Data of Satellite.

Sr. No.	Data Type	Source	Spatial Resolution	Time Period	Purpose
1	Landsat 7	USGS	30 m	2003	LULC 2003
2	Landsat 8	USGS	30 m	2013	LULC 2013
3	Sentinel-2	USGS	10 m	2023	LULC 2023
4	DEM	ALOS PALSAR SRTM	12.5 m 30 m	2007 2000	Slope, aspect, and hill shade

2.2.3. Digital Elevation Model

The digital elevation model (DEM) was obtained from https://search.asf.alaska.edu (accessed on 23 December 2023 and 28 April 2024) and https://earthexplorer.usgs.gov (accessed on 23 December 2023 and 28 April 2024). A DEM is a data set that can store information about continuously changing variables, such as soil depth, elevation, and water present in the soil. In this study, the DEM obtained from the Alaska website is an ALOS PALSAR data set with a resolution of 12.5 m. This ALOS PALSAR DEM was clipped in ARCGIS to fit the study areas of Multan and Dera Ghazi Khan only. For Bahawalpur, the ALOS PALSAR data set did not consist of all the required images; hence the USGS website was utilized to obtain the Shuttle Radar Topographic Mission (SRTM) data set having a 30 m resolution. The DEM helps to make further derived data sets such as slope, aspect, and hill shade. These derived data sets were used for making LULC maps in the QGIS MOLUSCE plug-in with improved accuracy. The cell size and coordinates of these variables must match those of the LULC maps. For the years 2003 and 2013, Sentinel satellite data were not available. Only Landsat 7 and Landsat 8 data were available, which provided a spatial resolution of 30 m. In 2023, Landsat 9 and Sentinel data were available. Sentinel data provide a spatial resolution of 10 m. The high resolution helps to identify finer-scale features in land cover during sample collection in supervised classification more accurately. Due to the high resolution of Sentinel, it was chosen to prepare the classified map for the year 2023. However, to maintain compatibility with earlier maps, the prepared map from Sentinel data was adjusted to match the 30 m resolution of Landsat data. Detailed information about satellite data is presented below in Table 1.

2.2.4. Driving Factors

Factors such as slope, aspect, hill shade, distance from the road, and distance from the river are variables used in MOLUSCE to predict LULC maps accurately. These factors were derived in ArcGIS using DEM data. MOLUSCE relies on driving factors to simulate land use changes. Inaccurate driving factors may cause unrealistic clustering of certain land uses, poor model performance in predicting real-world scenarios, and failure to address critical environmental problems. Slope data show that steeper slopes are unsuitable for agriculture and urban areas, so in the future, they will not contain such things. However, natural vegetation may grow here. Human activities with enhanced changing patterns will be observed in areas with less slope and elevation [1]. The aspect tells us about the face of the slope. The face, which receives more sunlight, produces more vegetation. Distance from road data show that areas near a road are likely to convert into urban and industrial development. The presence of a major road in an area with good resource management results in the expansion of that area [20]. Areas near the river will be fertile and suitable for vegetation, but they may be affected by floods. Hill shade is the shaded region that mostly consists of moist soil that generally does not have vegetation due to the projection of less sunlight. Soil with more moisture is less erodible by rainwater than dry and loose soil.

2.2.5. Population Data

Population and average annual growth rate (AAGR) data required for the study area were collected from the Pakistan Bureau of Statistics website, https://www.pbs.gov.pk (accessed on 12 June 2024), which were available in the form of census reports for 2017 and 2023. Previous census data for 1998, 1981, and 1972 were collected from the website https://citypopulation.de (accessed on 12 June 2024). Accurate population data are important for forecasting growth rate trends and using them for urban planning [6]. Population data for 2003 and 2013 were calculated using the linear interpolation method, represented by Equation (1).

$$P(t) = P_{initial} + \frac{(P_{final} - P_{initial})}{(T2 - T1)} \times (T - T1) \quad (1)$$

where:

P(t) = Population in required time period;
P_{final} = Population in final year;
$P_{initial}$ = Population in initial year;
T = Required time period;
T1 = Time period of initial year;
T2 = Time period of final year;
Population data for all the three districts are shown below in Table 2.

Table 2. Population of the Study Area.

District	1972	1981	1998	2003	2013	2017	2023
Multan	1,506,223	1,970,075	3,116,851	3,586,158	4,403,152	4,746,166	5,362,305
Bahawalpur	1,071,026	1,453,438	2,433,091	2,758,376	3,408,946	3,669,176	4,284,964
D.G. Khan	686,057	943,663	1,643,118	1,966,673	2,613,783	2,872,631	3,393,705

2.3. Data Processing

After collecting data, a set of techniques were applied for further processing of data to fulfill the research objectives, which are given below in Figure 2 in the form of a flow chart. Landsat 7, Landsat 8, and Sentinel satellite data were used to classify and prepare LULC maps for 2003, 2013, and 2023 using supervised classification in GEE. In the MOLUSCE tool, an extension of QGIS, classified LULC maps for 2003 and 2013 were given as input maps. MOLUSCE needs only the initial and final year data to simulate a map for the projected year based on historical data and observed changes and patterns of shifting from one class to another. The ten-year gap between 2003 and 2013 is sufficient to observe the changing pattern of converting classes from one to another. Driving factors, which were obtained from DEM, were also given as input to the MOLUSCE tool. The change detection process and transition potential modeling in MOLUSCE simulated the map for 2023. The 2023 simulated map, prepared by MOLUSCE, and an actual map created through supervised classification were validated within MOLUSCE to assess the model's performance with the given data. In the next step, the LULC map for the projected year 2033 was prepared within MOLUSCE. Finally, the prepared maps were analyzed.

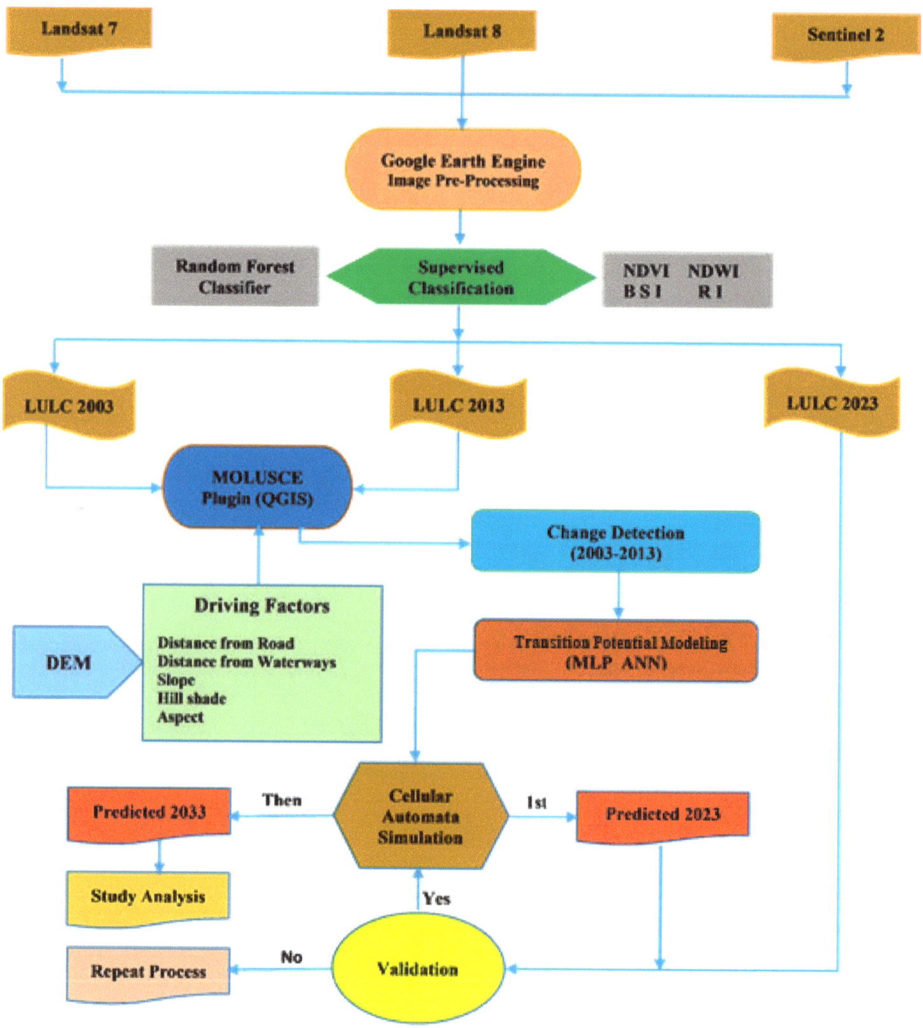

Figure 2. Methodology Flow Chart.

2.4. Combination of Bands

Satellite images in GEE comprise multiple bands that provide spectral data for distinguishing between classes. These bands act like filters on a camera, combining different colors to enhance visualization. These bands include red, green, blue, near-infrared (NIR), shortwave infrared (SWIR), and thermal infrared. In GEE, satellite images were clipped to fit the study area and shape files. Different color combinations were used for better visualization. Natural color comprises red, green, and blue bands that visualize the image as our eyes perceive it. To enhance vegetation features, a false color combination was used, which includes near-infrared, red, and green bands. Vegetation appears bright red when using a false color combination.

2.5. Indices

Different types of indices are helpful in GEE for distinguishing one class from another. They provide a better view of the image and help in sample collection for supervised classification. The indices used include the Normalized Difference Vegetation Index (NDVI), Normalized Difference Water Index (NDWI), Bare Soil Index (BSI), and Normalized Difference Built-up Index (NDBI). These are expressed below in Equations (2), (3), (4), and (5), respectively.

$$NDVI = \frac{(NIR - R)}{(NIR + R)} \tag{2}$$

$$NDWI = \frac{(Green - NIR)}{(Green + NIR)} \tag{3}$$

$$BSI = \frac{(SWIR + R) - (NIR + Blue)}{(SWIR + R) + (NIR + Blue)} \tag{4}$$

$$NDBI = \frac{(SWIR - NIR)}{(SWIR + R)} \tag{5}$$

where:

R = Red;
NIR = Near-infrared;
SWIR = Shortwave infrared.

2.6. Types of LULC Classification

There are two types of LULC classification:

1. Unsupervised classification.
2. Supervised classification.

Supervised Classification

Supervised classification is a technique used to convert image pixels into different land cover classes, resulting in a classified map. The image is classified by training a model to associate specific colors with land cover classes, based on visual interpretation and ground truthing data. In this study, supervised classification was used to classify maps. In GEE, supervised classification comprises three steps:

1. Sample collection;
2. Training a classifier;
3. Classifying the map.

Sample points representing location and class were collected in GEE by analyzing the satellite image and utilizing various indexes for better results. These samples were stored in a feature collection containing spectral information. A model was trained using the collected samples [21]. The random forest (RF) classifier, a well-known machine learning algorithm, was applied on the image to produce a classified map. The RF algorithm is the most reliable compared to others because it uses an object-based technique to produce accurate LULC maps for complex and large areas [14]. This technique was applied to the districts of Multan, Bahawalpur, and Dera Ghazi Khan to prepare LULC maps for 2003, 2013, and 2023.

2.7. Accuracy Assessment

The accuracy of a classified map indicates the quality and reliability of results, specifically how closely the classified map aligns with actual land cover. For accuracy assessment, a confusion matrix was developed in GEE that offers insights into the correct and incorrect samples across various classes. Overall accuracy and the kappa coefficient were calculated using Equations (6) and (7), which are given below:

$$Overall\ Accuracy = \frac{Total\ Number\ of\ correctly\ Classified\ Pixels\ (Diagonal)}{Total\ Number\ of\ Reference\ Pixels} \times 100 \quad (6)$$

$$Kappa\ Coefficient = \frac{(TS \times TCS) - \sum(Column\ Total\ \times\ Row\ Total)}{TS^2 - \sum(Column\ Total - Row\ Total)} \times 100 \quad (7)$$

where:

TS = Total sample;

TCS = Total correct sample.

Kappa coefficient is categorized as follows: <0 indicates no agreement, 0–0.2 slight agreement, 0.2–0.41 poor agreement, 0.41–0.6 moderate agreement, 0.6–0.8 significant agreement, and 0.8–1.0 almost perfect agreement [16]. A kappa value in the range of 0.6–0.8 is considered to indicate a strong agreement [22].

2.8. Prediction

Prediction of LULC for 2033 was conducted in QGIS using the MOLUSCE plug-in. MOLUSCE first calculates the changes in area and then predicts a future LULC map with ANN methods, specifically the CA and MLP algorithms. The CA-ANN model is a reliable tool widely used by researchers and planners for LULC predictions [2].

There are six stages of LULC prediction in MOLUSCE:

1. Data input;
2. Evaluation of correlation;
3. Change detection;
4. Transition potential modeling;
5. Cellular automata simulation;
6. Validation.

During data input, LULC maps for 2003, 2013, and 2023 were introduced into MOLUSCE, having the same classes, number of cells, and coordinates. LULC for 2003 was given as the initial data set, and LULC for 2013 was given as the final data set. Then derived factors such as slope, aspect, hill shade, distance from road, and distance from river were inserted as variables for accurate prediction. The data must contain the same geometry otherwise MOLUSCE cannot perform further steps. In the second step, evaluation of correlation, the derived factors were analyzed based on their Pearson correlation to determine which factors influence the changes. In the third step, the area of one class changing into another class was detected. In the fourth step of transition potential modeling, MOLUSCE uses the MLP algorithm, a specific type of ANN, containing different layers of interconnected neurons. The data are processed here, and the MLP algorithm places these data into a specific layer, with a specific class, where the specific data fall. For example, areas near roads are likely to fall into a built-up class in future. So, MLP, using a backpropagation training algorithm, gives us future change output data on the basis of input data. It compares the predicted map with the original one to identify errors during prediction. MLP identifies and resolves errors, creating an accurate map through iterations [23,24]. An ANN is a machine learning model that mimics the human brain's functioning and learns from data to make predictions. AI tools track new trends in a location over time and, after analyzing the data, predict future developments [23]. In the fifth step of cellular automata simulation, the algorithm automatically changes cells according to the influence of neighboring cells. For example, cells or pixels of bare soil surrounded by built-up areas may be influenced and converted into built-up cells or pixels. The LULC map for 2023 was predicted by simulation. In the last step, predicted LULC map for 2023 was validated within the MOLUSCE tool against the actual one prepared in GEE. The LULC map for 2033 was then predicted using a cellular automata simulation algorithm, with the number of iterations set to 2 and a 10-year interval.

3. Results and Discussion

This section focuses on analyzing the maps prepared from 2003 to 2023 for the three districts of South Punjab: Multan, Bahawalpur, and D.G. Khan, as well as examining the map for the projected year 2033 in this context. It also includes observing changing patterns from one class to another. The results of changes in areas across the three districts are compared with population growth rates.

3.1. LULC Maps from Google Earth Engine

The LULC maps were prepared in GEE using a supervised classification for 2003, 2013, and 2023 for three districts of Multan, Bahawalpur, and D.G. Khan, with an overall accuracy exceeding 90%. An accuracy of 85% or below indicates a significant difference between the classified map and the actual land cover. The future forecast will be unreliable if the map has an accuracy of less than 85% [25].

3.2. LULC Maps of Multan District

The LULC maps for 2003, 2013, and 2023 prepared using a supervised classification in GEE for Multan are shown in Figure 3 and consist of four classes: built-up, vegetation, water, and barren. After the maps were classified, they were reclassified in ArcGIS. The maps were processed in ArcGIS to calculate the area of each class in square kilometers. Detailed information about the area of each class, in square kilometers and as percentages, is provided in Tables 3 and 4. In 2003, the built-up, vegetation, water, and barren areas of Multan were 304.9 km^2, 3026 km^2, 26.7 km^2, and 290 km^2 which were 8.4%, 83%, 0.73%, and 8%, respectively, of the total area. The LULC map for 2013 revealed that the built-up, vegetation, water, and barren areas in Multan covered 459.9 km^2, 2815.4 km^2, 42.4 km^2, and 329.8 km^2, respectively, of the total area. Similarly, the LULC map for 2023 shows that the built-up, vegetation, water, and barren areas of Multan extended to 731.3 km^2, 2714.1 km^2, 38 km^2, and 164 km^2 which were 20.1%, 74.4%, 1.0%, and 4.5%, respectively, of the total area.

Table 3. Total Area of each LULC Class of Multan in km^2.

Class	Area in km^2				Temporal Area Changes in km^2		
	2003	2013	2023	2033	2003–2013	2013–2023	2023–2033
Built-up	304.79	459.9	731.3	932.25	155	271	201
Vegetation	3025.9	2815.4	2714.1	2552.9	−211	−101	−161
Water	26.7	42.3	37.95	33.59	15.7	−4.4	−4.36
Barren	290	329.8	164.05	128.7	39.8	−166	−35.4
Total	3647.4	3647.4	3647.4	3647.4	0	0	0

Table 4. Total Area of each LULC Class of Multan in Percentage.

Class	Area in %				Temporal Area Changes in %		
	2003	2013	2023	2033	2003–2013	2013–2023	2023–2033
Built-up	8.36	12.61	20.0	25.5	4.3	7.44	5.5
Vegetation	82.9	77.19	74.4	70	−5.76	−2.78	−4.4
Water	0.73	1.16	1.04	0.9	0.43	−0.12	−0.12
Barren	7.95	9.04	4.5	3.5	1.09	−4.54	−0.7
Total	100	100	100	100	0	0	0

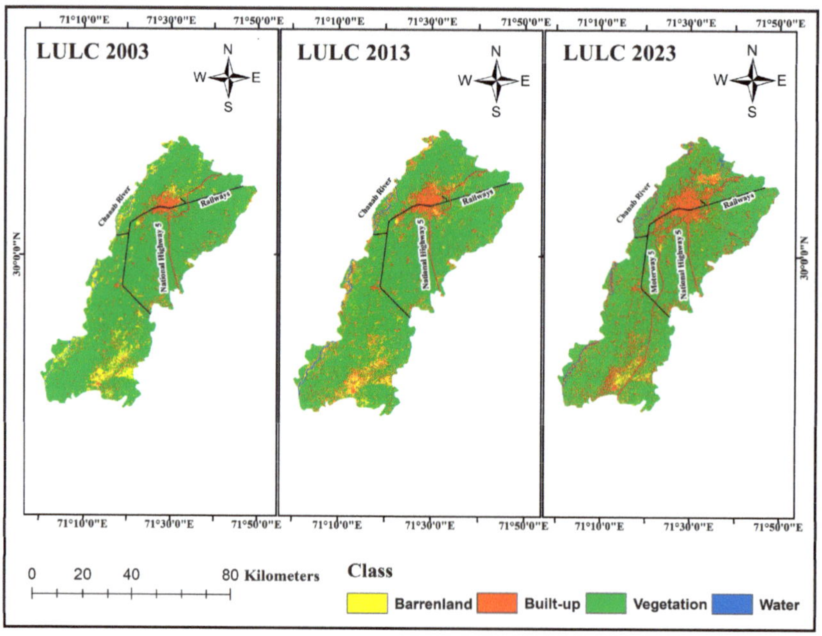

Figure 3. LULC maps of 2003, 2013, and 2023 for Multan district.

3.2.1. Driving Factors for Multan

The driving factors such as slope, aspect, distance from road, and distance from waterways were prepared in ArcGIS. These factors helped to determine which land cover classes might convert into another in future. Accurate LULC maps for the projected year could be prepared by considering these factors during prediction using the MOLUSCE tool in QGIS. The driving factors for Multan are shown below in Figure 4.

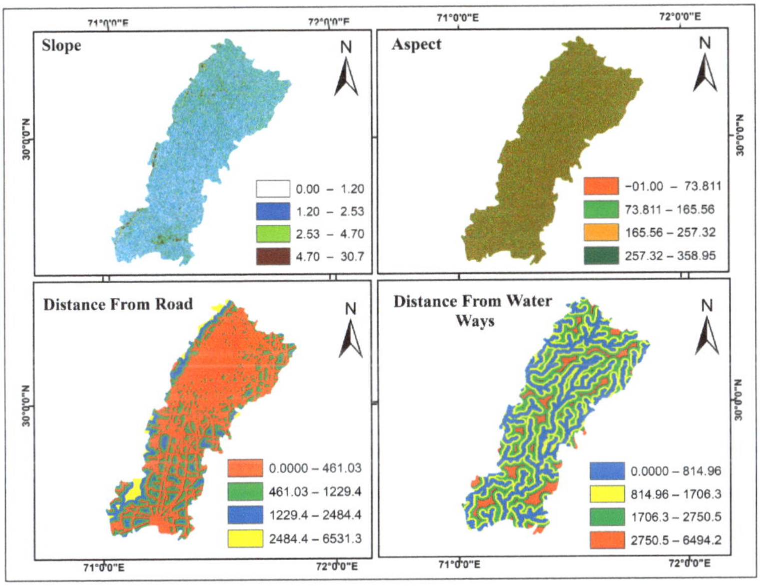

Figure 4. Driving Factors for Multan.

3.2.2. Projected LULC Map of Multan District

LULC map prediction helps to plan urban development, control unplanned urban sprawl, overcome environmental impacts, and better manage sustainable agriculture practices. In this study, the LULC map of Multan District for 2033 was prepared using the MOLUSCE tool in QGIS as given below in Figure 5, with a model accuracy of 74.15% and an overall kappa of 0.35. More accuracy could be obtained by increasing the driving factors. The predicted area for 2033 of Multan District was calculated in ArcGIS. The results shows that the built-up, vegetation, water, and barren land will spread over 932.25 km^2 (25.56%), 2552.94 km^2 (69.99%), 33.59 km^2 (0.92%), and 128.67 km^2 (3.53%), respectively.

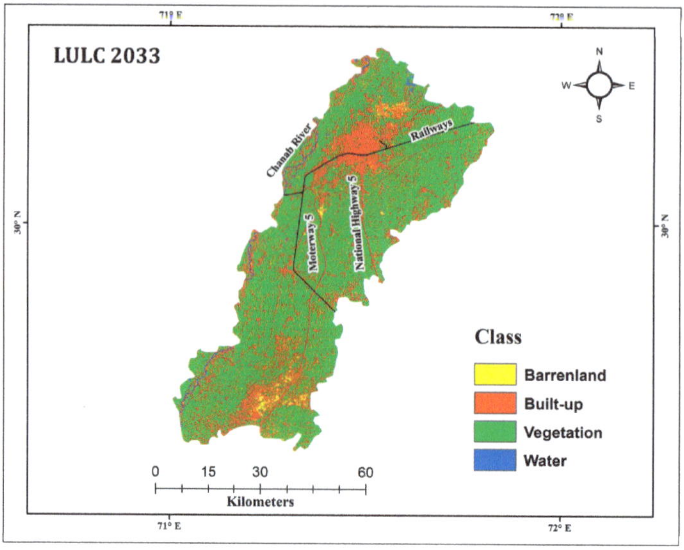

Figure 5. Predicted map for the year 2033.

3.2.3. Analysis of LULC Area of Multan District from 2000–2033

LULC analysis provides valuable information about the distribution of different classes in a specific location. LULC maps for 2003, 2013, and 2023 of Multan District were processed in ArcGIS. Areas with percentage changes for each class were calculated and are summarized below in Tables 3 and 4, respectively. The areas occupied by each class and their trends over time are shown below in Figures 6 and 7. Positive values in Tables 3 and 4 show an increasing trend whereas negative values indicate a decreasing trend for the respective classes.

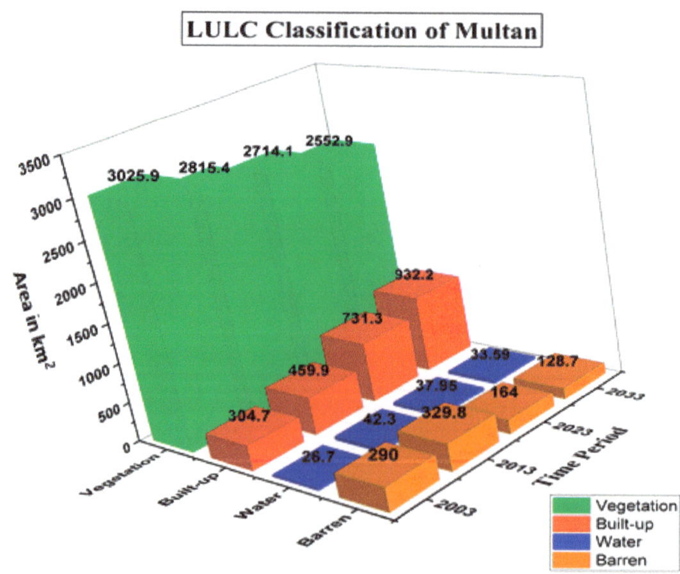

Figure 6. Graphical comparison of LULC Classes of Multan.

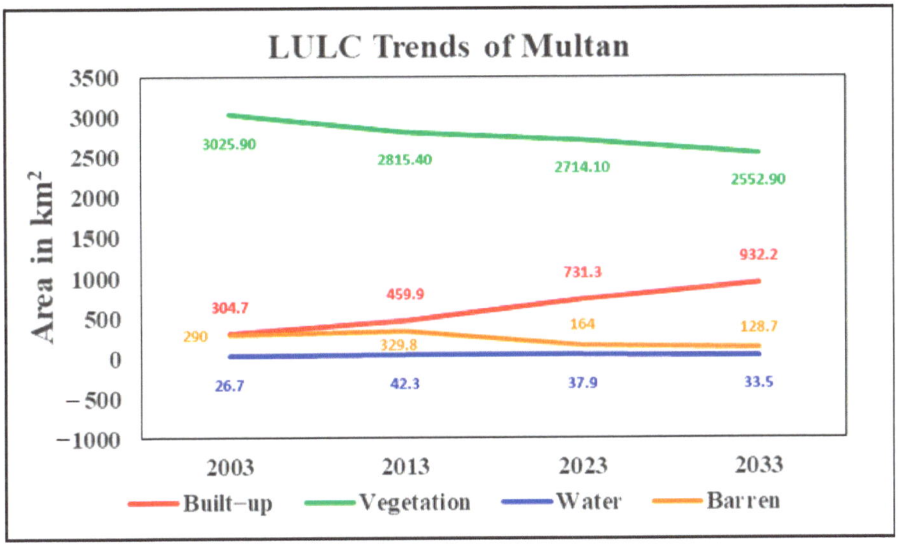

Figure 7. Comparison of LULC Classes of Multan in Trends.

Change detection from 2003 to 2013 showed a 155 km^2 (4.3%) increase in the built-up area due to city expansion. Vegetation decreased by 211 km^2 (5.76%), due to growth of housing and the development of new industrial zones. Barren land also increased by 39.8 km^2 (1.09%), partially due to agricultural land conversion and some areas cleared of trees for residential purposes. Surface water areas also increased by 0.43% due to higher rainfall and increased river water levels. Change detection between 2013 and 2023 shows a significant development in Multan, including industrialization and urbanization. Approximately 271 km^2 of land was converted to built-up land which is 7.4% of the total area. Vegetation has decreased by 101 km^2, or 2.78% of the total area, due to the conversion of orchards and crop fields into housing and developmental activities. Water-covered areas decreased due to lower river water levels and reduced rainfall. Barren land also decreased during this period to 166 km^2 (4.54% of the total area), with much of it converted into housing. A graphical representation of classes gaining and losing area within the specific time periods is given below in Figure 8.

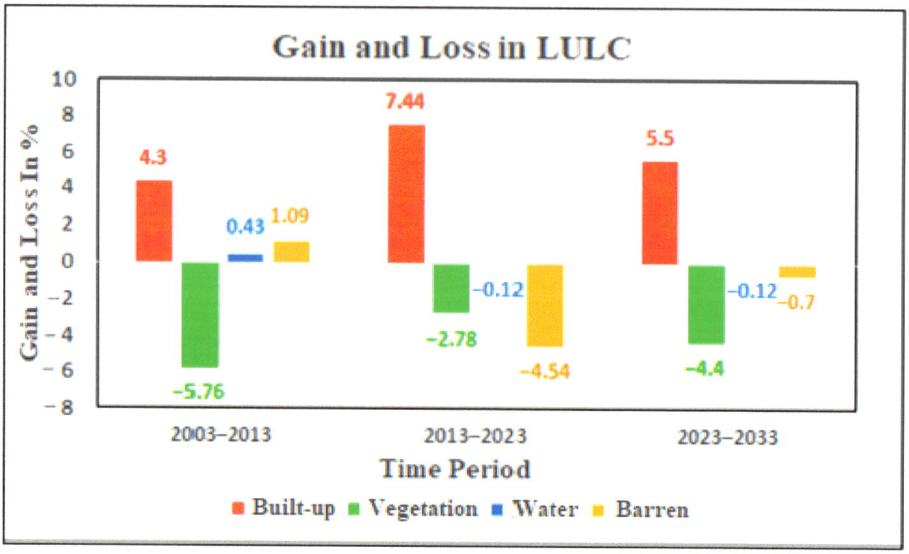

Figure 8. Gain and loss of LULC Classes of Multan in Percent.

3.3. LULC Maps of Bahawalpur District

The LULC maps of the district of Bahawalpur for 2003, 2013, and 2023 prepared using a supervised classification in GEE are shown below in Figure 9. This consists of four classes: built-up, vegetation, water, and barren. In 2003, the built-up, vegetation, water, and barren areas in the district of Bahawalpur covered 298.15 km^2, 4353.64 km^2, 13.62 km^2, and 17,688.14 km^2, respectively, which are 1.33%, 19.48%, 0.06%, and 79.13%, of the total area, respectively. The areas in square kilometers and their corresponding percentages are given below in Tables 5 and 6. The LULC for 2013, presented below in Figure 9 with Tables 5 and 6, shows that the built-up, vegetation, water, and barren areas of Bahawalpur District covered 830.29 km^2, 4056.16 km^2, 16.66 km^2, and 17,450.43 km^2 which are 3.71%, 18.15%, 0.075%, and 78.07% of the total area, respectively. Similarly, LULC for 2023 represents that the built-up, vegetation, water, and barren areas in Bahawalpur were 1184.26 km^2, 4333.29 km^2, 52.12 km^2, and 16,783.87 km^2, accounting for 5.30%, 19.39%, 0.23%, and 75.08% of the total area, respectively. All the classified results are summarized below in Tables 5 and 6 for a complete analysis of Bahawalpur District's LULC classes.

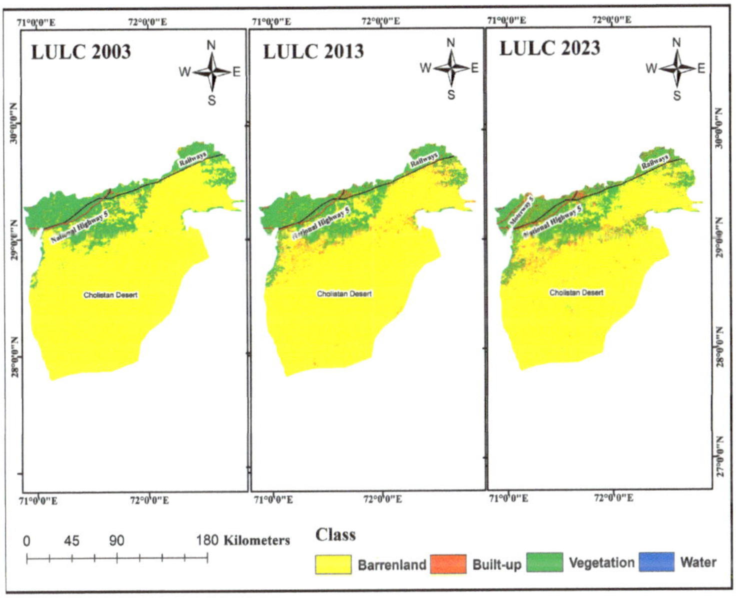

Figure 9. LULC Maps of 2003, 2013, and 2023 for Bahawalpur District.

Table 5. Area of each LULC class of Bahawalpur in km^2.

Class	Area in km^2				Temporal Area Changes in km^2		
	2003	2013	2023	2033	2003–2013	2013–2023	2023–2033
Built-up	298.2	830.3	1184.3	1296	532.14	353.97	111.62
Vegetation	4354	4056.2	4333.3	4390.3	−297.48	277.13	57.04
Water	13.6	16.7	52.1	55.6	3.04	35.46	3.63
Barren	17,688	17,450	16,784	16,612	−237.71	−666.56	−172.28
Total	22,354	22,354	22,354	22,354	0	0	0

Table 6. Area of each LULC class of Bahawalpur in Percentage.

Class	Area in %				Temporal Area Changes in %		
	2003	2013	2023	2033	2003–2013	2013–2023	2023–2033
Built-up	1.33	3.71	5.30	5.80	2.38	1.59	0.5
Vegetation	19.48	18.15	19.39	19.64	−1.33	1.24	0.25
Water	0.06	0.07	0.23	0.25	0.01	0.16	0.02
Barren	79.13	78.07	75.08	74.31	−1.06	−2.99	−0.77
Total	100	100	100	100	0	0	0

3.3.1. Driving Factors for Bahawalpur

Driving factors for the district of Bahawalpur such as slope, aspect, distance from roads, and distance from waterways or streams are presented below in Figure 10, which were prepared in the same way as prepared for Multan District using ArcGIS. Uniformity in the modeling approach was ensured by using the same driving factors in all three districts, and consistency is important for comparative analysis between them.

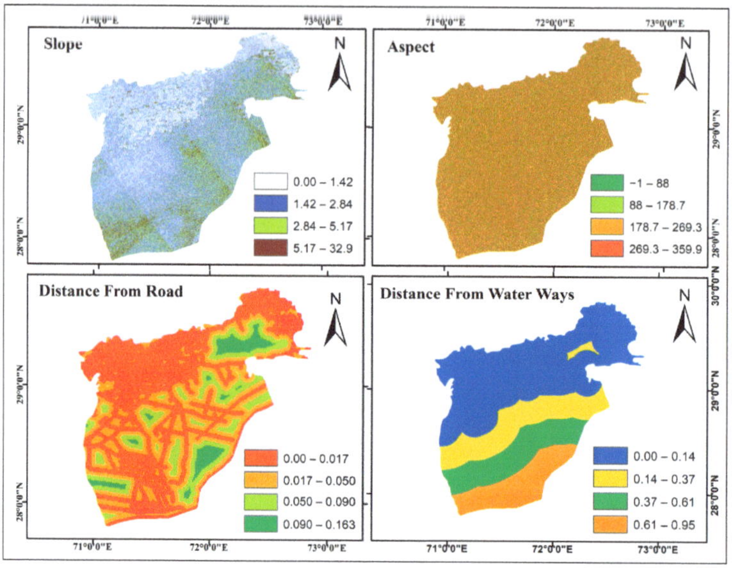

Figure 10. Driving Factors for Bahawalpur.

3.3.2. Projected LULC Map of Bahawalpur District

The Bahawalpur area projected map for 2033 is presented below in Figure 11. It was prepared in QGIS using the MOLUSCE plug-in with model accuracy of 89.13% and an overall kappa value of 0.70. Areas covered by each land use class are summarized below in Tables 5 and 6, which were calculated in ArcGIS. The result shows that the built-up, vegetation, water, and barren land areas are projected to expand to 1295.88 km^2 (5.80%), 4390.33 km^2 (19.64%), 55.75 km^2 (0.25%), and 16,611.59 km^2 (74.31%), respectively.

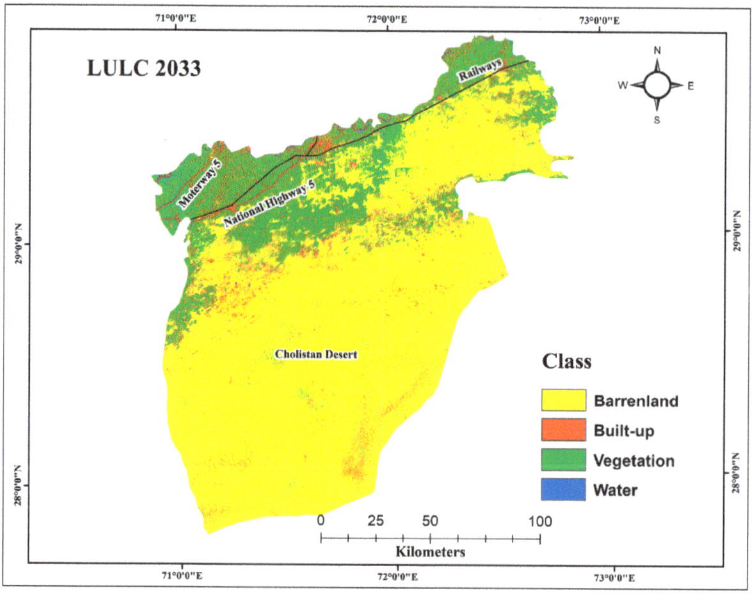

Figure 11. LULC 2033 of Bahawalpur.

3.3.3. Analysis of LULC Area of Bahawalpur District from 2003–2033

When the results of LULC for 2003 and for 2013 of the district of Bahawalpur were compared, there was a noticeable change found in the built-up area. In 2013, it increased to 532.14 km^2, which is 2.38% of the total area. The vegetation area is found to decrease to 297.48 km^2, which is 1.33% of the total area. The water area is found to increase to 3.04 km^2, which is 0.01% of the total area.

The barren area decreased by 237.71 km^2, which is 1.06% of the total area. By 2023, the built-up, vegetation, and water areas increased by 353.97 km^2 (1.59%), 277.13 km^2 (1.24%), and 35.46 km^2 (0.16%), respectively, as shown in Tables 5 and 6. Meanwhile, the barren land area decreased by 666.56 km^2 (1.06%) due to an increase in the built-up areas to accommodate housing requirements of the increased population. Meanwhile, vegetation areas decreased as they were converted into housing. A significant increase in vegetation was also observed in desert areas.

Barren land was decreased as areas were converted to built-up spaces and some regions were overtaken by vegetation. Orchard farming was introduced in the Thal Desert area using of drip irrigation systems. The area covered by each class in square kilometers and the percentage are described below in Tables 5 and 6. Graphical representations of data and trends in land use changes are shown in Figures 12 and 13, respectively.

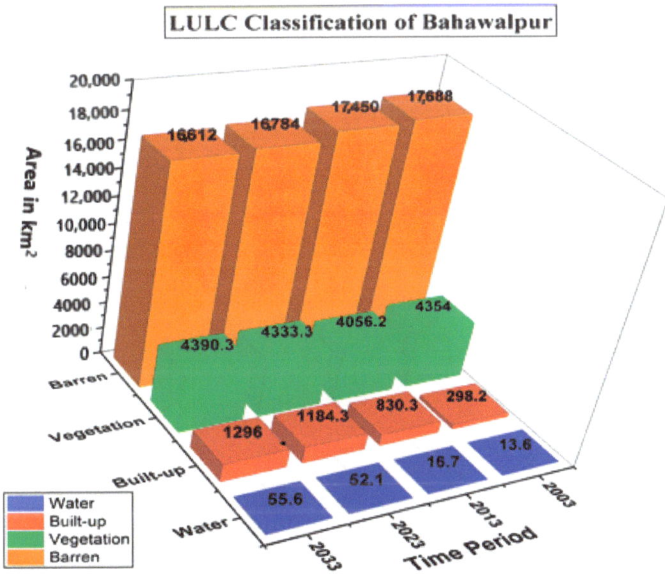

Figure 12. Graphical representation of LULC classes of Bahawalpur.

As shown in the gain and loss analysis in Figure 14, the built-up area increased by 2.38% and 1.59% from 2003–2013 and from 2013–2023, respectively. In the projected period of 2023–2033, it is expected to gain an additional 0.5% of the total area. The vegetation area decreased by 1.33% in the first decade but increased by 1.24% during the second one. In the projected years 2023–2033, it is anticipated to decrease again by 0.77%. On the other hand, the coverage area of water showed fluctuations, with an increase of 0.01% from 2003–2013 and 0.16% from 2013–2023. In the projected period of 2023–2033, it is expected to decrease again as summarized below in Figure 14.

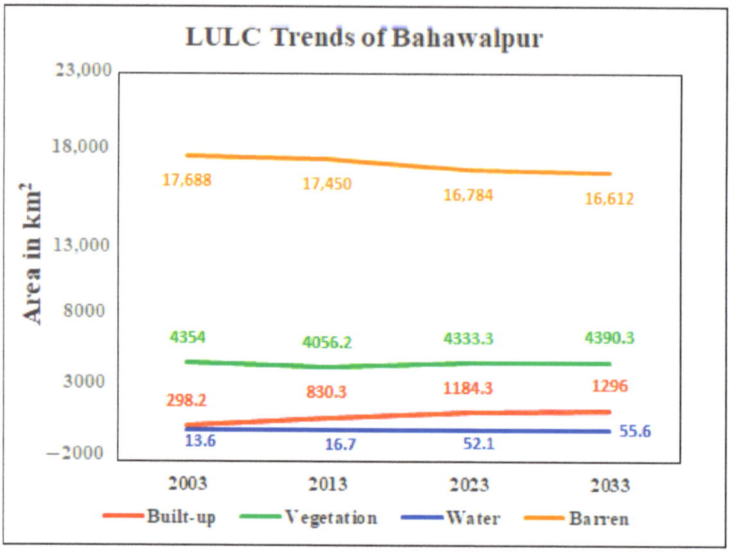

Figure 13. Trends of LULC classes of Bahawalpur.

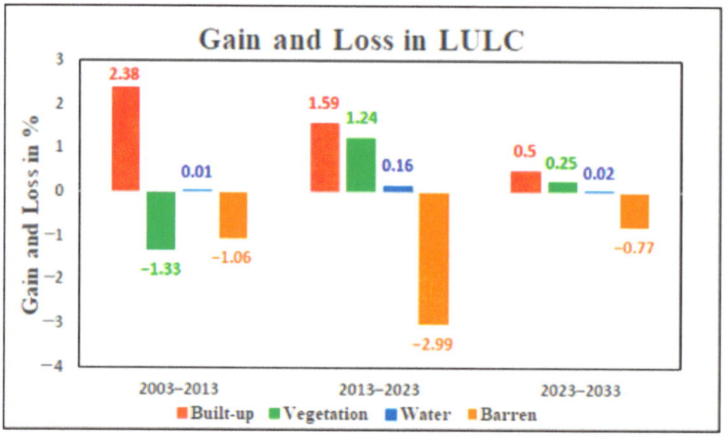

Figure 14. Gain and loss in percentage of Bahawalpur LULC classes.

3.4. LULC Maps of Dera Ghazi Khan District

The LULC maps for the years 2003, 2013, and 2023 for the district of D.G. Khan (Figure 15) were prepared in GEE using a supervised classification. The classified maps have five classes: built-up, vegetation, water, barren, and rock. Since the district of D.G. Khan also consists of rocky areas, rock was added in this study as a fifth class. It is observed that in 2003 the built-up, vegetation, water, barren, and rock areas of D.G. Khan District were spread over 367.81 km^2 (2.29%), 3618.28 km^2 (22.53%), 181.13 km^2 (1.13%), 5271.06 km^2 (32.82%), and 6623.26 km^2 (41.24%), respectively, as described below in Tables 7 and 8. Similarly, the LULC for 2013 (Tables 7 and 8) shows that the built-up, vegetation, water, barren, and rocky areas of D.G. Khan spanned over 624.91 km^2 (3.89%), 4717.08 km^2 (29.37%), 232.11 km^2 (1.45%), 3873.45 km^2 (24.12%), and 6613.99 km^2 (41.18%), respectively. The LULC for 2023 (Tables 7 and 8) reveals that the built-up, vegetation, water, barren, and rocky areas of D.G. Khan were 1325.01 km^2 (8.25%), 6261.42 km^2 (38.98%), 210.13 km^2 (1.31%), 2959.35 km^2 (18.43%), and 5305.63 km^2 (33.03%), respectively, of the total area.

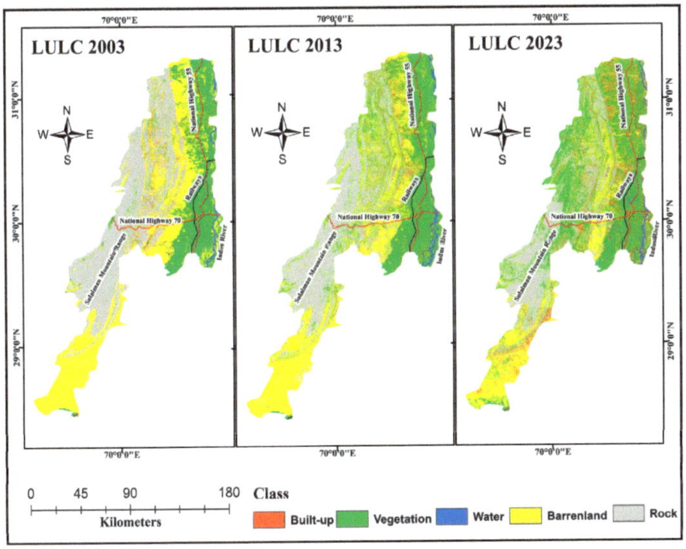

Figure 15. LULC maps of 2003, 2013, and 2023 for D.G. Khan.

Table 7. Area of each LULC class of D.G. Khan in km^2.

Class	Area in km^2				Temporal Area Changes in km^2		
	2003	2013	2023	2033	2003–2013	2013–2023	2023–2033
Built-up	367.81	624.91	1325	1960.5	257.1	700.1	635.5
Vegetation	3618.2	4717.1	6261.4	7181.9	1099	1544.3	920.5
Water	181.13	232.11	210.13	167.12	50.98	−21.98	−43.01
Barren	5271.1	3873.5	2959.4	1740	−1398	−914.1	−1219
Rock	6623.3	6613.8	5305.6	5011.9	−9.27	−1308	−293.7
Total	16,062	16,062	16,062	16,062	0	0	0

Table 8. Area of each LULC class of D.G. Khan in percentage.

Class	Area in %				Temporal Area Changes in %		
	2003	2013	2023	2033	2003–2013	2013–2023	2023–2033
Built-up	2.29	3.9	8.25	12.21	1.61	4.35	3.96
Vegetation	22.53	29.4	38.98	44.72	6.87	9.58	5.74
Water	1.13	1.45	1.31	1.04	0.32	−0.14	−0.27
Barren	32.82	24.12	18.43	10.83	−8.7	−5.69	−7.6
Rock	41.23	41.18	33.03	31.2	−0.05	−8.15	−1.83
Total	100	100	100	100	0	0	0

3.4.1. Driving Factors for D.G. Khan

The district of D.G. Khan also includes hilly areas and a hill shade factor was added along with other factors such as slope, aspect, distance from roads, and distance from waterways, as shown in Figure 16. These factors were added to enhance the prediction accuracy.

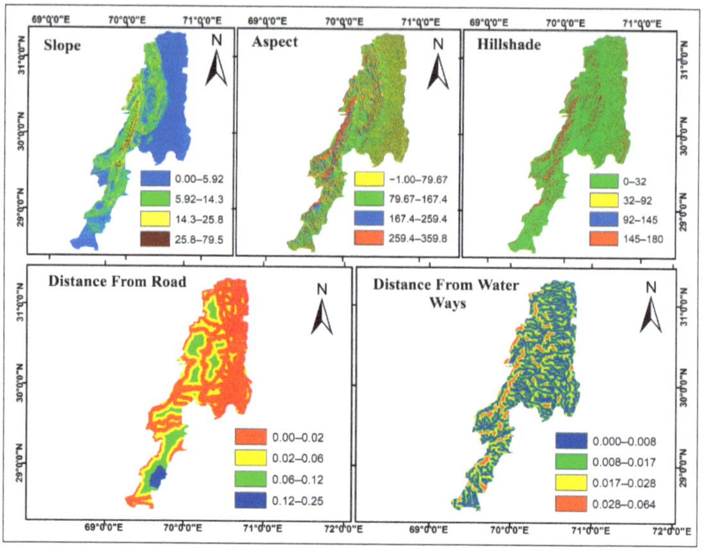

Figure 16. Driving factors for D.G. Khan.

3.4.2. Projected LULC Map of D.G. Khan District

The LULC map for 2033 (Figure 17) for D.G. Khan was predicted using MOLUSCE, achieving a correctness of 62.9% and an overall kappa value of 0.47. The data were further processed in ArcGIS to calculate the area covered by each class. By 2033, the built-up, vegetation, water, barren, and rocky areas are expected to expand over 1960.54 km^2 (12.21%), 7181.94 km^2 (44.72%), 167.12 km^2 (1.04%), 1740.04 km^2 (10.83%), and 5011.9 km^2 (31.20%), respectively. The areas for each class in square kilometers and percentages are presented in Tables 7 and 8, respectively.

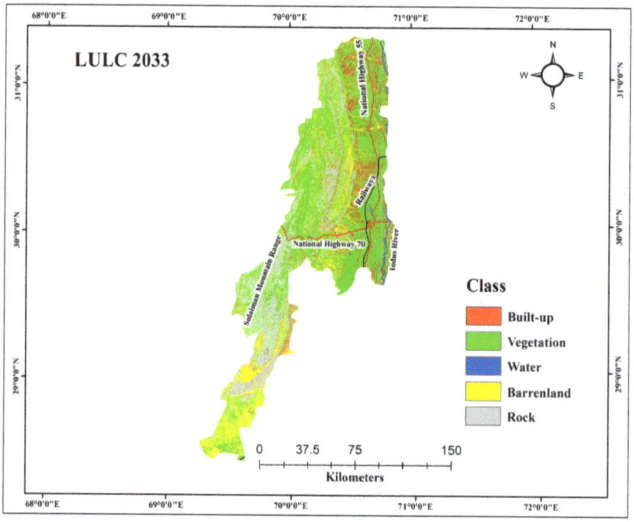

Figure 17. Predicted map of 2033 for D.G. Khan.

3.4.3. Analysis of LULC Area of D.G. Khan District from 2003–2033

An analysis of the LULC map is very important as it reveals the patterns of different classes of land use on the Earth's surface. It provides insight into urban growth and the impact of expansion on the environment. It aids in effective planning and policymaking to mitigate future environmental impacts. The analysis of the classified maps for D.G. Khan District, summarized below in Tables 7 and 8, shows a continuous increase in the built-up area from 2003 to 2023, with projections indicating that this trend persists in 2033, converting more than 4 times the area compared to 2003. This expansion in the built-up areas reflects the ongoing development and urbanization in the region. The vegetation area has also shown an increasing trend from 2003 to 2023, and this growth is expected to continue in 2033, driven by promotion of vegetation in rocky areas. The fluctuation in the water areas is attributed to variations in rainfall. The projected year result shows that, in the future, surface water will continue to decrease. The decrease in barren land is due to an increase in agriculture and built-up areas. Likewise, rocky areas are decreasing as vegetation is being promoted in these areas, and the growing business of rock crushing for gravel extraction is also contributing to this decline. A graphical analysis of classes with their changing trends is also shown below in Figures 18 and 19 for better understanding.

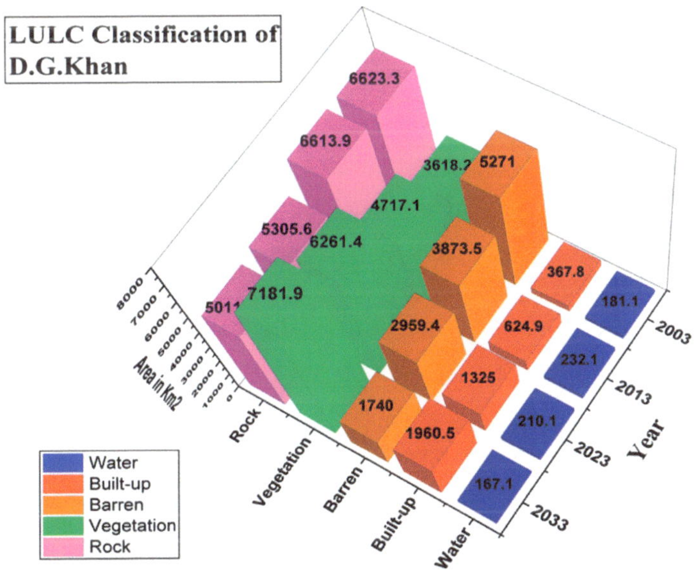

Figure 18. Graphical comparison of LULC classes of D.G. Khan.

The built-up area increased by 1.61% between 2003 and 2013 and by 4.35% from 2013 to 2023. In the projected years from 2023 to 2033, it is expected to gain an additional 3.96% of the area. The vegetation area also exhibited a positive trend, as shown in Figure 20, with an increase of 6.87% from 2003 to 2013 and 9.58% from 2013 to 2023. In the projected years 2023–2033, it is expected to gain an additional 5.74% of the area. Conversely, the barren, water, and rock areas have been steadily declining over time, and this trend is expected to continue in the projected period, with further decreases in the areas of these classes.

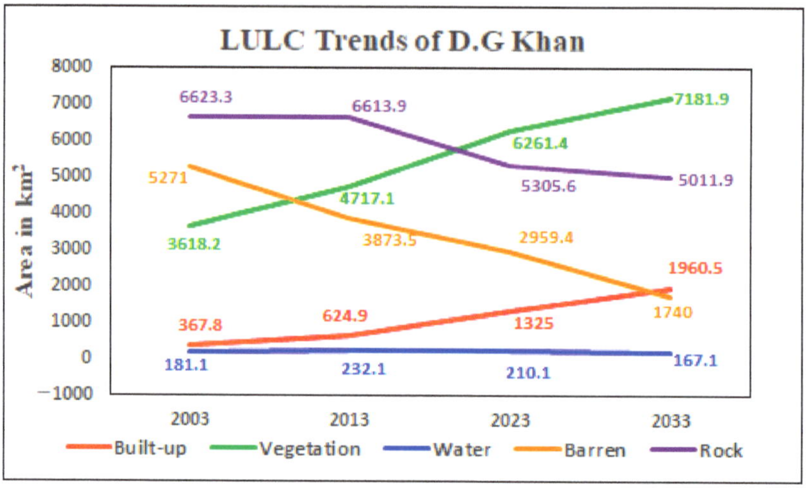

Figure 19. Trends of LULC classes in D.G. Khan.

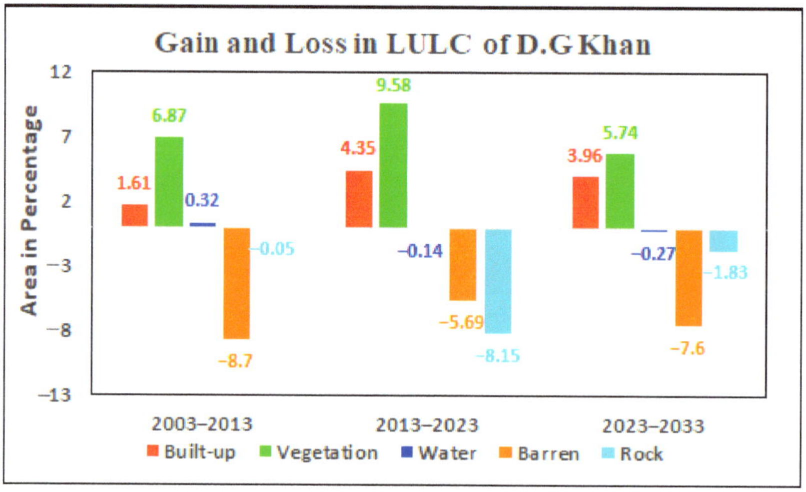

Figure 20. Gain and loss in percentage of LULC for D.G. Khan.

3.5. Projected Population in 2033

The projected population for 2033 was calculated by the compound annual growth rate method which is defined by Equation (8). The results of the method are summarized below in Table 9.

$$CAGR = (V_{final}/V_{initial})^{1/t} - 1 \tag{8}$$

where:

CAGR = Compound annual growth rate;
V_{final} = Final value;
$V_{initial}$ = Initial value;
t = Time in years.

Table 9. Population of study area.

District	1972	2003	2013	2017	2023	2033
Multan	1,506,223	3,586,158	4,403,152	4,746,166	5,362,305	6,564,063
Bahawalpur	1,071,026	2,758,376	3,408,946	3,669,176	4,284,964	5,527,616
D.G. Khan	686,057	2,237,170	2,649,371	2,872,631	3,393,705	4,480,548

3.6. *Comparative Analysis of Population and LULC Changes*

A comparative analysis of a city's population growth provided valuable insights into developmental activities in the form of urbanization and industrialization. It also highlights the environmental impacts resulting from changes in land use and land cover. The changes in growth rate and the percentage changes of other LULC classes were calculated by using Equation (9) as described below.

$$\text{Growth Rate} = (\text{Final Value} - \text{Initial Value})/\text{Initial Value} \times 100 \qquad (9)$$

3.6.1. Comparative Analysis of Population and LULC Changes in Multan

The percentage changes of all four land use classes of Multan were calculated using Equation (9), as described earlier, and results are presented in Table 10. To analyze how the land use classes of Multan District are affected by population growth and to predict the potential effect of future growth, the population growth rate was compared with the percentage changes in these classes. The study reveals that the population growth rates from 2003 to 2013 and from 2013 to 2023 were 22.7% and 21.8%, respectively. In the projected period from 2023 to 2033, the population of Multan District is expected to increase by 22.4%, based on the compound annual growth model. It was observed that although the population continuously increased from 2003 to 2033, the growth rate declined with each subsequent decade. The growth rate from 2013 to 2023 was impacted by COVID-19, which lead to increased mortality, reduced fertility rates, and less migration to urban areas. However, in the projected period of 2023 to 2033, a slight increase in growth rate reflects that these effects are diminishing. The built-up areas in Multan District continuously increased, with percentage changes of 50.9% from 2003 to 2013 and 58.9% from 2013 to 2023. In the projected decade of 2023 to 2033, the percentage change will have reached 27.4%, which is decreasing but still more than the population growth rate. Urban saturation, limited land availability, increasing property values, and higher land transfer fees will decrease the development of new housing and restrict the expansion of the built-up area in the projected years from 2023 to 2033. The percentage change in the built-up area was highest from 2013 to 2023 and is expected to keep increasing, though at a lower rate than in previous decades. The vegetation area in Multan consistently decreased, with a reduction of 6.9% from 2003 to 2013. Its percentage decreased to 3.6% from 2013 to 2023 and will continue to decrease to 5.9% in the projected years 2023 to 2033. The continuous reduction of vegetation cover poses an alarming situation for Multan District. Similarly, a percentage change of the other two classes is summarized below in Table 10.

The graphical analysis of the change rates in the district of Multan, comparing population growth with other land use classes, is presented in Figures 21 and 22 which are given below. Additionally, the trends illustrating the comparison between population growth and the changes in other classes are given below in Figure 23.

Table 10. Population growth rate and land use change analysis of Multan.

Time Period	Growth Rate Change in Percentage				
	Population	Built-Up	Vegetation	Water	Barren
2003–2013	22.7	50.9	−6.9	58.6	13.7
2013–2023	21.8	58.9	−3.6	−10.4	−50.3
2023–2033	22.4	27.5	−5.9	−11.95	−21.6

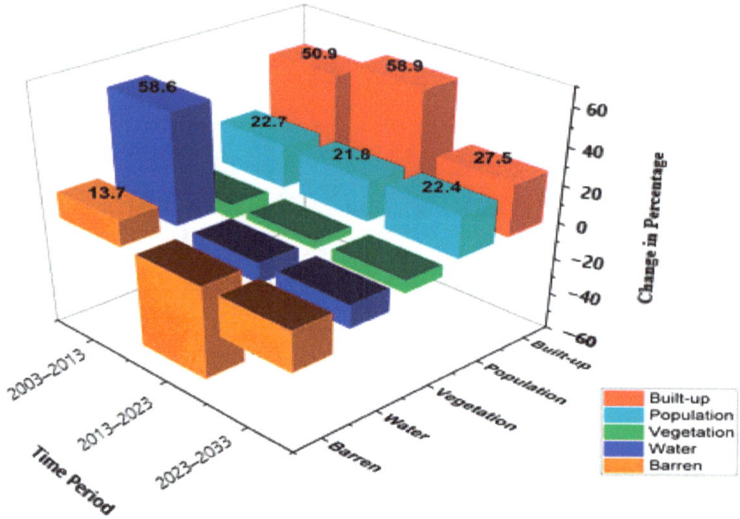

Figure 21. Analysis in 3D format.

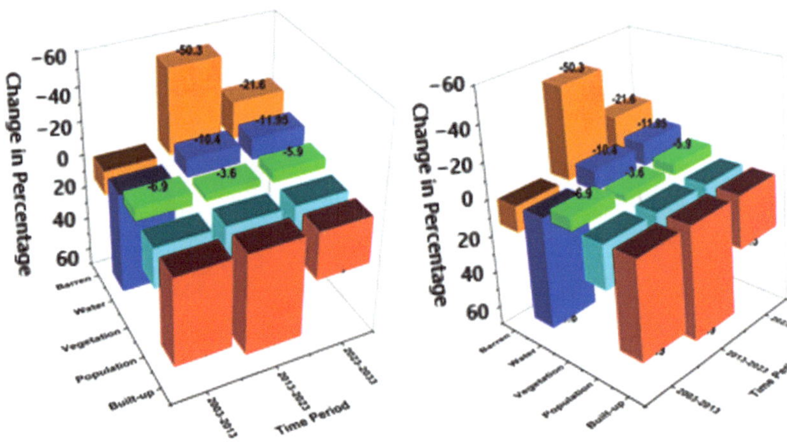

Figure 22. Population growth rate and land use change analysis of Multan.

The area of barren land initially increased between 2003 and 2013 at a rate of 58.6%, indicating the conversion of agricultural land into barren land for developmental activities. However, a sharp decrease was observed from 2013 to 2023 as the barren area was reduced at a rate of 101%, which was converted into built-up areas, and it will continue to decrease by 27.6% from the projected years 2023 to 2033. In the years 2003–2013, the areas covered in surface water increased by 13.7%, which is less than the population growth rate. They decreased more sharply from 2013 to 2023, at 50.3%, against the increase in population growth rate. In the projected years 2023 to 2033, the rate (−21.6%) again has a decreasing trend. The overall observation is that the population and built-up area are continuously increasing. Vegetation and barren areas are trending to shift into built-up areas. Although the population is increasing, the rate of population growth is less than the built-up expansion, which means an increase in urban sprawl and trends of unplanned urbanization. The vegetation area is more than all the other land cover classes, but its continuous reduction is trending towards development and urbanization. The percentage change of areas covered with surface water has a decreasing trend, and for an increasing population, the demand for water will be more in the future. To cover the demand of the population and to maintain agricultural sustainability, some serious steps need to be taken.

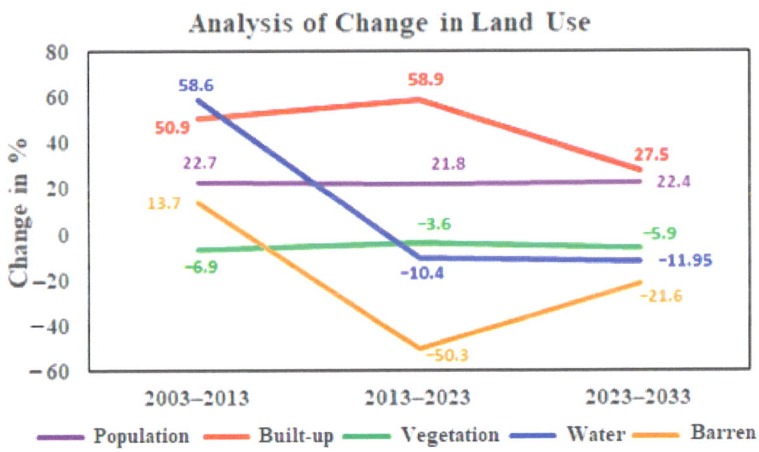

Figure 23. Changing trend in land use of Multan.

3.6.2. Comparative Analysis of Population and LULC Change in Bahawalpur

The study shows that the population growth rates are 23.6%, 25.7%, and 29% for the periods of 2003–2013, 2013–2023, and 2023–2033, respectively. It is observed that the built-up area percentage change has an increasing trend of 178.4% from 2003 to 2013, which is more than 7 times the population growth rate, reflecting massive development and urban sprawl. In the years 2013 to 2023, although the built-up change percentage (42%) decreased from the previous one, it was still more than the population growth rate. In the projected years 2023 to 2033, the built-up area rate (9.42%) appears to be lower than the population growth rate. The growth rate of areas covered with vegetation is less than the population growth rate. Initially, the vegetation percentage decreased in the years 2003 to 2013, which reflects the conversion of agricultural land to built-up areas. Vegetation increased in the years 2013–2023 to 6.83%, which is again less than the population growth rate but reflects some improved agriculture practices. In the projected years 2023 to 2033, vegetation appears to be in a decreasing trend both in terms of population and in the previous period. Growth rate and graphical analysis with comparison between different classes are expressed below in Table 11 and Figures 24 and 25, respectively.

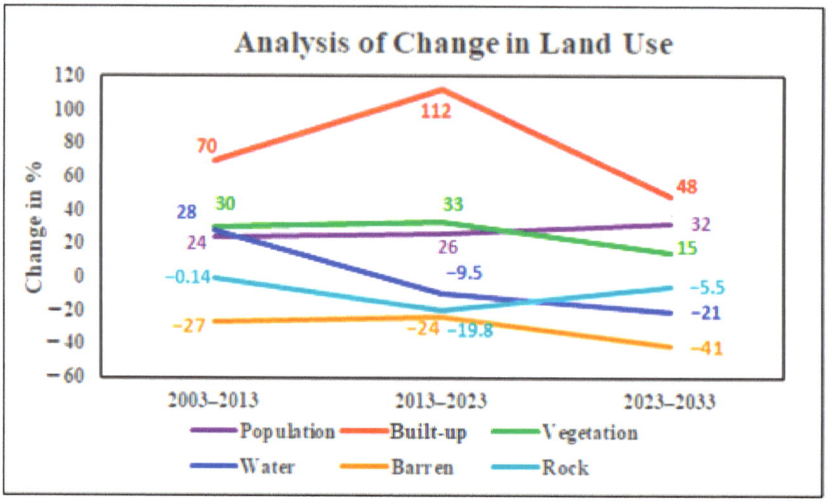

Figure 24. Trends in land use change of Bahawalpur.

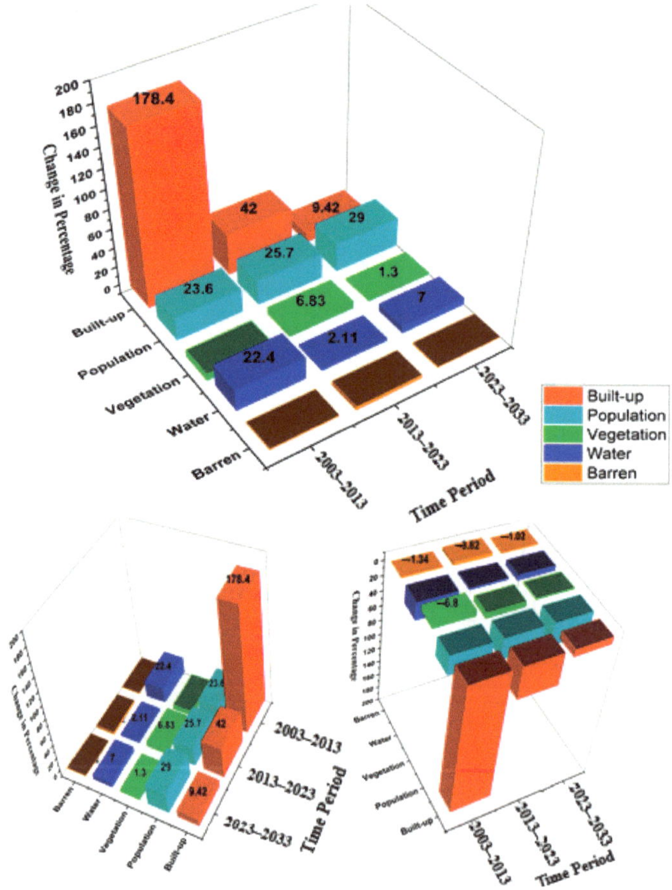

Figure 25. Population Growth rate and land use change analysis of Bahawalpur.

Table 11. Population growth rate and land use change analysis of Bahawalpur.

Time Period	Growth Rate Change in Percentage				
	Population	Built-Up	Vegetation	Water	Barren
2003–2013	23.6	178.4	−6.8	22.4	−1.34
2013–2023	25.7	42	6.83	2.11	−3.82
2023–2033	29	9.42	1.3	7	−1.02

Areas consisting of surface water reflect that in the years 2003 to 2013, the percentage change of water was 22.4%, which is near the population growth rate but, on the contrary, in the years 2013 to 2023, they decreased compared to the previous period. In the projected years 2023 to 2033, they appear to be increasing due to more rainfall. The barren area percentage change is in a decreasing trend as compared to the population growth rate and reflects that it is shifting towards built-up as well as vegetation areas.

3.6.3. Comparative Analysis of Population and LULC Change in D.G. Khan

A comparative analysis of population and LULC change in D.G. Khan is presented in Table 12. In the years 2003 to 2013, the population growth rate (18%) is less than the built-up (69%) as well as the vegetation (30%) percentage change, which indicates that both the built-up area and vegetation area are expanding more than the population growth rate. In the years 2013 to 2023, the built-up area expanded at a rate of 112%, which was four times greater than the population growth rate (28%). That sharp increase in the change in built-up areas is indicating massive development in the form of urban sprawl and industrialization. The vegetation percentage change (33%) is near the population growth rate but slightly high, which indicates plantations and improved green color. In the projected years 2023 to 2033, the built-up percentage change will be more due to the population growth rate, and more development in the form of urban sprawl will be observed. The change in the percentage of vegetation area will be less compared to the previous change, which will affect agriculture and climate. Proper planning and management are needed to improve vegetation in the future. This could be carried out by improving agriculture practices, promoting plantations, and decreasing the built-up area in the form of controlling urban sprawl.

Table 12. Population growth rate and land use change analysis of D.G. Khan.

Time Period	Growth Rate Change in Percentage					
	Population	Built-Up	Vegetation	Water	Barren	Rock
2003–2013	24	70	30	28	−27	−0.14
2013–2023	26	112	33	−9.5	−24	−19.8
2023–2033	32	48	15	−21	−41	−5.5

Graphical analysis of the population compared to percentage change in other classes is described below in Figure 26. The trend analysis of change is also presented below in Figure 27. Positive values show an increasing trend of a class, while on the other hand negative values show a decreasing trend.

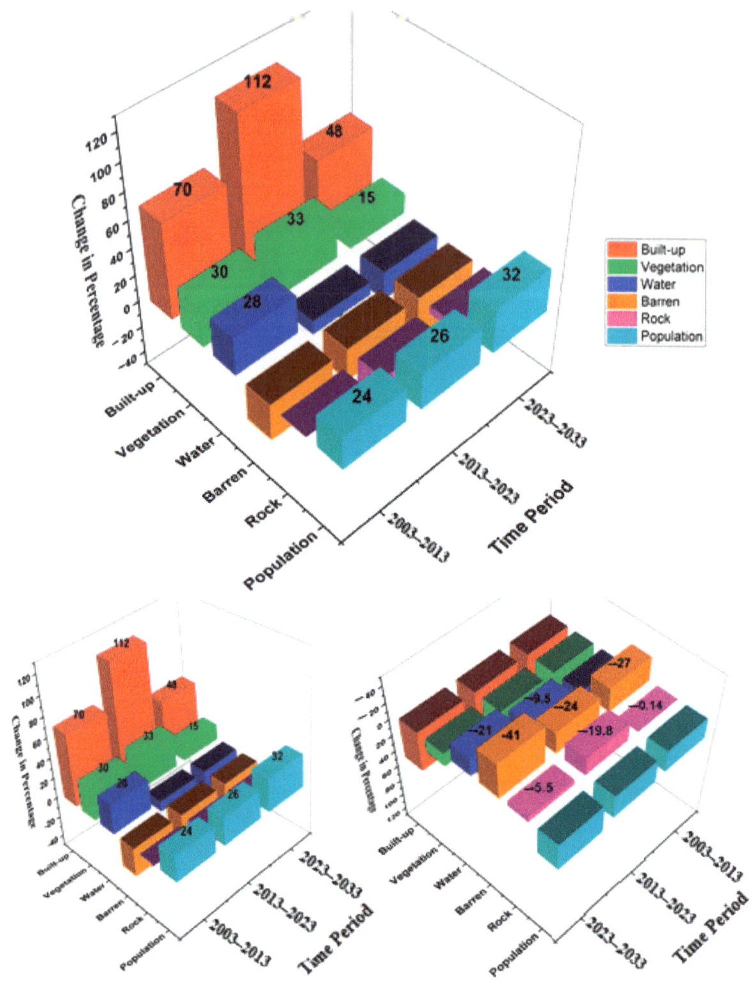

Figure 26. Population growth rate and land use change analysis of D.G. Khan.

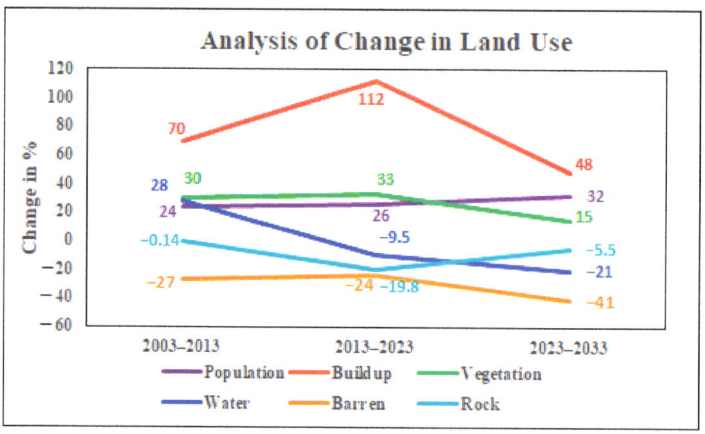

Figure 27. Trends in land use change in D.G. Khan.

3.7. Comparative Analysis of Population

The population growth rate increased for all three districts from 2003–2013, as shown in Figure 28 below. The population growth rate of D.G. Khan was the highest compared to Multan and Bahawalpur in this period. The population growth rate of Bahawalpur and D.G. Khan is continuously increasing while the population growth rate of Multan is decreasing from 2013–2023. In the projected years 2023–2033, the population growth rate of all three districts will be have an increasing trend.

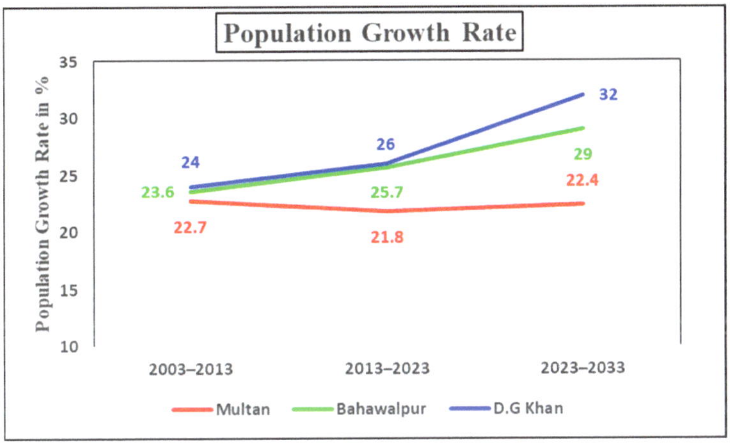

Figure 28. Population growth rate comparison of Multan, Bahawalpur, and D.G. Khan.

3.8. Comparative Analysis of Land Use Classes of Study Areas

The changes in the percentage of built-up, vegetation, water, and barren areas of all three districts were compared, and they are summarized below in Figure 29. The results showed that between the years 2003 and 2013, there was the highest development in built-up areas of Bahawalpur compared to Multan and D.G. Khan. From 2013–2023, the built-up area of Multan and D.G. Khan showed an increasing trend due to rapid urbanization during this period, migration from rural to urban areas, and lower property values. At the same time, Bahawalpur's built-up area again showed a decreasing trend during this period. From 2013 to 2023, the vegetation area of all three districts showed inflexion points as before 2013, particularly in 2010, there was a flood in the study area, which played a critical role in recharging the groundwater and improving the soil moisture level. These conditions were favorable for agriculture practices and vegetation growth in the years 2013 to 2023. Additionally, in the project "Punjab Irrigated Agriculture Improvement Project", the government also provided subsidies to farmers for the installation of drip irrigation, land development, the concrete lining of water courses, and purchasing laser land levelers for the promotion of crop per drop boosted vegetation cover. In the projected years 2023–2033, the expansion of the built-up areas in all three districts is decreasing due to high property values and skyrocketing material prices. Figure 29 reflects those areas consisting of surface water of Multan, which had the highest increasing trend compared to Bahawalpur and D.G. Khan from 2003–2013. This increment was due to floods and rainfall. The surface water of all three districts decreased between the years 2013 and 2023. In the projected years 2023–2033, areas consisting of surface water of Multan and D.G. Khan districts are decreasing while Bahawalpur is showing an increasing trend.

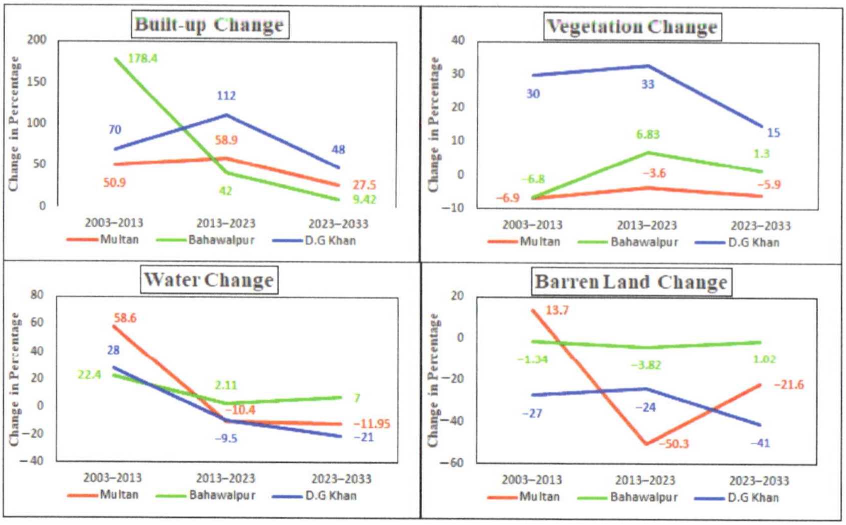

Figure 29. Comparison of percentage change in land use classes.

Vegetation areas of Multan and Bahawalpur decreased from 2003–2013, while vegetation areas of D.G. Khan increased in this period. Vegetation areas between the years 2013 and 2023 showed an increasing trend, while in the projected years 2023–2033, they showed decreasing trends for all three districts. The decrease in vegetation areas is due to shifting toward built-up areas. Barren land increased most significantly in Multan between 2003 and 2013. This increasing trend is due to cutting down trees and orchards while leaving land vacant for the construction of housing. The decreasing trend in barren areas showed that vacant areas were used for construction as well as vegetation. Fluctuation in surface water is due to fluctuations in rainfall.

4. Discussion

4.1. Important Findings

This study demonstrated and highlighted the significant trend and transformation in LULC of three districts of South Punjab: Multan, Bahawalpur and Dera Ghazi Khan, with classified maps for the last twenty years with ten-year intervals: 2003, 2013, and 2023. This long interval of ten years between adjacent classified maps helped to observe physical changes on the maps that are physically not observable while using short intervals. Additionally, satellite images were readily available for these years to ensure consistency and reliability in analysis. To ensure consistency and minimize the impact of seasonal variations on the analysis, satellite imagery for a period of six months (1st January to 30th June) of each year, i.e., 2003, 2013, and 2023, was used. GEE was used to calculate the NDVI values of all the images, and then average annual values were calculated for each year. Rapid urbanization was observed in Multan from 2013–2023 as conversion to built-up areas was higher in this decade than the previous one as well as the projected decade, which is 7.39%. However, the built-up area of Multan has a continuously increasing trend. The vegetation area of Multan is significantly decreasing from 82.9% to 70% between the years 2003 and 2033, which is an alarming situation for Multan to maintain its natural environment. In Bahawalpur, the built-up area modestly increased from 1.33% in 2003 to 5.80% in 2033 with relatively stable vegetation cover having a slight increase, from 19.48% to 19.64%. In Dera Ghazi Khan, the built-up area dramatically increased from 2.29% to

12.21%, which reflects rapid urbanization. Its vegetation area is also increasing from 22.53% in 2003 to 44.72% in 2033, which reflects the use of improved agricultural practices.

4.2. Limitations of the Study

In terms of vegetation, this study did not consider the details of further classification of vegetation. In built-up areas, details about buildings, industries, schools, and other infrastructure are missing. Similarly, the class of water could be further divided into surface and groundwater, but in this study only surface water was under consideration. The derived factors included in this study were confined to slope, aspect, distance from road, and distance from waterways. A more accurate map for the projected years could be obtained by using some more derived factors in MOLUSCE during prediction. MOLUSCE predicts projected years' maps on the basis of historical data and observed patterns. It does not guarantee that future predictions will be a hundred percent accurate. There will be uncertainty in future predictions as factors like government policies regarding land use, climate change, and unexpected population growth could highly influence land use and land cover. The LULC map classified by MOLUSCE could be compared with the classified map of a specific year to know the percentage of correctness. For projected years, with the absence of actual classified data, MOLUSCE could not directly find the percentage of correctness. Once classified data for projected years become available, they can be used to directly validate and find the percentage of correctness of the classified map.

4.3. Implications of the Findings

The findings of this study highlight the need for urban planning to control unplanned urban sprawl with sustainable agriculture practices to promote environmental sustainability.

5. Conclusions

Continuous LULC monitoring is crucial for optimal land use, urban planning, and environmental sustainability. This study has demonstrated the preparation of up-to-date LULC maps for the years 2003 to 2023 with ten-year intervals of Multan, Bahawalpur, and D.G. Khan districts and predicted future changes up to 2033 with RS and GIS techniques based on AI. This study demonstrated and highlighted the significant trend and transformation in LULC of three districts of South Punjab. The result of the study illustrated that, in Multan, the built-up area is continuously increasing and will increase from 8.36% in 2003 to 25.56% in 2033, while the vegetation area is continuously decreasing from 82.9% in 2003 to 70% 2033, indicating a loss of green spaces. The population has grown from 1.5 million in 1972 to 5.36 million in 2023 and will reach 6.56 million in 2033 which correlates with built-up areas. The rapid increase in population shifts land towards urban uses, and it is a threat to its agriculture sustainability. Results of Bahawalpur District revealed that built-up areas expanded modestly by 1.33% in 2003 to 5.80% in 2033. Vegetation cover remained relatively stable, with a slight increase from 19.48% in 2003 to 19.64% in 2033. Waterbody areas continuously increased from 0.06% to 0.25%, while, on the other hand, barren areas notably decreased from 79.13% to 74.31%. The population of Bahawalpur increased from 1.07 million in 1972 to 4.28 million in 2023 and will reach 5.52 million in 2033. The relatively small increase in built-up areas compared to the population highlights urban density. Results of D.G. Khan District showed that built-up areas dramatically increased from 2.29% in 2003 to 12.21% in 2033, with an indication of significant urbanization. Vegetation areas will increase from 22.53% in 2003 to 44.72% in 2033, suggesting afforestation and better land management practices. There was a steep decline in rock and barren areas, from 41.23% in 2003 to 31.2% in 2033 and 32.82% in 2003 to 10.83% in 2033, respectively. The population increased from 686,057 in 2003 to 3,393,705 in 2023, and in the projected year, it will reach 4,480,548, which indicates rapid urbanization and population growth. The findings revealed unplanned urban sprawl exceeding the population growth rate. The analysis revealed that urban expansion is replacing agricultural areas, contributing to climate change, inflation, and future food security challenges.

Author Contributions: Conceptualization, T.A. and M.S.; methodology, T.A., M.S. and R.A.; software, T.A.; validation, M.S., R.A., M.A.I.B. and I.A.; formal analysis, T.A., M.S., H.U.F. and M.U.A.; investigation, T.A. and M.U.A.; resources, M.S. and R.A.; data curation, T.A.; writing—original draft preparation, T.A., R.A. and M.S.; writing—review and editing, M.A.I.B., I.A., H.U.F. and M.U.A.; visualization, T.A. and R.A.; supervision, M.S. and R.A.; project administration, M.S., M.A.I.B., H.U.F. and M.U.A. All authors have read and agreed to the published version of the manuscript.

Funding: This research received no external funding.

Data Availability Statement: The original contributions presented in the study are included in the article, further inquiries can be directed to the co-corresponding author (M.S.).

Conflicts of Interest: The authors declare no conflicts of interest.

References

1. Muhammad, R.; Zhang, W.; Abbas, Z.; Guo, F.; Gwiazdzinski, L. Spatiotemporal Change Analysis and Prediction of Future Land Use and Land Cover Changes Using QGIS MOLUSCE Plugin and Remote Sensing Big Data: A Case Study of Linyi, China. *Land* **2022**, *11*, 419. [CrossRef]
2. Kondum, F.A.; Rowshon, M.K.; Luqman, C.A.; Hasfalina, C.M.; Zakari, M.D. Change analyses and prediction of land use and land cover changes in Bernam River Basin, Malaysia. *Remote Sens. Appl. Soc. Environ.* **2024**, *36*, 101281. [CrossRef]
3. Roy, S.K.; Alam, T.; Mojumder, P.; Mondal, I.; Al Kafy, A.; Dutta, M.; Ferdous, N.; Al Mamun, A.; Mahtab, S.B. Dynamic assessment and prediction of land use alterations influence on ecosystem service value: A pathway to environmental sustainability. *Environ. Sustain. Indic.* **2024**, *21*, 100319. [CrossRef]
4. Baghel, S.; Kothari, M.K.; Tripathi, M.P.; Singh, P.K.; Bhakar, S.R.; Dave, V.; Jain, S.K. Spatiotemporal LULC change detection and future prediction for the Mand catchment using MOLUSCE tool. *Environ. Earth Sci.* **2024**, *83*, 1–15. [CrossRef]
5. Rawat, J.S.; Kumar, M. Monitoring land use/cover change using remote sensing and GIS techniques: A case study of Hawalbagh block, district Almora, Uttarakhand, India. *Egypt. J. Remote Sens. Sp. Sci.* **2015**, *18*, 77–84. [CrossRef]
6. Beshir, S.; Moges, A.; Dananto, M. Trend analysis, past dynamics and future prediction of land use and land cover change in upper Wabe-Shebele river basin. *Heliyon* **2023**, *9*, e19128. [CrossRef]
7. Al Kafy, A.; Dey, N.N.; Al Rakib, A.; Rahaman, Z.A.; Nasher, N.M.R.; Bhatt, A. Modeling the relationship between land use/land cover and land surface temperature in Dhaka, Bangladesh using CA-ANN algorithm. *Environ. Chall.* **2021**, *4*, 100190. [CrossRef]
8. Mondal, M.S.; Sharma, N.; Garg, P.K.; Kappas, M. Statistical independence test and validation of CA Markov land use land cover (LULC) prediction results. *Egypt. J. Remote Sens. Sp. Sci.* **2016**, *19*, 259–272. [CrossRef]
9. Hashim, M.; Rahman, A.; Qasim, M.; Farooq, M.U.; Muneer, S.; Ahmed, Z. The spatio-temporal analysis of land use land cover changes in Multan city, Pakistan. *Nat. Appl. Sci. Int. J.* **2023**, *4*, 120–134. [CrossRef] [PubMed]
10. Rana, D.; Kumari, M.; Kumari, R. Land Use and Land Coverage Analysis with Google Earth Engine and Change Detection in the Sonipat District of the Haryana State in India. *Eng. Proc.* **2022**, *27*, 85. [CrossRef]
11. Al-Dousari, A.E.; Mishra, A.; Singh, S. Land use land cover change detection and urban sprawl prediction for Kuwait metropolitan region, using multi-layer perceptron neural networks (MLPNN). *Egypt. J. Remote Sens. Sp. Sci.* **2023**, *26*, 381–392. [CrossRef]
12. Guan, D.J.; Li, H.F.; Inohae, T.; Su, W.; Nagaie, T.; Hokao, K. Modeling urban land use change by the integration of cellular automaton and Markov model. *Ecol. Modell.* **2011**, *222*, 3761–3772. [CrossRef]
13. Jain, M. Future land use and land cover simulations with cellular automata-based artificial neural network: A case study over Delhi megacity (India). *Heliyon* **2024**, *10*, e34662. [CrossRef]
14. Alshari, E.A.; Abdulkareem, M.B.; Gawali, B.W. Classification of land use/land cover using artificial intelligence (ANN-RF). *Front. Artif. Intell.* **2022**, *5*, 964279. [CrossRef] [PubMed]
15. Wang, S.W.; Munkhnasan, L.; Lee, W.K. Land use and land cover change detection and prediction in Bhutan's high altitude city of Thimphu, using cellular automata and Markov chain. *Environ. Chall.* **2021**, *2*, 100017. [CrossRef]
16. Albano, R.; Sole, A.; Adamowski, J.; Perrone, A.; Inam, A. Using FloodRisk GIS freeware for uncertainty analysis of direct economic flood damages in Italy. *Int. J. Appl. Earth Obs. Geoinf.* **2018**, *73*, 220–229. [CrossRef]
17. Saharkar, U.R.; Shahapure, S.S. Identification of Land Use Land Cover (LULC) Changes Using Geo-Spatial Tools for Indrayani River Basin. *SSRG Int. J. Civ. Eng.* **2024**, *11*, 65–78. [CrossRef]

18. Sari, I.L.; Weston, C.J.; Newnham, G.J.; Volkova, L. Land cover modelling for tropical forest vulnerability prediction in Kalimantan, Indonesia. *Remote Sens. Appl. Soc. Environ.* **2023**, *32*, 101003. [CrossRef]
19. Nahib, I.; Turmudi, T.; Windiastuti, R.; Suryanta, J.; Dewi, R.S.; Lestari, S. Comparing of Land Change Modeler and Geomod Modeling for the Assessment of Deforestation (Case Study: Forest Area at Poso Regency, Central Sulawesi Province). *Int. J. Adv. Eng. Manag. Sci.* **2018**, *4*, 597–607. [CrossRef]
20. Munthali, M.G.; Davis, N.; Adeola, A.M.; Botai, J.O.; Kamwi, J.M.; Chisale, H.L.W.; Orimoogunje, O.O.I. Local perception of drivers of Land-Use and Land- Cover change dynamics across Dedza district, Central Malawi region. *Sustain.* **2019**, *11*, 832. [CrossRef]
21. Olaode, A.; Naghdy, G.; Todd, C. Unsupervised Classification of Images: A Review. *Int. J. Image Process.* **2014**, *8*, 325–342.
22. Ramadan, G.F.; Hidayati, I.N. Prediction and Simulation of Land Use and Land Cover Changes Using Open Source QGIS. A Case Study of Purwokerto, Central Java, Indonesia. *Indones. J. Geogr.* **2022**, *54*, 344–351. [CrossRef]
23. Notarangelo, N.M.; Hirano, K.; Albano, R.; Sole, A. Transfer Learning with Convolutional Neural Networks for Rainfall Detection in Single Images. *Water* **2021**, *13*, 588. [CrossRef]
24. Salahudin, H.; Shoaib, M.; Albano, R.; Inam Baig, M.A.; Hammad, M.; Raza, A.; Akhtar, A.; Ali, M.U. Using Ensembles of Machine Learning Techniques to Predict Reference Evapotranspiration (ET_0) Using Limited Meteorological Data. *Hydrology* **2023**, *10*, 169. [CrossRef]
25. Dammag, A.Q.; Jian, D.; Cong, G.; Derhem, B.Q.; Latif, H.Z. Predicting spatio-temporal land use/land cover changes and their drivers forces based on a cellular automated Markov model in Ibb City, Yemen. *Geocarto Int.* **2023**, *38*, 2268059. [CrossRef]

land

Article

Exploring the Impact of Climate Variables and Scenario Simulation on Ecosystem Service Value Profits and Losses in China

Dahai Yu [1] **and Chang You** [2,*]

[1] Faculty of Engineering and IT, University of Melbourne, Melbourne, VIC 3052, Australia; dahaiy@student.unimelb.edu.au

[2] School of Ethnology and Sociology, Minzu University of China, Beijing 100081, China

* Correspondence: 22400130@muc.edu.cn; Tel.: +86-18535003403

Abstract: Ecosystem restoration can yield multiple benefits, and the quantitative accounting of ecosystem service value (ESV) profits and losses is of significant importance to the economic benefits of ecosystem restoration. This study reveals the dynamic impacts of climate change on ESVs by analyzing the effects of climate variables on ESV profits and losses across different periods and scenarios. The research findings are as follows: (1) From 1990 to 2020, and extending to simulated projections for 2030, China's ESV exhibits a high distribution pattern in the southern regions. In 2030, under the natural development scenario (NDS), the southwestern region shows a coexistence of high and low ESVs. Under the ecological protection scenario (EPS), ESV in the southwestern region increases, whereas under the urban development scenario (UDS), ESV in the southwest decreases. (2) In both the NDS and UDS, the trends in ESV profits and losses continue from 2010 to 2020. Under the EPS, there is a significant increase in ESV in the southwestern region. The largest contributors to ESV loss are the conversion of grassland to unused land and forest to farmland. The southwestern region shows the most significant spatial differences in ESV profits and losses, with an increase in ESV profits in the northeastern region. In contrast, other regions show no significant spatial differences in ESV profits and losses. (3) From 1990 to 2000, Bio13 (the precipitation of the wettest month) and Bio12 (annual precipitation) had a significant positive impact on ESV profits and losses, indicating that increased precipitation promotes the functioning of ESVs. This study indicates that fluctuations in precipitation and temperature are significant climate factors influencing the value of ESV. Due to climate change, precipitation patterns and temperature swings are now key determinants of ESV changes. By carefully studying ESV profits and losses and their driving factors, this research can serve as the scientific basis for ecosystem restoration and management strategies.

Keywords: deep learning; profits and losses; climate change; scenario simulation

1. Introduction

Ecosystem services (ESs) serve as an essential link between human society and the natural environment, acting as a catalyst to promote sustainable development [1]. One key indicator used to assess ecosystem health and function is its ecosystem service value (ESV) [2]. ESV refers to all tangible and intangible benefits that ecosystems provide to human society, such as cleaner air and water purification systems, recreation opportunities, and recreational services [3]. Their supply is essential to economic and societal progress [4]. Climate change and pollution have greatly diminished ES supplies; human activities now exceed ecosystem carrying capacities [5], necessitating the quantification of ESV profits and losses as an essential step toward effective ecosystem restoration and management [6]. Environmental degradation, pollution, climate change, and ecological degradation have had a devastating impact on ESV's profits and losses, with climate change being particularly significant [7]. Climate variables such as precipitation and temperature exert considerable influence on the functionality of ESs [8]. Climate change can lead to notable changes in ESV across different periods and scenarios [9]. For instance, increased precipitation may enhance ecosystem functionality by boosting water resource availability and promoting plant growth, thereby improving both productivity and stability in an ecosystem [10]. Conversely, temperature fluctuations can negatively impact ESV by increasing plant transpiration and soil moisture evaporation rates, reducing water resource availability [11]. Climate change is widely recognized as a major determinant of ESV changes, as the stability and variability of precipitation and temperature are crucial factors [12]. Moreover, the effects of climate change are not limited to these variables; other environmental factors, such as wind speed and humidity levels, also play a role [13]. Extreme weather events like flooding, drought, and heatwaves have downstream and lasting impacts on ecosystems, significantly altering their structure and function and subsequently affecting ESV [14]. For example, flooding can cause soil erosion and deteriorate water quality, threatening purification services; droughts reduce water supply, negatively impacting agricultural production and drinking water availability; heat waves affect both human health and biodiversity. These events, driven by climate change, necessitate further research to develop effective adaptation and mitigation strategies [15].

To better comprehend and predict the impact of climate change on ESV, many scholars have conducted extensive research [16]. Watson et al. (2020) used general circulation models to simulate 20 climate scenarios with projected shifts in Koppen–Geiger climate classifications. They combined global ESVs with 20 climate scenarios and utilized these global ESVs as indicators to track ESV distribution during transitions among climate categories [17]. Yang et al. (2021) also explored this subject matter extensively [18]. Wilkes et al. (2024) assessed climate change's impacts on biodiversity and ecosystem services in China, concluding that climate change is having a substantial effect on productivity, species interactions, biological invasions in northern agropastoral ecotones, as well as fragile ecosystems of China [19]. Asmus et al. (2019) created a framework to investigate species' vulnerability to climate change, their roles in food webs, contributions to ESs, and overall persistence despite climate-induced species loss [20]. Leal Filho et al. (2021) studied the effects of climate change on African ESV through case studies from different African countries in order to gain clarity on its destructive impact [21]. Furthermore, Zhang et al. (2022) created a network between climate forcings and ecological systems in order to analyze how climate change is driving change within ecologies [22]. Sharma et al. (2023) pointed out that climate change and biodiversity loss pose significant systemic risks to humanity and proposed an integrated approach to addressing climate change through the circular economy [23]. Rahim et al. (2024) studied potential ESs for climate change adaptation [24]. However, these studies mainly focus on climate scenario simulations and biodiversity assessments on a global or specific regional scale, lacking detailed dynamic analysis of ESV profits and losses in China. In contrast, our study not only reveals the dynamic impacts of climate change on ESV within China but also analyzes the key roles of specific climate variables, such as precipitation and temperature fluctuations on ESV. We found that the precipitation of the wettest month (Bio13) and annual precipitation (Bio12) had a significant positive impact on ESV profits and losses during the period from 1990 to 2000, while temperature seasonality (Bio4) showed a negative impact. These findings are consistent with global trends observed in previous studies but provide more specific regional data. Additionally, we discovered that in future scenario simulations, ESV significantly increases under the ecological protection scenario (EPS), whereas it decreases under the urban development scenario (UDS). These findings align with those of Wilkes et al. (2024) in Northern China but further detail the spatiotemporal variations under different scenarios [19]. The innovation of this study lies in its detailed analysis of the specific impacts of climate change on ESV profits and losses across different regions and scenarios within China, filling a gap in previous research that lacked detailed regional analyses. By employing various scenario simulation methods, our study not only validates the conclusions of previous research but also provides more localized scientific evidence. This evidence offers more targeted guidance for future ecosystem restoration and management strategies. The results of our study will provide policymakers with scientific evidence to help them formulate more effective climate change adaptation and mitigation strategies, promoting ecological sustainability.

While studies based on historical data can provide a theoretical foundation for current land planning and ecosystem management, predicting future ESV changes based on existing land use and analyzing the impact of climate factors on ESV changes can offer early warnings. This can prompt us to preemptively mitigate potential future risks and help us take more robust actions in the future. Our study employs several advanced methods, including the PLUS (Patch-Generating Land Use Simulation) model, LSTM (Long Short-Term Memory) model, and SHAP (Shapley Additive Explanations) method, to dynamically analyze the impact of climate change on historical and future ESV profits and losses. The PLUS model combines the random forest algorithm and Markov Chain model, enabling the precise capture of the spatial heterogeneity and temporal dynamics of LUCs. Lu et al. (2022) utilized the PLUS model to simulate future land spatial patterns under different development scenarios in the lower Yellow River, revealing the spatiotemporal differentiation of ESV and providing references for optimizing land use and ecological protection [25]. Zhu et al. (2024) used the PLUS model to analyze the driving factors of ecosystem function changes, identifying gross domestic product (GDP) and population density as the main social factors influencing wetland area changes [26]. Zhang et al. (2024) integrated the InVEST-PLUS models to evaluate and predict changes in ecosystem carbon storage under different land use scenarios in the Chengdu metropolitan area of Central and Western China by 2050 [27]. Evidently, applying the PLUS model not only simulates LUCs under different future scenarios but also provides data support for formulating rational land use planning. However, these studies primarily focus on simulating LUCs without deeply exploring the specific impacts of climate variables such as precipitation and temperature fluctuations on ESV. Our study not only employs the PLUS model but also integrates the LSTM model and SHAP method to offer a more comprehensive analysis of the dynamic impact of climate change on ESV, thereby filling this research gap.

In handling time series data, the LSTM model excels. LSTM is a special recurrent neural network that effectively addresses the gradient vanishing and exploding problems in traditional RNNs for long time-series data by introducing forget gates, input gates, and output gates [28]. Our study utilizes the LSTM model to analyze the time series data of climate variables, providing more accurate dynamic analysis by predicting ESV changes under future climate scenarios. Deng et al. (2024) constructed four LSTM-based models, LSTM, Stack-LSTM, Bi-LSTM, and CNN-LSTM, to assess hydrological changes under future climate change [29]. Yang et al. (2023) explored the feasibility of improving flow simulation performance in the tropical Kelantan River Basin (KRB) of Peninsular Malaysia by combining the hydrological model SWAT (Soil and Water Assessment Tool) with the deep learning model Bi-LSTM (Bidirectional Long Short-Term Memory) [30]. To interpret the outputs of the LSTM model [31], this study employs the SHAP method. SHAP can decompose the model's output into contributions from each input feature, revealing the independent impact and interactions of each feature on the prediction results [32]. In this study, the SHAP method is used to interpret the outputs of the LSTM model, helping to identify the independent impacts and interactions of climate variables on ESV changes. Huang et al. (2024) transformed four meteorological features—precipitation, temperature, relative humidity, and wind speed—into six hydrological features and used an LSTM-SHAP composite model to analyze the potential relationships between feature variables and the target variable (streamflow), predicting daily streamflow in the Sandu River Basin (SRB) [33]. Sun et al. (2024) established a regionally applicable multi-output LSTM prediction model to determine the performance of semi-rigid asphalt pavement and conducted a stochastic analysis of the model based on SHAP, providing references for preventative maintenance decision making for China's highway network [34]. By comprehensively employing these methods, we not only verify the conclusions of previous research but also provide more localized data support, revealing regional differences and the specific impacts of climate factors on ESV, which are rarely seen in the existing literature [35]. The application of the PLUS model [36], LSTM model [37], and SHAP method [38] allows for a comprehensive and systematic simulation and interpretation of the impacts of climate change on ESV profits and losses. The PLUS model is used to simulate LUCs, the LSTM model predicts the dynamic impacts of climate variables on ESV, and the SHAP method interprets the model's prediction results. The combined application of these methods not only validates prior research conclusions but also provides localized data support, offering more targeted guidance for future ecosystem restoration and management strategies. The uniqueness of this study lies in its detailed analysis of the specific impacts of climate change on ESV profits and losses across different regions and scenarios within China, filling a gap in regional detailed analysis. Using multi-scenario simulation, this study provides scientific evidence for policymakers to help formulate more effective climate change adaptation and mitigation strategies, promoting ecological sustainability.

By simulating ESV profits and losses under different climate scenarios from 1990 to 2030, this study aims to better predict the potential impacts of future climate change on ESV profits and losses. The research will address the following scientific questions: (1) What is the spatiotemporal heterogeneity of ESV profits and losses in China under different times and scenarios? (2) Which climate factors have the greatest impact on ESV under different climate scenarios? (3) How can multi-scenario climate change simulations provide scientific evidence for future ecosystem restoration and management? By gaining an in-depth understanding of the impact of climate change on ESV profits and losses, this study can provide more precise guidance for ecosystem restoration and management. It will provide policymakers with scientific evidence to facilitate the formulation of more effective climate change adaptation and mitigation strategies. The research results not only reveal the dynamic impacts of climate change on ESV profits and losses but also offer scientific evidence for future ecosystem management and policymaking. The ultimate goal is to provide pathways for achieving ecological sustainability.

2. Materials and Methods

2.1. Study Area

China is a vast country (Figure 1) with a land area of 9.6 million square kilometers. The country spans several climatic zones from north to south, including subrigid, temperate, warm temperate, subtropical, and tropical zones. China's climate types vary widely, ranging from the subrigid monsoon climate in the northeast to the tropical monsoon climate in the south. These regions exhibit significant climatic differences, with uneven distributions of precipitation and temperature. China's ecosystems include forests, grasslands, wetlands, deserts, and farmlands that serve essential functions like water conservation, soil retention, biodiversity protection, and climate regulation [39]. Over recent years, climate change's impact on China's ecosystems has become ever more obvious. Alterations to precipitation patterns and more frequent extreme weather events threaten their stability and sustainability, endangering their very existence as ES. Human activities, including urbanization, agricultural expansion, and industrial development, have had profound effects on ecosystems, leading to habitat fragmentation, biodiversity loss and degradation in ecosystem functions. This research seeks to reveal the dynamic impacts of climate change on ESV profits and losses and provide scientific evidence for ecosystem restoration and management.

Figure 1. Overview of the study area.

2.2. Data Sources

We carefully selected high-quality data from reputable sources (Table 1). The land use data were reclassified into 6 categories (farmland, forest, grassland, water, built-up land, and unused land) using ArcGIS Desktop 10.8. This ensures consistency and comparability across different periods (1990, 2000, 2010, and 2020). The agricultural statistics data are standardized to ensure consistency and accuracy, including grain production, prices, and planting areas for the years 1990, 2000, 2010, and 2020. Digital elevation model (DEM) data were processed to generate slope data, unified to 1 km resolution to match other datasets. Annual precipitation and the average temperature data were resampled to 1 km resolution to ensure consistency across different climate variables. Population density and regional average GDP data were also resampled to 1 km resolution to ensure consistency. Vector data for rivers, roads, and railways were used to calculate the Euclidean distance from each 1 km × 1 km grid center to the nearest river, road, and railway using ArcGIS 10.8. This method provides a robust representation of distances, which is crucial for accurately analyzing spatial relationships.

This study analyzed the impact of climate variables on ESV profits and losses and applied the SHAP method to reveal the nonlinear effects of climate variables on ESV profits and losses changes. To ensure the stability and reliability of the analysis, we utilized high-resolution climate data from WorldClim (https://worldclim.org/data/worldclim21.html (accessed on 20 January 2024)), provided in raster format with 1 km resolution (Supplementary Material S1). To avoid the influence of multicollinearity, we calculated the correlation coefficients between variables and set a threshold to exclude highly correlated variables. Additionally, we removed variables with low contribution rankings to simplify the model and enhance its interpretability and predictive capability. We ultimately retained the following climate variables: Bio2 (mean diurnal range), Bio4 (temperature seasonality), Bio12 (annual precipitation), Bio13 (the precipitation of the wettest month), Bio15 (precipitation seasonality), Bio16 (the precipitation of the wettest quarter), Bio18 (the precipitation of the warmest quarter), and Bio19 (the precipitation of the coldest quarter).

Table 1. Details of the data and its pre-processing steps.

Data Type	Data Source	Format	Resolution	Time Range	Pre-Processing Steps
Land use data	Chinese Academy of Sciences Resource and Environment Science Data Center (http://www.resdc.cn/ (accessed on 20 January 2024))	Raster	1 km	1990, 2000, 2010, 2020	Reclassified into six categories: farmland, forest, grassland, water, built-up land, and unused land using ArcGIS 10.8.
Agricultural statistics data	China Rural Statistical Yearbook (https://www.stats.gov.cn/ (accessed on 20 January 2024))	Table	-	1990, 2000, 2010, 2020	Standardized data processing, including grain production, prices, and planting areas.
Digital elevation model	http://www.gscloud.cn/ (accessed on 20 January 2024)	Raster	1 km	-	Generated slope data, unified resolution to 1 km.
Annual precipitation	http://www.geodata.cn/ (accessed on 20 January 2024)	Raster	1 km	-	Resampled data, unified resolution to 1 km.
Annual mean temperature	http://www.geodata.cn/ (accessed on 20 January 2024)	Raster	1 km	-	Resampled data, unified resolution to 1 km.
Population density	http://www.resdc.cn/ (accessed on 20 January 2024)	Raster	1 km	-	Resampled data, unified resolution to 1 km.
Regional average gross domestic product	http://www.resdc.cn/ (accessed on 20 January 2024)	Raster	1 km	-	Resampled data, unified resolution to 1 km.
River data	http://www.resdc.cn/ (accessed on 20 January 2024)	Vector	-	-	-
Major road data	http://www.resdc.cn/ (accessed on 20 January 2024)	Vector	-	-	-
Railway data	http://www.resdc.cn/ (accessed on 20 January 2024)	Vector	-	-	-
Distance to rivers, major roads, and railways	-	Raster	1 km	-	Calculated Euclidean distance using ArcGIS 10.8 based on 1 km × 1 km grid to determine the distance from each grid centroid to the nearest river, road, and railway.

2.3. Calculation of the Ecological Contribution of LUC to ESV Profits and Losses

We adopt the standard equivalent factor value calculation method proposed by Xie Gaodi et al. [40,41], using the net profit of grain production per unit area of farmland as the *ESV* of one standard equivalent factor. Rice, wheat, corn, and soybeans are the four main grains in China [42]. Considering the fluctuations in the value and planting area of these four grains across different years, we use the average net profit of these four grains from 1990 to 2020 as the standard equivalent factor value in this study. The calculation formula is as follows [43]:

$$V_a = \frac{1}{7}\sum_{i=1}^{n} \frac{a_i p_i q_i}{A} (i = 1, 2, \cdots, n) \tag{1}$$

$$VE_{ij} = C_{ij} V_a \ (i,\ j = 1, 2, \cdots, n) \tag{2}$$

$$ESV = \sum A_k E_k \tag{3}$$

where V_a represents the economic value of crops per unit area in China; i represents the type of crop, p_i represents the current year's price for the ith crop; q_i represents the yield per unit area of the jth crop; a_i represents the total planting area of the ith crop; and A represents the total planting area of the four types of crops. VE_{ij} represents the ESV coefficient of the jth ecosystem service function contained in the ith ecosystem; C_{ij} represents the economic value of the jth service in the ith ecosystem relative to per unit area of farmland; V_a represents the economic value generated by crops per unit area; ESV represents the total ESV; A_k represents the area of the kth land type; and E_k represents the ESV corresponding to per unit area of the kth land type. The revised ESVs are shown in Table 2.

Table 2. Revised ecosystem service value of different land use types (USD/hm^2).

Function	Farmland	Forest	Grassland	Water	Built-Up Land	Unused Land
FP	175.78	40.17	37.12	69.47	0.00	0.80
RM	38.97	92.27	54.62	38.71	0.00	2.39
WS	−207.60	47.72	30.23	691.47	0.00	1.59
GR	141.58	303.45	191.96	151.13	0.00	10.34
CR	73.97	907.95	507.47	340.96	0.00	7.95
ED	21.48	266.06	167.57	493.68	0.00	32.61
HA	237.83	594.17	371.72	7084.40	0.00	19.09
SC	82.72	369.47	233.85	171.81	0.00	11.93
NC	24.66	28.24	18.03	13.26	0.00	0.80
BD	27.04	336.46	212.64	553.07	0.00	11.14
AL	11.93	147.55	93.86	355.81	0.00	4.77
Aggregate	628.37	3133.49	1919.04	9963.76	0.00	103.40

Note: FP: food production; RM: raw material production; WS: water supply; GR: gas regulation; CR: climate regulation; ED: environment depuration; HA: hydrological adjusting; SC: soil conservation; NC: nutrient cycle maintenance; BD: biodiversity; AL: esthetic landscape.

Furthermore, the following formulas are used to calculate the profits and losses of ESV and the ecological contribution of LUC to these profits and losses. The specific formulas are as follows [44,45]:

$$LP_{ik} = \sum A_{ij} \times \left(VC_{ik} - VC'_{ik}\right) \quad (4)$$

$$EL_{ij} = \frac{\left(VC_{ik} - VC'_{ik}\right) \times A_{ij}}{\sum_{i=1}^{n} \sum_{j=1}^{n} \left[\left(VC_{ik} - VC'_{ik}\right) \times A_{ij}\right]} \times 100\% \quad (5)$$

where LP_{ik} is the value profit and loss of ESV after Class i land is converted to Class k land. A_{ij} represents the area of Class i land converted into Class j land. $VC_{ik} - VC'_{ik}$ is the ESV equivalent of the kth ESV of the ith land use type before and after the land use transition, respectively.

2.4. Multi-Scenario Land Use Projections for 2030

The PLUS model is an LUC simulation tool that combines the random forest algorithm and Markov Chain, capable of accurately capturing the spatial heterogeneity of land use changes [46,47]. This method is not only suitable for predicting future LUCs but also provides scientific evidence for land use planning under different scenarios [36]. Specifically (Figure 2), the PLUS model first extracts land expansion information from two phases of land use data and generates development probabilities for various land use types using the random forest algorithm combined with driving factors (such as distance to major rivers, railways, and main roads, elevation, slope, per capita GDP, population density, potential evapotranspiration, and annual average rainfall). To select the driving factors reasonably, this study, based on existing research and prior knowledge, chose the 7 most significant driving factors from over 20 potential factors. These driving factors and their processing and application in the PLUS model are shown in Table 3. Then, the Markov Chain is used to predict the total number of pixels for future land use, serving as a constraint for land use layout. Finally, in the CARS (Conversion of Land Use and its Effects at Small Regional Scales) module [48], we set the total number of future pixels, cost matrix, neighborhood weight, and related parameters to simulate and predict the spatial distribution of various land use types.

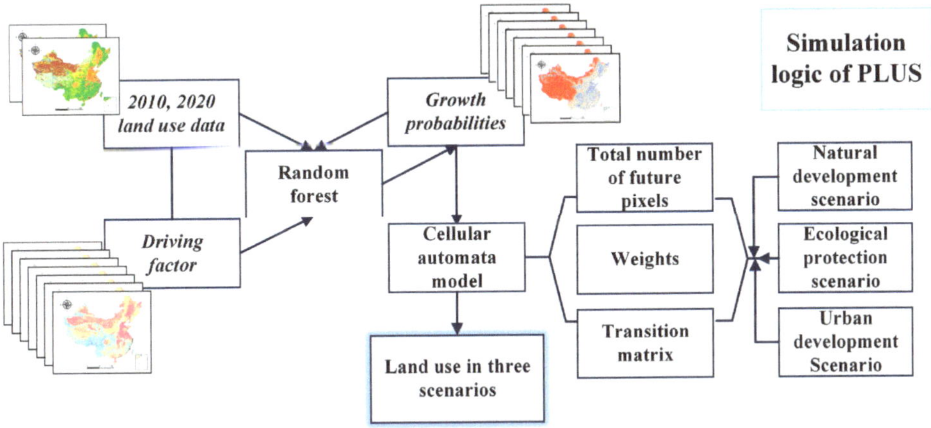

Figure 2. Multi-scenario prediction logic of the Patch-Generating Land Use Simulation model.

Table 3. Rationality of driver selection and its application in the Patch-Generating Land Use Simulation model.

Driving Factor	Rationale	Processing and Application in PLUS Model
Population density	Population density is a key driver of urbanization and land use change; high-density areas experience significant land changes [49,50].	Input to random forest algorithm to generate development probabilities for different land use types.
Regional average gross domestic product	Economic development level directly influences land use; areas with high GDP are more likely to undergo land development [51,52].	Input to random forest algorithm to generate development probabilities for different land use types.
Digital elevation model	Elevation affects the distribution of land types; low elevation areas are more suitable for agriculture and construction [53,54].	Input to random forest algorithm to generate development probabilities for different land use types.
Slope	Slope is a crucial determinant of hydrological characteristics, affecting land usability [55,56].	Data input to random forest algorithm to generate development probabilities for different land use types.
Annual precipitation	Precipitation affects ecosystem functions and biodiversity; high precipitation areas are usually forests and grasslands [57].	Data input to random forest algorithm to generate development probabilities for different land use types.
Distance to major rivers	Proximity to rivers influences water resource availability; areas close to rivers are suitable for agriculture and construction [58].	Used in random forest algorithm to generate land development probabilities, simulating dependence on rivers.
Distance to major roads	Proximity to roads affects accessibility and economic activity; areas with high traffic accessibility are easier to develop [59].	Used in random forest algorithm to generate land development probabilities, simulating the impact of transportation accessibility.

To validate the accuracy of the model, we used the 2020 land use data as a basis and employed the PLUS model to simulate the land use situation in 2030. Additionally, we simulated the 2020 land use situation using the 2010 land use data and compared the simulation results with the actual data from 2020. The results showed a Kappa value of 0.9734 and an FOM value of 0.9218, indicating that the model has very high simulation accuracy. This provides a solid foundation for future land use predictions in this study. In this study, we set up three scenarios (natural development scenario [NDS], ecological protection scenario [EPS], and urban development scenario [UDS]), each representing different land use strategies and environmental planning. The detailed settings of each scenario are provided in Supplementary Table S2. In the conversion cost matrix, a value of 1 indicates permitted conversion, while a value of 0 indicates restricted conversion (Table 4). All scenario predictions use the same driving factors to ensure the comparability and scientific validity of the results.

Table 4. Conversion cost matrix for multi-scenario land projections.

Type	Natural Development Scenario						Ecological Protection Scenario						Urban Development Scenario					
	I	II	III	IV	V	VI	I	II	III	IV	V	VI	I	II	III	IV	V	VI
I	1	1	1	1	1	1	1	0	0	0	0	0	1	1	1	1	1	1
II	1	1	1	1	1	1	0	1	0	0	0	0	1	1	1	1	1	1
III	1	1	1	1	1	1	1	1	1	1	1	1	1	1	1	1	1	1
IV	0	0	0	1	0	0	0	0	0	1	0	0	0	0	0	1	0	0
V	1	1	1	1	1	1	1	1	1	1	1	1	0	0	0	0	1	0
VI	1	1	1	1	1	1	1	1	1	1	1	1	1	1	1	1	1	1

Note: I, II, III, IV, V, and VI represent farmland, forest, grassland, water, built-up land and unused land respectively.

2.5. Deep Learning Simulation of the Impact of Climate Variables on ESV Profits and Losses

In this study, we used the LSTM (Long Short-Term Memory) model to simulate the impact of climate variables on ESV profits and losses. LSTM combines three gating mechanisms, forget gate, input gate, and output gate, which effectively regulate the flow of information and prevent the issues of gradient explosion and vanishing, thereby modeling time series features more accurately [60]. We applied the LSTM model to the climate variables in China, using 19 climate variables from the Shared Socioeconomic Pathway (SSP) 126 scenario for the period 2021–2040, to capture their impact on ESV profits and losses. The SSP126 scenario represents a low greenhouse gas emission pathway aligned with sustainable development goals. These 19 climate variables, including temperature and precipitation, were downloaded from the WorldClim database, and previously described variables were utilized in this analysis. In the implementation process, we used Python 3.8 along with its deep learning libraries TensorFlow and Keras to build and train the LSTM model [61]. The input for the LSTM model consists of climate variable data, and the output is the ESV profits and losses. The model is trained using past data through supervised learning, optimizing the model parameters by minimizing the error between the predicted values and the actual values. To help readers better understand the architecture of the LSTM model, Figure 3 illustrates the schematic of the LSTM neural network we used, including the specific operations of the forget gate, input gate, and output gate.

To better interpret the outputs of our LSTM model, we utilized the SHAP method [62]. It decomposes feature importance into individual effects and interactions with other features, thus providing explanations for black-box models [63]. SHAP is locally accurate and consistent, making it suitable for interpreting various machine learning models, including LSTM. In this study, we combined the LSTM model with the SHAP method (LSTM-SHAP) to reveal the nonlinear relationships between climate drivers and ESV profits and losses. Through the LSTM-SHAP method [34], we were able to identify the independent effects of individual climate variables on ESV changes and analyze their interactions.

In terms of implementation, we used Python 3.8 and the SHAP library [64], combined with the LSTM model, to predict the contribution proportions of climate-driving factors (Figure 4). The specific steps are as follows: First, the time series data of climate variables are input into the LSTM model. The LSTM model performs temporal prediction on these climate variables and outputs the contribution proportions of each climate-driving factor to ESV profits and losses. Then, we use the SHAP toolkit to calculate the Shapley values for each input variable, a process based on the *SHAP* formula:

$$SHAP(x_i) = \sum_{S \subseteq N \setminus \{i\}} \frac{|S|!(n-|S|-1)!}{n!} \left[f\left(S \bigcup \{i\}\right) - f(S) \right] \tag{6}$$

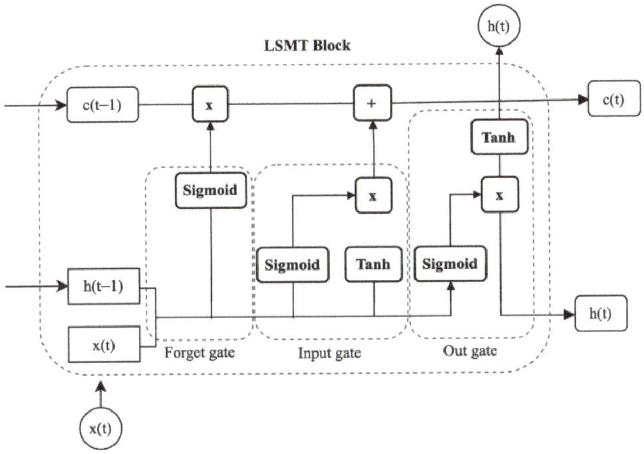

Figure 3. Schematic diagram of Long Short-Term Memory neural network architecture. Notes: x(t) represents the input features at time step t, while h(t) is the hidden state (output) at the same time step. c(t) is the cell state at time t, corresponding to c(t−1) and h(t−1) from the previous time step. The forget gate controls which information is discarded, the input gate determines which new information is added, and the output gate manages the output to the hidden state. The sigmoid function compresses inputs to [0, 1] for information retention, while the tanh function generates outputs within [−1, 1]. The addition operation (+) combines signals from different sources.

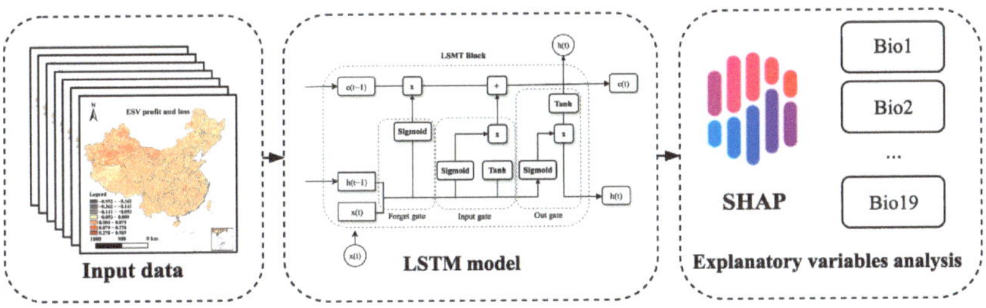

Figure 4. Flow of Shapley Additive Explanations method for interpreting Long Short-Term Memory model outputs.

Here, x_i represents each climate variable, S is a subset of the remaining variables in the model, and $f(S)$ is the predicted value of the model output for the subset of variables. n indicates the size of the current subset or the number of specific variables. N denotes the total number of all variables. This formula calculates the marginal contribution of each climate variable to the changes in ESV.

3. Results

3.1. Analysis of Driving Factors in Scenario Forecasting of LUC in China

Based on the existing research results and prior knowledge, we selected 7 driving factors from more than 20 possible factors. The importance of these seven driving factors is shown in Figure 5. By comparing the contribution of driving factors to various LUCs, we found differences in the main factors affecting various LUCs (Figure 6). For farmland and built-up land, population density is the most important factor driving LUC. The high population density means more frequent human activities, which promote the development of urbanization, the utilization of farmland, and the conversion of farmland into built-up land. For grassland and forests, rainfall is the most important transformation driver. Natural resource-rich areas are often accompanied by adequate rainfall, which is crucial in maintaining grassland and forest biodiversity, soil health, and landscape ecological value.

The slope is the main driving factor of water transformation, which is more conducive to forming lakes and wetlands in low-lying areas. GDP mainly drives the transformation of unused land. Economic development promotes the development of unused land and turns it into built-up land.

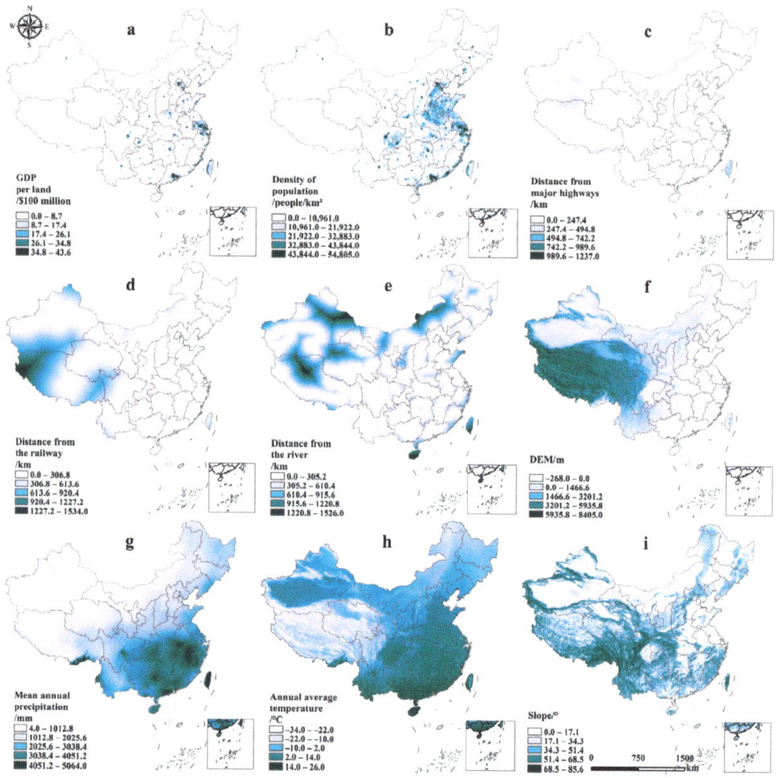

Figure 5. Drivers of multi-scenario modeling of 2030 land use projections. Notes: (**a**) GDP per land. (**b**) density of population. (**c**) distance from major highways. (**d**) distance from the railway. (**e**) distance from the river. (**f**) DEM (digital elevation model). (**g**) mean annual precipitation. (**h**) annual average temperature. (**i**) slope.

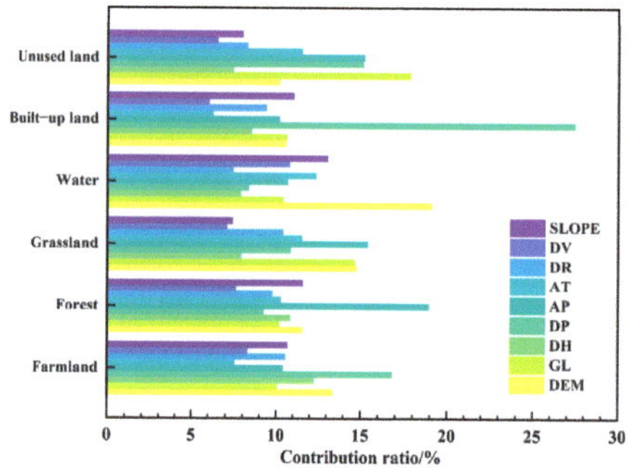

Figure 6. Contribution of driving factors affecting land use change. Note: DH: distance from major highways; DR: distance from the railway; DV: distance from the river; GL: GDP per land; DP: density of population; AT: annual average temperature; AP: mean annual precipitation. DEM: digital elevation model.

3.2. Multi-Scenario Simulation of LUC in China from 1990 to 2030

According to the spatial–temporal pattern evolution of land use (Figure 7), farmland is mainly distributed in the eastern and northeastern regions of China. The forest is mainly distributed in the south and southeast of China, and a large amount of unused land, including sandy and bare land, is widely distributed in the northwest of China. Grassland, built-up land, and water areas are dispersed. From 1990 to 2020, unused land significantly expanded in Central and Eastern China, and built-up land showed an aggregation trend in Eastern China. In contrast, the changing trend of other types of land was not obvious.

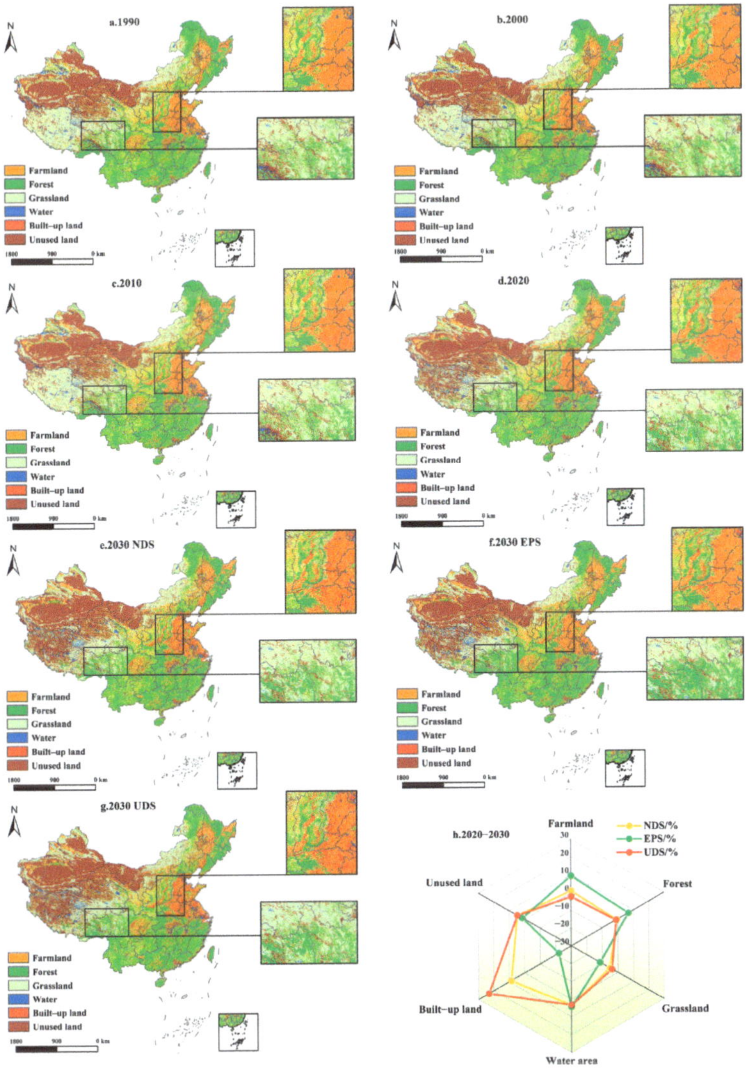

Figure 7. Multi-scenario simulation of land use change. Note: (**a–d**) represent the spatial distribution of land use in China for the years 1990, 2000, 2010, and 2020, respectively. (**e–g**) depict the spatial distribution of land use under the natural development scenario, ecological protection scenario, and urban development scenario for the year 2030. (**h**) shows the proportions of land use increase or decrease under different scenarios from 2020 to 2030.

Compared with the information shown in Figure 7d, clear temporal and spatial changes occur in all types of land under different scenarios in 2030 (Figure 7e). Under the NDS, the density of unused land increased, and grassland decreased in Southwest China. The amount of built-up land increased in the eastern region, but there was no obvious spatiotemporal variation in other directions. This suggests that under the NDS, China's urban construction is likely to achieve further development in 2030. In the EPS (Figure 7f), the forest density in Southwest China increased, the built-up land density in Eastern China was lower than that in the NDS, and there was no significant variation in other directions. In the UDS (Figure 7g), the density of unused land increased, and the density of grassland decreased in Southwest China. The density of built-up land in Eastern China increased further than in other scenarios. There was no significant variation in other directions.

In the different scenarios, while different types of land use showed spatial differentiation, the amount of land use also significantly changed (Figure 7h). Under the NDS, except for the relatively high increased proportion of built-up land (8.58%), the changes in other types of land were relatively balanced. Under the EPS, the rates of increase in farmland (9.48%) and forest (6.84%) were the highest, followed by the decreased rates of built-up land (−22.05%) and grassland (−11.4%), whereas the water area did not significantly change. The ecological protection policy mainly protects farmland, forest, and water, which indicates that the scenario simulation is consistent with the actual situation. Under the UDS, built-up land (23.11%) increased by a large proportion, and the rest of the land changed similarly to the natural development scenario.

3.3. Multi-Scenario Simulation of ESV in China from 1990 to 2030

China's ESV generally shows a trend of "high in the south and low in the north". The ESVs ranged from USD 21.9 to USD 37.1 million for the southern region and from USD 0 to USD 21.9 million for most of the northern region (Figure 8). In most regions of the northwest, ESV < USD 10.9 million, and in the southwest, ESV > USD 10.9 million. There is a significant difference in the total ESV between the east and the northeast. From 1990 to 2020, the areas with ESV > USD 37.1 million gradually increased in Southwest China, and the areas with ESV < USD 10.9 million also increased in Southwest China, resulting in the obvious polarization of ESV in Southwest China. This spatial variation can be partly attributed to terrain and geological factors, as regions with higher elevations and more complex geological structures often provide diverse habitats and richer biodiversity, leading to higher ESV. Conversely, flatter and more homogeneous terrains with less geological variation tend to have lower ESV. The changing trend of ESV in other regions is not clear.

Based on the land use data for 2030 under different scenarios after the forecast, the corresponding ESVs were calculated. Compared to the 2020 ESV (Figure 8d), the 2030 NDS shows an increase in the area of the ESV distribution of USD 0–10.9 million in the Southwest and a denser distribution of USD 37.1–64.9 million. The ESVs in the Southwest showed a high–low value differentiation pattern, whereas there was no significant variation in other directions. In the EPS, ESVs of USD 21.9 to USD 37.1 million show an expanding trend in South, Southwest, Northwest, Central, and Eastern China. Compared with the EPS, the ESV (USD 10.9–21.9 million) under the UDS increases in South, Southwest, Northwest, Central, East, and Northeast China, and the ESV (USD 21.9–37.1 million) decreases significantly in the distribution area. This reduction refers to the decrease in ESV compared to 2020, which is typically associated with changes in land use, such as conversion to non-vegetated areas, thereby diminishing the ecosystem's ability to provide services.

From 2020 to 2030 (Figure 8h), the ESV changes in all land use types under the NDS were relatively balanced. The ESVs of farmland (0.17%), built-up land (4.71%), and unused land (3.29%) slightly increased, and the ESVs of water area (−6.21%), grassland (−6.21%), and forest (−1.55%) moderately decreased. Under the EPS, the ESV of the forest (5.77%) and farmland (8.35%) significantly increased, while the ESV of built-up land (−24.83%) and water area (−12.31%) significantly decreased. Under the UDS, the ESV of built-up land (18.72%) significantly increased, and the ESV of other land showed a small change in range.

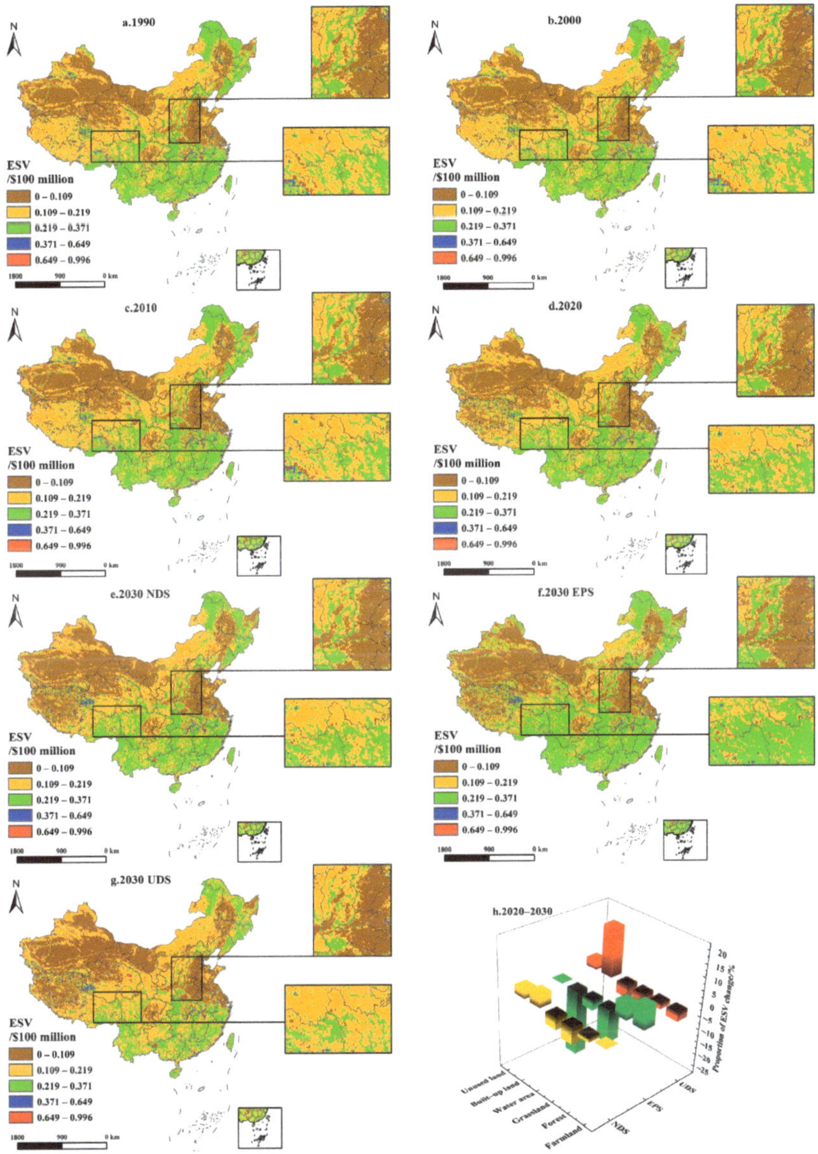

Figure 8. Spatial distribution of ecosystem service value in China from 1990 to 2030. Note: (**a**–**d**) represent the spatial distribution of ecosystem service values (ESVs) in China for the years 1990, 2000, 2010, and 2020, respectively. (**e**–**g**) depict the spatial distribution of ESV under the natural development scenario, ecological protection scenario, and urban development scenario for the year 2030. (**h**) shows the proportion of changes in ESV for different land use types from 2020 to 2030.

3.4. Multi-Scenario Simulation of ESV Profit and Loss

From 1990 to 2000, the ESV profits and losses in Northwest China were relatively concentrated in the range of USD 0 to USD 0.079 billion, while in other regions, the distribution of ESV profits and losses was more uniform, primarily fluctuating within the range of USD −0.053 to USD 0 billion and USD 0 to USD 0.079 billion (Figure 9).

During the period from 2000 to 2010, there were significant ESV profits and losses in the northwest and eastern regions of China, with a decrease from the USD 0 to USD 0.079 billion range to the USD −0.053 to USD 0 billion range and a slight increase within the USD 0 to USD 0.079 billion range in the southwest region. From 2010 to 2020, some areas in Southwest and Northwest China experienced more severe fluctuations in ESV profits and losses. This reduction refers to the decrease in ESV compared to the year 2020, which typically corresponds to changes in land use, such as conversion to non-vegetated areas, thereby diminishing the capacity of ecosystems to provide services. The decrease in ESV can be attributed to increased developmental pressures such as urbanization and industrialization, leading to habitat loss and degradation and consequently affecting the provision of ecosystem services. Conversely, areas that saw an increase in ESV may have benefited from conservation efforts and ecological restoration. Other regions did not show significant changes. From 2020 to 2030, under the NDS and UDS, the trends in ESV profits and losses continued from the 2010 to 2020 period. Under the EPS, certain areas in Southwest China experienced a more noticeable increase in ESV, suggesting that the EPS is more conducive to promoting the growth of ESV.

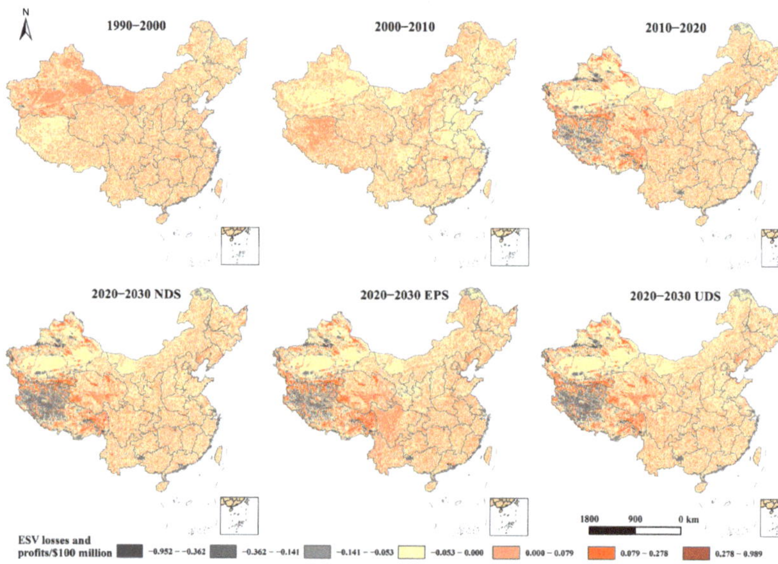

Figure 9. Spatial change in ecosystem service value profit and loss from 1990 to 2030.

3.5. Multi-Scenario Simulation of the Effect of LUC on ESV Profit and Loss

According to Formula (5), the contribution degree of each LUC to ESV was calculated (Figure 10). This study only analyzed the LUC for which the contribution degree was greater than 0.1%. From 1990 to 2020, the conversion of farmland to forest (11.94%), unused land to grassland (7.20%), and farmland to water area (6.23%) contributed the most to ESV profit. The conversion of forest to farmland (−12.46%), grassland to unused land (−7.18%), and forest to grassland (−5.74%) had the highest contribution to ESV loss. From 2000 to 2010, the conversion of farmland to waters (15.37%) and unused land to waters (9.04%) showed an ESV profit contribution that was the highest. The conversion from water area to unused land (−9.00%), from water area to farmland (−6.27%), and from grassland to farmland (−6.06%) had the highest contribution to ESV loss. From 2010 to 2020, the contribution of farmland to forest (9.02%), unused land to grassland (8.39%), and grassland to water area (6.14%) to ESV profit was the highest. The grassland conversion to unused land (−13.42%), forest conversion to farmland (−9.37%), and water conversion to unused land (−6.14%) contributed the most to ESV loss.

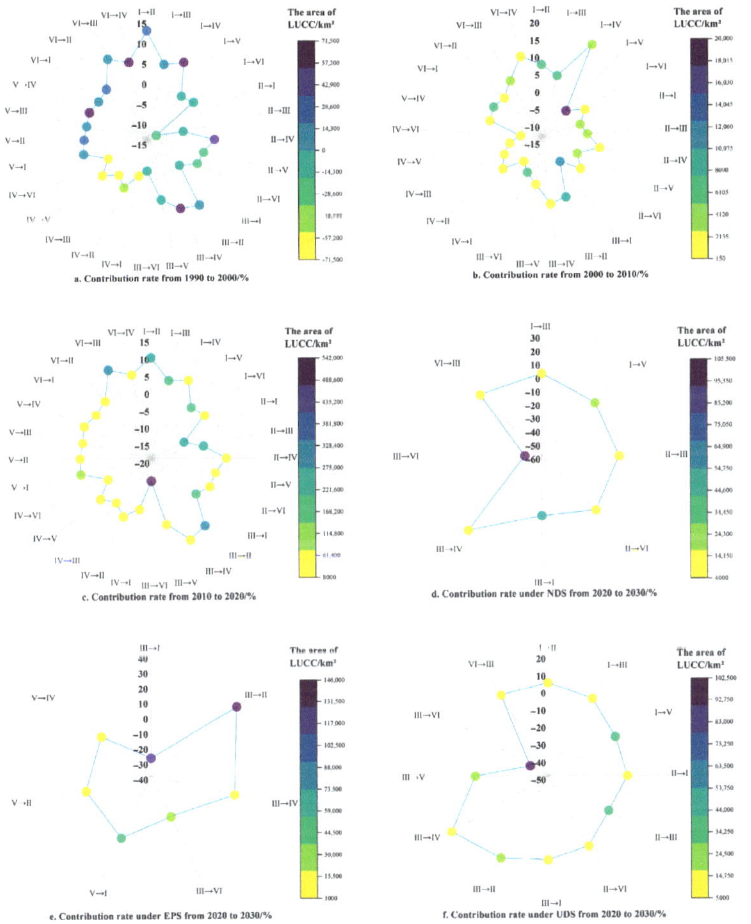

Figure 10. Contribution rate of land use change to ecosystem service value change under multiple scenarios. Note: I, II, III, IV, V and VI represent farmland, forest, grassland, water area, built-up land, and unused land, respectively. I→II indicates conversion of farmland to forest, and other transitions follow similarly.

Under the NDS, grassland shifted to unused land (105,435 km^2), with the highest contribution to ESV loss (−47.55%) and ESV reduction at USD 1815.60. In addition, converting grassland to farmland contributed more to ESV loss (−15.53%), reducing the ESV by USD 1290.70. The transformation of grassland to water area improved the ESV (17.66%), whereas the transformation contribution of other land groups was lower. Under the EPS, the transitions that contributed the most to ESV profit were grassland to forest (32.59%), grassland to water area (16.61%), and built-up land to farmland (5.56%). At the same time, the conversion of grassland to farmland of 119,020 km^2 (−28.27%) and grassland to unused land of 307,333.3 km^2 (−10.27%) resulted in ESV losses of USD 1290.70 and USD 1815.60, respectively. The contribution of other transformations to ESV change was low. In the UDS, the contribution of grassland to unused land (−38.37%), forest to grassland (−9.54%), and grassland to built-up land (−7.95%) to ESV loss was the highest. Grassland to water (14.27%), grassland to forest (4.74%), and unused land to grassland (3.90%) had the highest contribution to ESV profit. The above-mentioned analysis indicates that grassland transformation is crucial in regulating ESV.

3.6. Impact of Climate Variables on ESV Profits and Losses

The impact of climate variables on ESV profits and losses exhibits significant differences across various periods and scenarios (Figure 11). During the 1990–2000 period, Bio13 (the precipitation of the wettest month) and Bio12 (annual precipitation) had a significant positive impact on ESV profits and losses. However, the negative impact of Bio4 (temperature seasonality) indicated that temperature fluctuations weakened ecosystem stability, with notable ESV losses during years with large temperature variations. In the 2000–2010 period, the negative impact of Bio15 (precipitation seasonality) gradually became apparent. At the same time, the negative effect of Bio2 (the mean diurnal range) also became significant. Moving into the 2010–2020 period, the negative influence of Bio4 (temperature seasonality) continued to intensify, while precipitation-related variables such as Bio12 (annual precipitation) and Bio13 (the precipitation of the wettest month) continued to have a positive impact on ESV. The variable Bio13 (the precipitation of the wettest month) may indeed depend on which specific month is the wettest and the amount of precipitation received during that month, which influences whether there is a gain or decrease in ESV. For the 2020–2030 forecast under different scenarios, Bio13 and Bio12 maintained a strong positive impact under the NDS. However, under EPS, while the positive impact of precipitation on ESV remained, temperature-related variables, especially Bio4 and Bio16 (the precipitation of the wettest quarter), exhibited stronger negative and positive impacts, respectively. In the UDS, the negative impact of Bio4 (temperature seasonality) was the most pronounced. All these climate variables (Bio1 to Bio19) are derived from the WorldClim database, which provides high-resolution global climate data (Table S1). Specifically, Bio15 refers to precipitation seasonality, which is particularly important in regions with monsoonal climates.

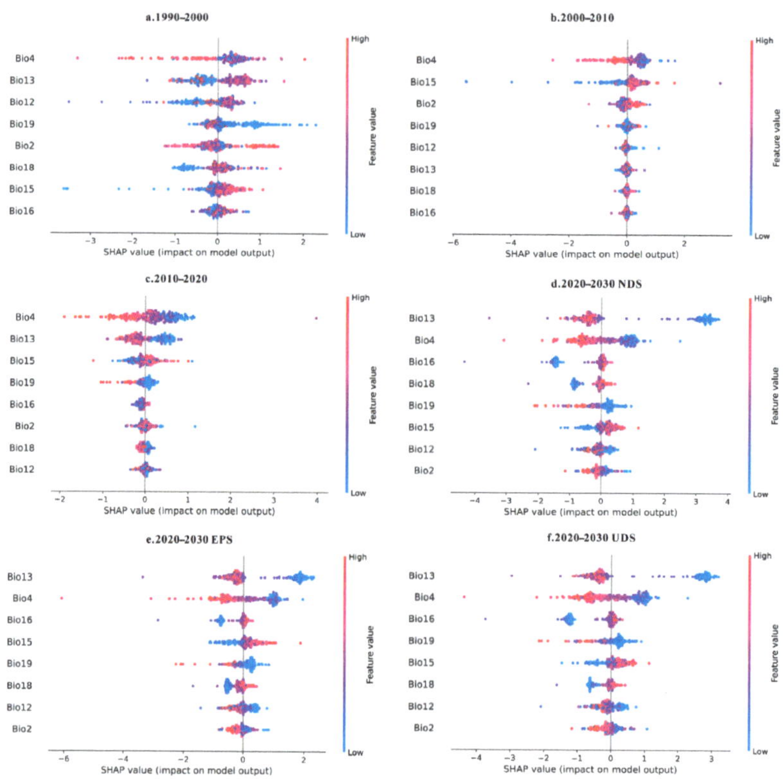

Figure 11. Impact of climate variables on ecosystem service value profits and losses.

4. Discussion

4.1. Ecological Impacts of LUC on ESV Under Multiple Scenarios

Similar to the results of prior studies at different scales, this study confirms that LUC causes ESV changes at the national scale; where LUC may be related to human activities and changes in natural resources, different land types can provide a certain amount of ESV, and a specific ESV can be affected by human activities and changes in natural resources [10]. The human activity factor is mainly exhibited by the increased development of land with accelerated urbanization, and more land will be transformed into land for construction [1]. In addition, the over-exploitation of land by humans causes land degradation, resulting in the desertification and sanding of land, eventually forming unused land; overgrazing by herders also leads to grassland degradation. Unlike prior studies [13], this study found that grasslands are critical in regulating ESV changes. In particular, the conversion of grasslands to unused land from 1990 to 2000 and 2010 to 2020 caused a significant decrease in ESV in Southwest China, which indicates that grasslands are more advantageous for maintaining soil health and ensuring a series of other ecological functions, such as biodiversity, which also indicates that the reasons for ESV changes are complex [11]. In contrast, other studies indicate that the ecosystem services of forests and wetlands also play a crucial role in maintaining ecological stability [65]. This study complements the role of grasslands in ecosystem maintenance, particularly their critical importance under specific regional environmental pressures. This finding highlights the significance of grassland conservation and management, especially in the context of rapid urbanization.

From 1990 to 2020, the total ESV in most regions of China was in the range of USD 0–37.1 million, and the improvement in ESV during the study period was most sensitive to the conversion of farmland to forests, grasslands, and watersheds and the conversion of grasslands to watersheds. However, the deterioration of ESV was most sensitive to the transformation of forests to grasslands and grasslands and watersheds to unused lands, which shows that forests, grasslands, and watersheds play an integral role in environmental purification. This is consistent with some studies [66–68], which emphasize the necessity of the rational use and protection of natural resources. Thus, forests, grasslands, and watersheds impact environmental purification, soil health maintenance, biodiversity enhancement, and hydrological regulation [45]. Therefore, in future ecological planning, certain amounts of forests, grasslands, and watersheds need to be included in the scope of protection to ensure the stability of the total ESV.

Grasslands play an instrumental role in ecological restoration, and in this context, the government needs to strengthen the management of grassland ecology and improve ecological protection policies. Government departments can try to link farmers' income, grassland ecological protection, and ecological industry development so that social, ecological, and economic benefits are linked, and the three can develop synergistically; the government can also go deep into the grassroots to help farmers establish new concepts and learn new methods of ecological restoration and ecological management, adopt production methods that match environmental capacity and ecological carrying capacity, and realize the harmonious coexistence of herders and ecology. The government should improve relevant laws, define the scope of use of grasslands, prohibit herders from using the grasslands in the governance area, and impose economic or criminal penalties on violators.

4.2. Spatiotemporal Characteristics of ESV Profit and Loss

Based on the land use assessment and prediction of the spatial and temporal dynamics of ESV loss and profit in China, there is a scientific basis for the optimal allocation and effective use of future national land resources to guide ecological conservation and regional economic development [69]. From 1990 to 2000, economic development activities in Northwest China were relatively limited. The natural environment in this region is quite fragile, but due to the relatively small impact of human activities, the change in ESV was not significant. Between 2000 and 2010, economic development in Northwest China and the eastern regions accelerated, with urban expansion and industrialization leading to the degradation of the natural environment. Particularly in the eastern region, as the economic center of China, urbanization and industrialization exerted significant pressure on the ecological environment, with ESV losses decreasing from USD 0–0.079 billion to USD −0.053–0 billion. The slight increase in ESV within the USD 0–0.079 billion range in the Southwest may be attributed to the implementation of ecological protection policies and the impact of ecological restoration projects. From an economic perspective, it is projected that by 2030, under the NDS, the ESV in the southwest region could generate economic benefits of approximately USD 37.1 million to USD 10.9 million, while under the EPS, with the protection of forests and grasslands, the ESV is expected to increase by about USD 10 million. This increase in economic benefits not only reflects the direct contributions of ESV but also helps to indirectly promote local economic development and green industries such as ecotourism. In contrast to the findings of Ai et al. (2024), who emphasized the impact of population density on ESV [42], this study reveals the direct threat that rapid economic development poses to ESVs, providing a new perspective for understanding the dynamic relationship between urbanization and ecology. Moreover, in areas experiencing rapid urbanization, increased environmental pressure may lead to a loss of ESV activities indeed exerting pressure on ESVs. In this study, the fluctuations in ESV in the southwest region align with ecological protection policies, further validating the timeliness of these policies in influencing environmental changes. From 2010 to 2020, parts of Southwest and Northwest China experienced more intense fluctuations in ESV, which may be related to China's increased development efforts in the western regions, such as the western development strategy, and the increased vulnerability of the ecological environment in this region under the impact of climate change. This study reveals that under high pressure, especially from urbanization and agricultural expansion, ESV is expected to fluctuate significantly. The fluctuations in ESV in the southwest region during this period indicate that potential economic losses in the future could increase further, with projected losses reaching USD 21 million. Consistent with the findings of Carbone et al. (2022) [70], our results emphasize the interaction between economic policies and environmental policies, highlighting the long-term impacts of climate change on ecological services, which provides practical cases for local governance. The trends in ESV changes under the three different predicted scenarios (NDS, EPS, and UDS) for 2020–2030 are similar to those from 2010 to 2020, which may indicate that the development trends and environmental policies of the past decade continue to have an influence. Under the EPS, certain areas in Southwest China exhibited a more noticeable increase in ESV; this result is consistent with the conclusions of Boonman et al. (2023) [71], emphasizing the effectiveness of ecological protection measures. It is anticipated that by 2030, the economic benefits under the EPS will further increase, with the growth in ESV potentially bringing an additional economic value of approximately USD 15 million to the region.

From 1990 to 2020, the conversion of farmland to forest in Northern, Central, and Southwestern China provided the maximum contribution to ESV profit, while the conversion of forest to farmland in the same region provided the greatest contribution to ESV loss, indicating that the policy of returning farmland to forest in China plays an important role in ESV improvement. Thus, the implementation of this policy needs to be maintained in the future. Consistent with the views of Li et al. (2019) [72], this study further confirms the effectiveness of the Grain for Green policy and indicates that this policy needs to be continued in the future to ensure the stability of ecosystem services. The conversion of unused land to grassland and grassland to unused land in Northwest and Southwest China was second in terms of contribution to both ESV profit and loss, indicating that the role of grassland in maintaining ESV is second only to that of forest; thus, the protection of grassland needs to be strengthened. In 2020–2030, under the NDS, EPS, and UDS, grassland conversion to cropland in Northwest and Southwest China and grassland conversion to unused land in East and Central China all contributed larger ESV losses, while grassland conversion to water and forest in Southwest China both contribute to larger ESV profits, indicating that the development of farmland and unused land depletes ESV. However, forest, grassland, and water all improve ESV. In the future, China must insist on reforestation, grass conversion, and the rational use of water resources to achieve greater ESV profits [73].

In this study, the PLUS model was used for multi-scenario prediction, which can intuitively reflect the LUC and ESV under different scenarios, and the research results were found to be reliable after testing. However, in selecting driving factors, government decisions, human factors, and social development were not considered. In contrast to the study by Li et al. (2022) [74], our research highlights the impact of more complex social and economic changes on ecosystem services. Therefore, the effect of LUC on ESV in the context of multiple complex relationships should be further analyzed in subsequent studies, and the land use data can be further subdivided to improve the accuracy of the data and analysis. In studying the spatiotemporal dynamic changes in ESV profit and loss in China, the quantitative accuracy of these figures is still insufficient. Although ESV evaluation based on a remote sensing model has strong computational power and a scientific nature, it cannot accurately reflect the spatiotemporal characteristics of ESV profit and loss in China only via numerical and formula simulation. In the future, it is essential to combine the measured data from the study area with continuous debugging and simulations to improve the accuracy of determining the actual levels of ESV profit and loss.

4.3. Impact of Climate Change on ESV Profits and Losses

The impact of climate variables on ESV profits and losses shows significant differences across different time periods and scenarios. We found that precipitation and temperature fluctuations are key climate factors affecting ESV profits and losses. Specifically, during the 1990–2000 period, Bio13 (the precipitation of the wettest month) and Bio12 (annual precipitation) had a significant positive impact on ESV profits and losses, indicating that increased precipitation helps enhance ES functions. However, the negative impact of Bio4 (temperature seasonality) suggests that temperature fluctuations weaken ecosystem stability, particularly in years with large temperature variations, resulting in notable ESV losses. These results reflect the different mechanisms through which precipitation and temperature fluctuations influence ESVs. Increased precipitation, especially annual precipitation and precipitation during the wettest month, can provide sufficient water resources, promoting plant growth and biodiversity and thereby enhancing ES functions. Conversely, intensified temperature fluctuations may lead to stress responses in ecosystems, disrupt ecological balance, and reduce ESV. Consistent with the findings of Austhof et al. (2024) [75], this study further reveals the nuanced impact of temperature seasonality on specific ecosystem service functions, which has not been thoroughly explored in previous research. Compared to the study by Swenson et al. (2023) [76], these findings emphasize the fundamental role of precipitation in sustaining ecosystem services, while also highlighting the adverse effects of temperature fluctuations on service functions.

From 2000 to 2010, Bio15 (precipitation seasonality) gradually manifested negative impacts, signaling that its uneven distribution put undue stress on ESV. Changes in precipitation seasonality can cause water shortages during certain times of the year that fail to meet biological community needs and disrupt the service functions of ecosystems, thereby diminishing their service functions. Simultaneously, the negative impact of Bio2 (the mean diurnal range) also became significant, suggesting that under extreme climate conditions, increased day–night temperature differences are similarly detrimental to the maintenance of ES functions. Large diurnal temperature variations can affect plants' physiological processes, such as photosynthesis and respiration. Although temperatures remain above freezing and day length is unchanged, significant temperature fluctuations can cause thermal stress on plants. This increased physiological stress on plants may impact their growth cycles and overall health, reducing productivity and stability within the ecosystem. Compared to the study by Yao et al. (2022) [77], this research provides more detailed insights into the specific impacts of diurnal temperature variation on the profits and losses of ESVs, revealing the high sensitivity of ecosystem services to changes in temperature fluctuations.

Entering the 2010–2020 period, the negative effect of Bio4 (temperature seasonality) continued to intensify, while precipitation-related variables such as Bio12 (annual precipitation) and Bio13 (the precipitation of the wettest month) continued to have a positive impact on ESV. Bio4, which represents temperature seasonality, is calculated as the standard deviation of monthly temperature averages multiplied by 100. This metric indicates the degree of temperature variation throughout the year. Increased temperature seasonality means greater fluctuations in temperature between seasons, which can lead to environmental stress on ecosystems. Precipitation-related variables such as Bio12 (annual precipitation) and Bio13 (the precipitation of the wettest month) continued to have a positive impact on ESV. During this period, the challenges posed by climate fluctuations to ecosystems intensified, and precipitation stability became a crucial factor in maintaining ESV. The worsening negative impact of temperature seasonality on ecosystems may be related to the increased frequency of extreme weather events caused by climate change. These events put significant pressure on ecosystems, leading to decreased service functions. Through the analysis of long-term data series, this study confirms the universal importance of precipitation stability on ESV across different climatic contexts, providing a new reference perspective for research on similar ecosystems globally.

Future scenario projections indicate that from 2020 to 2030, under the NDS, Bio13 and Bio12 will continue to have a strong positive impact, especially in years with abundant precipitation, significantly enhancing ESs. However, under the EPS, although the positive effect of precipitation on ESV remains, temperature-related variables, particularly Bio4 and Bio16 (the precipitation of the wettest quarter), show stronger negative and positive impacts, respectively. Lastly, under the UDS, the negative impact of Bio4 (temperature seasonality) is the most pronounced. Urbanization exacerbates temperature fluctuations, further weakening ESV, while the influence of precipitation-related variables decreases. Compared to traditional climate model predictions, this study incorporates human activity factors across multiple scenarios, such as different urbanization pathways, providing a more comprehensive analysis of the response mechanisms of ESV. Overall, precipitation and temperature fluctuations are key climatic factors affecting ESV profits and losses. Over time, the intensification of climate change and human activities has made the stability of precipitation and the variability of temperature crucial determinants of ESV changes. Future climate change projections suggest that the impact of climate variables on ESV under different scenarios carries a certain degree of uncertainty, necessitating further research and monitoring to formulate effective ecosystem protection and management strategies. The multi-period and multi-scenario analysis approach of this study not only validates existing theories but also reveals new impact pathways, providing innovative analytical perspectives for future evaluations of ESVs. Understanding how climate change impacts ESVs will enable us to more accurately anticipate and respond to future impacts, providing a scientific basis for protecting and managing ecosystems. This study highlights the significance of precipitation stability and temperature variability for maintaining ESVs, providing insights for climate adaptation strategies in the future.

Although this study deeply analyzed the impact of 19 climate variables on ESV profits and losses across different regions and scenarios in China using the LSTM-SHAP combined model, it still has certain limitations. Currently, the data we can access are limited to the 19 climate factors officially designated for the SSP126 scenario from 2021 to 2040, which may restrict our in-depth analysis of other potentially critical climate variables. Using these forecast-based data in analysis and modeling might introduce potential biases; therefore, decision makers should fully consider this when making relevant management and policy decisions. Future research will strive to overcome this limitation. We plan to introduce more climate factors, such as wind speed and humidity, which will help provide a more comprehensive understanding of the impact of climate change on ESV profits and losses. Additionally, we will integrate more scenario simulation data to further refine and enrich the research content. These efforts will contribute to providing more comprehensive and accurate scientific evidence for ecosystem restoration and management strategies. In this context, we also recognize that using the distance from the center of a 1 km grid to the nearest roads, rivers, and railways as an indicator in this study may introduce some uncertainties. While the 1 km grid scale offers a balance between environmental feature coverage and data processing operability in geographical and ecological research, it can still be inadequate in accurately representing certain regional characteristics. Specifically, although the distance from the grid cell center to the nearest roads, rivers, and railways simplifies complex spatial analyses and maintains consistency in overall analyses, the presence of multiple transportation and water networks within a 1 km unit and the relatively large area of the unit mean that the center point distance might not fully reflect the spatial characteristics of all features within the unit. For example, the actual roads or water bodies in a 1 km unit might be closer to the boundary of the area, and this spatial approximation might reduce the precision in depicting more complex relationships between features. In future research, considering higher resolution data could be a potential solution to reduce this uncertainty. Specifically, for each 1 km grid cell, we could use more detailed spatial analysis methods (such as area-based or multipoint approaches) to assess the actual impact of roads and water bodies. These methods, by integrating proximity from multiple locations or utilizing diverse distance calculations, can enhance the understanding of ecosystem services within specific regions. This improvement not only helps enhance data representativeness but might also improve the responsiveness and accuracy of analyses concerning ecosystem changes and impacts.

5. Conclusions

This study employed multi-scenario simulations to examine the profits and losses of ESV in China from 1990 to 2030, with particular attention paid to climate change's influence. The primary findings are as follows: Over time, China's ESV displays an overall high trend in the south and low trend in the north, with ESV levels significantly higher in southern regions than northern ones from 1990 to 2030. From 1990 to 2020, however, southwest regions show distinct polarization in ESV performance between different areas, with significant increases and decreases over this timeframe. Under the NDS, ESV changes in various land uses are relatively balanced. In the EPS, the ESV of forests and farmland increases significantly, while the ESV of built-up land and water decreases significantly. Under the UDS, the ESV of built-up land increases significantly, with smaller changes in other land use types. The observed variations in climate variables impacted ESV profits and losses. During the 1990–2000 period, precipitation had a significant positive impact on ESV, while temperature seasonality had a negative impact. While these findings suggest potential implications related to climate variation, the study's scope was limited to analyzing observed data and did not comprehensively assess climate change impacts. Further research is needed to draw more definitive conclusions regarding the long-term effects of climate change on ESV. From 2000 to 2010, the negative impacts of precipitation seasonality and the mean diurnal range on ESV gradually became apparent. From 2010 to 2020, the negative effect of temperature seasonality continued to intensify, while the positive impacts of precipitation-related variables persisted. Future scenario projections indicate that under the NDS, the positive impact of precipitation on ESV remains significant; in the EPS, the impacts of precipitation and temperature variables are more complex; and in the UDS, the negative effect of temperature seasonality is most pronounced. Converting farmland to forest, converting unused land to grassland, and converting farmland to water contribute the most to ESV profits, while converting forest to farmland, grassland to unused land, and forest to grassland contribute the most to ESV losses. Under future scenarios, grassland transformation plays a crucial role in regulating ESV. This paper provides new insights into the dynamic changes in ESV and their driving factors by comprehensively considering the dual impacts of climate and land use. This not only enriches existing research but also provides important theoretical support for practical policymaking. In future research, the further exploration of the interactions between different regions and long-term trend changes will be of significant importance for achieving sustainable development goals nationwide.

Based on the above research conclusions, the following policy recommendations are proposed to enhance the ESV and promote sustainable development in China: (1) Consider long-term climate impacts in regional planning: when formulating regional development plans, fully consider the long-term impacts of climate change and implement comprehensive measures to address extreme climate events and mitigate the adverse effects of temperature seasonality and precipitation seasonality changes on ecosystem. (2) Strengthen water resource management: in areas with significant precipitation seasonality changes, enhance water resource management by building reservoirs, improving rainwater collection, and increasing water use efficiency. (3) Integrate ESV into economic policies: When formulating economic development policies, use ESV as an important indicator. Encourage green GDP accounting and consider environmental benefits when evaluating economic performance. Actively promote green industries such as ecological agriculture and ecotourism, encourage businesses to utilize more eco-friendly production methods, reduce environmental damage through regional ESV, and enhance environmental education. (4) Raise public awareness on ESV/sustainable development issues. Encourage and assist public participation in ecological protection activities. (5) Consider climate change in land use and ecological policies: Integrating climate change factors into land use planning and ecological protection policies to enhance ecosystem resilience and adaptability while mitigating the negative effects of extreme climate events on ecosystems can help us better navigate climate change's complexities and drive the sustainable development of ecosystems in China. By adhering to this recommendation, we can navigate its complexity for sustainable ecosystem development in China.

Supplementary Materials: The following supporting information can be downloaded at: https://www.mdpi.com/article/10.3390/land13111924/s1, Table S1: Climate variables and the data sources. Table S2: Scenario setting for the PLUS model.

Author Contributions: Conceptualization, methodology, software, formal analysis, data curation and writing—original draft, D.Y.; writing—review and editing, methodology, software, formal analysis, data curation, visualization, supervision, project administration, funding acquisition, C.Y. All authors have read and agreed to the published version of the manuscript.

Funding: This research was funded by the National Key Research and Development Program of China (2022YFF1303001).

Data Availability Statement: Data available on request due to restrictions. The data presented in this study are available on request from the corresponding author. The data are not publicly available due to privacy and anonymity.

Conflicts of Interest: The authors declare that they have no known competing financial interests or personal relationships that could have appeared to influence the work reported in this paper.

References

1. Pan, H.Z.; Zhang, L.; Cong, C.; Deal, B.; Wang, Y.T. A dynamic and spatially explicit modeling approach to identify the ecosystem service implications of complex urban systems interactions. *Ecol. Indic.* **2019**, *102*, 426–436. [CrossRef]
2. Gu, L.; Gong, Z.W.; Du, Y.X. Evolution characteristics and simulation prediction of forest and grass landscape fragmentation based on the "Grain for Green" projects on the. *Ecol. Indic.* **2021**, *131*, 13. [CrossRef]
3. Hou, Y.Z.; Zhao, W.W.; Liu, Y.X.; Yang, S.Q.; Hu, X.P.; Cherubini, F. Relationships of multiple landscape services and their influencing factors on the Qinghai-Tibet Plateau. *Landsc. Ecol.* **2021**, *36*, 1987–2005. [CrossRef]
4. Morgado, R.; Ribeiro, P.F.; Santos, J.L.; Rego, F.; Beja, P.; Moreira, F. Drivers of irrigated olive grove expansion in Mediterranean landscapes and associated biodiversity impacts. *Landsc. Urban Plan.* **2022**, *225*, 11. [CrossRef]
5. Meng, Q.X.; Zhang, L.K.; Wei, H.J.; Cai, E.X.; Xue, D.; Liu, M.X. Linking Ecosystem Service Supply-Demand Risks and Regional Spatial Management in the Yihe River Basin, Central China. *Land* **2021**, *10*, 26. [CrossRef]
6. Chen, W.X.; Zeng, J.; Li, N. Change in land-use structure due to urbanisation in China. *J. Clean Prod.* **2021**, *321*, 14. [CrossRef]

7. de Mello, K.; Taniwaki, R.H.; de Paula, F.R.; Valente, R.A.; Randhir, T.O.; Macedo, D.R.; Leal, C.G.; Rodrigues, C.B.; Hughes, R.M. Multiscale land use impacts on water quality: Assessment, planning, and future perspectives in Brazil. *J. Environ. Manag.* **2020**, *270*, 16. [CrossRef]
8. Kefalas, G.; Kalogirou, S.; Poirazidis, K.; Lorilla, R.S. Landscape transition in Mediterranean islands: The case of Ionian islands, Greece 1985–2015. *Landsc. Urban Plan.* **2019**, *191*, 19. [CrossRef]
9. Mansour, S.; Al-Belushi, M.; Al-Awadhi, T. Monitoring land use and land cover changes in the mountainous cities of Oman using GIS and CA-Markov modelling techniques. *Land Use Policy* **2020**, *91*, 15. [CrossRef]
10. Osborne, P.E.; Alvares Sanches, T. Quantifying how landscape composition and configuration affect urban land surface temperatures using machine learning and neutral landscapes. *Comput. Environ. Urban Syst.* **2019**, *76*, 80–90. [CrossRef]
11. Zheng, L.; Wang, Y.; Li, J.F. Quantifying the spatial impact of landscape fragmentation on habitat quality: A multi-temporal dimensional comparison between the Yangtze River Economic Belt and Yellow River Basin of China. *Land Use Policy* **2023**, *125*, 16. [CrossRef]
12. Chen, X.; Yu, L.; Du, Z.R.; Xu, Y.D.; Zhao, J.Y.; Zhao, H.L.; Zhang, G.L.; Peng, D.L.; Gong, P. Distribution of ecological restoration projects associated with land use and land cover change in China and their ecological impacts. *Sci. Total Environ.* **2022**, *825*, 15. [CrossRef]
13. Wang, Z.Y.; Lechner, A.M.; Yang, Y.J.; Baumgartl, T.; Wu, J.S. Mapping the cumulative impacts of long-term mining disturbance and progressive rehabilitation on ecosystem services. *Sci. Total Environ.* **2020**, *717*, 15. [CrossRef]
14. Marull, J.; Cunfer, G.; Sylvester, K.; Tello, E. A landscape ecology assessment of land-use change on the Great Plains-Denver (CO, USA) metropolitan edge. *Reg. Environ. Chang.* **2018**, *18*, 1765–1782. [CrossRef]
15. Shifaw, E.; Sha, J.M.; Li, X.M.; Bao, Z.C.; Zhou, Z.L. An insight into land-cover changes and their impacts on ecosystem services before and after the implementation of a comprehensive experimental zone plan in Pingtan island, China. *Land Use Policy* **2019**, *82*, 631–642. [CrossRef]
16. Jin, Z.H.; Xiong, C.S.; Luan, Q.L.; Wang, F. Dynamic Evolutionary Analysis of Land Use/Cover and Ecosystem Service Values on Hainan Island. *Int. J. Environ. Res. Public Health* **2023**, *20*, 18. [CrossRef]
17. Watson, L.; Straatsma, M.W.; Wanders, N.; Verstegen, J.A.; de Jong, S.M.; Karssenberg, D. Global ecosystem service values in climate class transitions. *Environ. Res. Lett.* **2020**, *15*, 15. [CrossRef]
18. Yang, H.J.; Gou, X.H.; Yin, D.C. Response of Biodiversity, Ecosystems, and Ecosystem Services to Climate Change in China: A Review. *Ecologies* **2021**, *2*, 19. [CrossRef]
19. Wilkes, L.N.; Barner, A.K.; Keyes, A.A.; Morton, D.; Byrnes, J.E.K.; Dee, L.E. Quantifying co-extinctions and ecosystem service vulnerability in coastal ecosystems experiencing climate warming. *Glob. Chang. Biol.* **2024**, *30*, 17. [CrossRef]
20. Asmus, M.L.; Nicolodi, J.; Anello, L.S.; Gianuca, K. The risk to lose ecosystem services due to climate change: A South American case. *Ecol. Eng.* **2019**, *130*, 233–241. [CrossRef]
21. Leal, W.; Azeiteiro, U.M.; Balogun, A.L.; Setti, A.F.F.; Mucova, S.A.R.; Ayal, D.; Totin, E.; Lydia, A.M.; Kalaba, F.K.; Oguge, N.O. The influence of ecosystems services depletion to climate change adaptation efforts in Africa. *Sci. Total Environ.* **2021**, *779*, 13.
22. Zhang, Y.; Wu, T.; Song, C.S.; Hein, L.; Shi, F.Q.; Han, M.C.; Ouyang, Z.Y. Influences of climate change and land use change on the interactions of ecosystem services in China's Xijiang River Basin. *Ecosyst. Serv.* **2022**, *58*, 9. [CrossRef]
23. Sharma, R.; Malaviya, P. Ecosystem services and climate action from a circular bioeconomy perspective. *Renew. Sust. Energ. Rev.* **2023**, *175*, 16. [CrossRef]
24. Rahim, N.H.; Razak, S.A.; Chang, X.; Saun, F.C.; Khan, M.N.; Hamzah, S.N.; Rohman, F.; Ali, B.; Kaplan, A.; Iqbal, M.; et al. Review Ecosystem Services by Urban Forest (UF) towards Climate Change Adaptation: A Review. *Pol. J. Environ. Stud.* **2024**, *33*, 3503–3513. [CrossRef]
25. Lu, C.; Qi, X.; Zheng, Z.S.; Jia, K. PLUS-Model Based Multi-Scenario Land Space Simulation of the Lower Yellow River Region and Its Ecological Effects. *Sustainability* **2022**, *14*, 17. [CrossRef]
26. Zhu, X.L.; Qie, R.Q.; Luo, C.; Zhang, W.Q. Assessment and Driving Factors of Wetland Ecosystem Service Function in Northeast China Based on InVEST-PLUS Model. *Water* **2024**, *16*, 19. [CrossRef]
27. Zhang, Y.; Liao, X.Y.; Sun, D.Q. A Coupled InVEST-PLUS Model for the Spatiotemporal Evolution of Ecosystem Carbon Storage and Multi-Scenario Prediction Analysis. *Land* **2024**, *13*, 24. [CrossRef]
28. Wu, Z.J.; Cui, N.B.; Zhang, W.J.; Liu, C.W.; Jin, X.L.; Gong, D.Z.; Xing, L.W.; Zhao, L.; Wen, S.L.; Yang, Y.N. Estimating soil moisture content in citrus orchards using multi-temporal sentinel-1A data-based LSTM and PSO-LSTM models. *J. Hydrol.* **2024**, *637*, 14. [CrossRef]
29. Deng, C.; Yin, X.; Zou, J.C.; Wang, M.M.; Hou, Y.K. Assessment of the impact of climate change on streamflow of Ganjiang River catchment via LSTM-based models. *J. Hydrol.-Reg. Stud.* **2024**, *52*, 15. [CrossRef]
30. Yang, S.; Tan, M.L.; Song, Q.X.; He, J.; Yao, N.; Li, X.G.; Yang, X.Y. Coupling SWAT and Bi-LSTM for improving daily-scale hydro-climatic simulation and climate change impact assessment in a tropical river basin. *J. Environ. Manag.* **2023**, *330*, 13. [CrossRef]
31. Wang, C.; Sheng, Q.Q.; Zhu, Z.L. Exploring Ecological Quality and Its Driving Factors in Diqing Prefecture, China, Based on Annual Remote Sensing Ecological Index and Multi-Source Data. *Land* **2024**, *13*, 19. [CrossRef]
32. Airiken, M.; Li, S.C. The Dynamic Monitoring and Driving Forces Analysis of Ecological Environment Quality in the Tibetan Plateau Based on the Google Earth Engine. *Remote Sens.* **2024**, *16*, 15. [CrossRef]

33. Huang, F.Y.; Zhang, X.Y. A new interpretable streamflow prediction approach based on SWAT-BiLSTM and SHAP. *Environ. Sc Pollut. Res.* **2024**, *31*, 23896–23908. [CrossRef]
34. Sun, X.; Wang, H.L.; Mei, S.L. Explainable highway performance degradation prediction model based on LSTM. *Adv. Eng. Inform* **2024**, *61*, 12. [CrossRef]
35. Chen, H.; Huang, S.H.; Xu, Y.P.; Teegavarapu, R.S.V.; Guo, Y.X.; Nie, H.; Xie, H.W.; Zhang, L.Q. River ecological flow earl warning forecasting using baseflow separation and machine learning in the Jiaojiang River Basin, Southeast China. *Sci. Tota Environ.* **2023**, *882*, 14. [CrossRef]
36. Xu, X.L.; Wang, S.Y.; Rong, W.Z. Construction of ecological network in Suzhou based on the PLUS and MSPA models. *Ecol. Indic* **2023**, *154*, 15. [CrossRef]
37. Ding, W.F.; Sun, H.H. Prediction of PM2.5 concentration based on the weighted RF-LSTM model. *Earth Sci. Inform.* **2023**, *16* 3023–3037. [CrossRef]
38. Yang, Y.L.; Shi, M.C.; Liu, B.J.; Yi, Y.; Wang, J.Y.; Zhao, H.Y. Contribution of ecological restoration projects to long-term changes ir PM2.5. *Ecol. Indic.* **2024**, *159*, 13. [CrossRef]
39. Gong, J.; Li, J.Y.; Yang, J.X.; Li, S.C.; Tang, W.W. Land Use and Land Cover Change in the Qinghai Lake Region of the Tibetar Plateau and Its Impact on Ecosystem Services. *Int. J. Environ. Res. Public Health* **2017**, *14*, 21. [CrossRef]
40. Costanza, R.; de Groot, R.; Sutton, P.; van der Ploeg, S.; Anderson, S.J.; Kubiszewski, I.; Farber, S.; Turner, R.K. Changes in the global value of ecosystem services. *Glob. Environ. Chang. Hum. Policy Dimens.* **2014**, *26*, 152–158. [CrossRef]
41. Xie, G.D.; Zhang, C.X.; Zhen, L.; Zhang, L.M. Dynamic changes in the value of China's ecosystem services. *Ecosyst. Serv.* **2017**, *26* 146–154. [CrossRef]
42. Ai, M.S.; Chen, X.; Yu, Q. Spatial correlation analysis between human disturbance intensity (HDI) and ecosystem services value (ESV) in the Chengdu-Chongqing urban agglomeration. *Ecol. Indic.* **2024**, *158*, 18. [CrossRef]
43. Li, J.; Qiu, J.; Amani-Beni, M.; Wang, Y.Y.; Yang, M.; Chen, J.W. A Modified Equivalent Factor Method Evaluation Model Based or Land Use Changes in Tianfu New Area. *Land* **2023**, *12*, 22. [CrossRef]
44. Chang, Y.X.; Zou, T.H.; Yoshino, K.; Luo, S.Z.; Zhou, S.G. Ecological policy benefit valuation based on public feedback: Forest ecosystem services in Wuyishan nature reserve, China. *Sci. Total Environ.* **2019**, *673*, 622–630. [CrossRef]
45. Popak, A.E.; Markwith, S.H. Economic Valuation of Bee Pollination Services for Passion Fruit (Malpighiales: Passifloraceae Cultivation on Smallholding Farms in Sao Paulo, Brazil, Using the Avoided Cost Method. *J. Econ. Entomol.* **2019**, *112*, 2049–2054 [CrossRef]
46. Pelorosso, R.; Apollonio, C.; Rocchini, D.; Petroselli, A. Effects of Land Use-Land Cover Thematic Resolution on Environmental Evaluations. *Remote Sens.* **2021**, *13*, 26. [CrossRef]
47. Mugiraneza, T.; Ban, Y.F.; Haas, J. Urban land cover dynamics and their impact on ecosystem services in Kigali, Rwanda using multi-temporal Landsat data. *Remote Sens. Appl.-Soc. Environ.* **2019**, *13*, 234–246. [CrossRef]
48. Bidone, F. Driving governance beyond ecological modernization: REDD plus and the Amazon Fund. *Environ. Policy Gov.* **2022** *32*, 110–121. [CrossRef]
49. Moore, C.M.; Catella, S.A.; Abbott, K.C. Population dynamics of mutualism and intraspecific density dependence: How θ-logistic density dependence affects mutualistic positive feedback. *Ecol. Model.* **2018**, *368*, 191–197. [CrossRef]
50. Moller, S.G.; Rajan, S.; Moller-Hansen, S.; Kragholm, K.; Ringgren, K.B.; Folke, F.; Hansen, C.M.; Lippert, F.K.; Kober, L.; Gislason G.; et al. Pre-hospital factors and survival after out-of-hospital cardiac arrest according to population density, a nationwide study *Resusc. Plus* **2020**, *4*, 9. [CrossRef]
51. Hontecillas-Prieto, L.; García-Domínguez, D.J.; Palazón-Carrión, N.; García-Sancho, A.M.; Nogales-Fernández, E.; Jiménez-Cortegana, C.; Sánchez-León, M.L.; Silva-Romeiro, S.; Flores-Campos, R.; Carnicero-González, F.; et al. CD8+NKs as a potential biomarker of complete response and survival with lenalidomide plus R-GDP in the R2-GDP-GOTEL trial in recurrent/refractory diffuse large B cell lymphoma. *Front. Immunol.* **2024**, *15*, 10. [CrossRef] [PubMed]
52. Luo, R.; He, D.M. The dynamic impact of land use change on ecosystem services as the fast GDP growth in Guiyang city. *Ecol. Indic.* **2023**, *157*, 12. [CrossRef]
53. Hu, H.; Zhang, G.H.; Ao, J.F.; Wang, C.L.; Kang, R.H.; Wu, Y.L. Multi-information PointNet plus plus fusion method for DEM construction from airborne LiDAR data. *Geocarto Int.* **2023**, *38*, 18. [CrossRef]
54. Jiang, M.J.; Lu, Y.X.; Wang, H.N.; Chen, Y.R. Multi-field coupling analysis of mechanical responses in methane hydrate exploitation with a practical numerical approach combining T plus H with DEM. *Comput. Geotech.* **2024**, *166*, 18. [CrossRef]
55. Tu, W.F.; Li, L.P.; Li, S.C.; Zhou, Z.Q.; Lei, T.; Sun, S.Q. VC plus plus Software Development and Stability Analysis of a Slope Based on the Slice-Free Method. *Geotech. Geol. Eng.* **2020**, *38*, 1311–1322. [CrossRef]
56. Reheman, R.; Kasimu, A.; Duolaiti, X.; Wei, B.H.; Zhao, Y.Y. Research on the Change in Prediction of Water Production in Urban Agglomerations on the Northern Slopes of the Tianshan Mountains Based on the InVEST-PLUS Model. *Water* **2023**, *15*, 19 [CrossRef]
57. Nazemi, R.; Daryasafar, A.; Bazyari, A.; Najafi, S.A.S.; Ashoori, S. Modeling asphaltene precipitation in live crude oil using cubic plus association (CPA) equation of state. *Pet. Sci. Technol.* **2020**, *38*, 257–265. [CrossRef]
58. Kaus, A.; Michalski, S.; Hänfling, B.; Karthe, D.; Borchardt, D.; Durka, W. Fish conservation in the land of steppe and sky: Evolutionarily significant units of threatened salmonid species in Mongolia mirror major river basins. *Ecol. Evol.* **2019**, *9*, 3416–3433. [CrossRef]

59. Braziene, A.; Vencloviene, J.; Tamosiunas, A.; Dedele, A.; Luksiene, D.; Radisauskas, R. The influence of proximity to city parks and major roads on the development of arterial hypertension. *Scand. J. Public Health* **2018**, *46*, 667–674. [CrossRef]
60. Assadhan, B.; Bashaiwth, A.; Binsalleeh, H. Enhancing Explanation of LSTM-Based DDoS Attack Classification Using SHAP With Pattern Dependency. *IEEE Access* **2024**, *12*, 90707–90725. [CrossRef]
61. Song, H.L.; Li, Y.T.; Zou, X.F.; Hu, P.; Liu, T.B. Elite male table tennis matches diagnosis using SHAP and a hybrid LSTM-BPNN algorithm. *Sci. Rep.* **2023**, *13*, 17. [CrossRef] [PubMed]
62. Li, R.J.; Feng, K.L.; An, T.; Cheng, P.J.; Wei, L.C.; Zhao, Z.H.; Xu, X.Y.; Zhu, L. Enhanced Insights into Effluent Prediction in Wastewater Treatment Plants: Comprehensive Deep Learning Model Explanation Based on SHAP. *ACS EST Water* **2024**, *4*, 1904–1915. [CrossRef]
63. Li, Q.X.; Ji, Y.J.; Zhu, M.R.; Zhu, X.Y.; Sun, L.J. Unsupervised feature selection using chronological fitting with Shapley Additive explanation (SHAP) for industrial time-series anomaly detection. *Appl. Soft. Comput.* **2024**, *155*, 18. [CrossRef]
64. Cruz-Victoria, J.C.; Netzahuatl-Muñoz, A.R.; Cristiani-Urbina, E. Long Short-Term Memory and Bidirectional Long Short-Term Memory Modeling and Prediction of Hexavalent and Total Chromium Removal Capacity Kinetics of *Cupressus lusitanica* Bark. *Sustainability* **2024**, *16*, 25. [CrossRef]
65. Aryal, K.; Ojha, B.R.; Maraseni, T. Perceived importance and economic valuation of ecosystem services in Ghodaghodi wetland of Nepal. *Land Use Policy* **2021**, *106*, 12. [CrossRef]
66. Liu, D.D.; Guo, S.L.; Liu, P.; Zou, H.; Hong, X.J. Rational Function Method for Allocating Water Resources in the Coupled Natural-Human Systems. *Water Resour. Manag.* **2019**, *33*, 57–73. [CrossRef]
67. Kane, D.S.; Pope, K.L.; Koupal, K.D.; Pegg, M.A.; Chizinski, C.J.; Kaemingk, M.A. Natural resource system size can be used for managing recreational use. *Ecol. Indic.* **2022**, *145*, 7. [CrossRef]
68. Zhou, Y.X.; Lu, B.; Jia, W.; Huang, C.J.; Ma, Y.Q. The conflict between natural resource use and welfare supply-Natural resources is a bless or a curse? *Resour. Policy* **2023**, *82*, 9. [CrossRef]
69. Vande Velde, K.; Huge, J.; Friess, D.A.; Koedam, N.; Dahdouh-Guebas, F. Stakeholder discourses on urban mangrove conservation and management. *Ocean Coast. Manag.* **2019**, *178*, 11. [CrossRef]
70. Carbone, J.C.; Bui, L.T.M.; Fullerton, D.; Paltsev, S.; Wing, I.S. When and How to Use Economy-Wide Models for Environmental Policy Analysis. *Annu. Rev. Resour. Econ.* **2022**, *14*, 447–465. [CrossRef]
71. Boonman, H.; Verstraten, P.; van der Weijde, A.H. Macroeconomic and environmental impacts of circular economy innovation policy. *Sustain. Prod. Consump.* **2023**, *35*, 216–228. [CrossRef]
72. Li, W.Q.; Zinda, J.A.; Zhang, Z.M. Does the "Returning Farmland to Forest Program" Drive Community-Level Changes in Landscape Patterns in China? *Forests* **2019**, *10*, 25. [CrossRef]
73. Jiang, H.Q.; Wu, W.J.; Wang, J.N.; Yang, W.S.; Gao, Y.M.; Duan, Y.; Ma, G.X.; Wu, C.S.; Shao, J.C. Mapping global value of terrestrial ecosystem services by countries. *Ecosyst. Serv.* **2021**, *52*, 10. [CrossRef]
74. Li, J.W.; Dong, S.C.; Li, Y.; Wang, Y.S.; Li, Z.H.; Li, F.J. Effects of land use change on ecosystem services in the China-Mongolia-Russia economic corridor. *J. Clean Prod.* **2022**, *360*, 14. [CrossRef]
75. Austhof, E.; Warner, S.; Helfrich, K.; Pogreba-Brown, K.; Brown, H.E.; Klimentidis, Y.C.; Walter, E.S.; Jervis, R.H.; White, A.E. Exploring the association of weather variability on Campylobacter—A systematic review. *Environ. Res.* **2024**, *252*, 11. [CrossRef]
76. Swenson, L.M. Can Simple Metrics Identify the Process(es) Driving Extreme Precipitation? *Climate* **2023**, *11*, 22. [CrossRef]
77. Yao, H.; Peng, H.J.; Hong, B.; Ding, H.W.; Hong, Y.T.; Zhu, Y.X.; Wang, J.; Cai, C. Seasonal and diurnal variation in ecosystem respiration and environmental controls from an alpine wetland in arid northwest China. *J. Plant Ecol.* **2022**, *15*, 933–946. [CrossRef]

land

MDPI

Article

Spatiotemporal Relationship Between Carbon Metabolism and Ecosystem Service Value in the Rural Production–Living–Ecological Space of Northeast China's Black Soil Region: A Case Study of Bin County

Yajie Shang [1,2], Yuanyuan Chen [1,2], Yalin Zhai [1,2] and Lei Wang [1,2,*]

1 College of Landscape Architecture, Northeast Forestry University, Harbin 150040, China; shangyajie0121@outlook.com (Y.S.); chenyuanyuan@nefu.edu.cn (Y.C.); zhaiyl_a@nefu.edu.cn (Y.Z.)
2 Key Lab for Garden Plant Germplasm Development & Landscape Eco-Restoration in Cold Regions of Heilongjiang Province, Harbin 150040, China
* Correspondence: wanglei@nefu.edu.cn

Abstract: Amid global climate challenges and an urgent need for ecological protection, the northeastern black soil region—one of the world's remaining "three major black soil regions"—confronts significant tensions between agricultural economic development and land ecological protection, threatening national food security. Based on the "production–ecology–life" (PLE) classification system, this study established a dual-dimensional evaluation for carbon metabolism and ESV in horizontal and vertical dimensions. The horizontal flow of carbon and ESV was traced across different ecosystems, while the spatial and temporal dynamics of carbon metabolism and ESV were analyzed vertically. Spatial autocorrelation analyses were employed to examine the interaction patterns between carbon metabolism and ESV. The findings reveal that (1) cropland production space remains the dominant spatial type, exhibiting fluctuating patterns in the size of other spatial types, with a notable reduction in water ecological space. (2) From 2000 to 2020, high-value carbon metabolism density areas were primarily concentrated in the central region, while low-value areas gradually decreased in size. Cropland production space and urban living space served as key compartments and dominant pathways for carbon flow transfer in the two periods, respectively. (3) The total ecosystem service value (ESV) showed a downward trend, decreasing by CNY 1.432 billion from 2000 to 2020. The spatial distribution pattern indicates high values in the center and northwest, contrasting with lower values in the southeast. The flow of ecological value from forest ecological space to cropland production space represents the main loss pathway. (4) A significant negative correlation exists between carbon metabolism density and ESV, with areas of high correlation predominantly centered around cropland production space. This study provides a scientific foundation for addressing the challenges facing the black soil region, achieving synergistic resource use in pursuit of carbon neutrality, and constructing a more low-carbon and sustainable spatial pattern.

Keywords: production–living–ecological space; carbon metabolism; ecosystem service value; spatial and temporal relationships; northeastern black soil area

1. Introduction

Carbon emissions are currently one of the most significant contributors to global climate change, leading to increasing environmental concerns [1]. Tracking regional carbon inflows and outflows from a metabolic perspective can effectively quantify the strength of carbon cycle flows between compartments of regional ecosystems, thereby identifying key nodes for carbon reduction and enhancement [2]. As significant carbon reservoirs within the carbon cycle, terrestrial ecosystems are crucial in combating global climate change and achieving regional carbon neutrality. These ecosystems provide goods and services to humans directly and indirectly, and their value is reflected in ecosystem service values (ESVs) [3].

Imbalances in carbon metabolism not only directly contribute to climate change but also indirectly affect ecosystem services through various pathways. For instance, excessive carbon emissions exacerbate climate change and disrupt climate patterns, which, in turn, destabilize ecosystems, reduce carbon sequestration capacity, and weaken the effectiveness of carbon sinks, ultimately diminishing the value of ecosystem services [4]. Conversely, healthy ecosystems can effectively absorb carbon dioxide and enhance carbon storage capacity, thereby mitigating the impacts of climate change. The economic value of their carbon sequestration services can be quantified through the ecosystem service value (ESV) [5]. It is important to note that the relationship between carbon metabolism and ecosystem service value is not limited to carbon sequestration; it is also closely linked to the evolution of PLE space. Urbanization and changes in agricultural land use are often accompanied by alterations in land cover and ecosystem types, which directly influence the sources and intensity of carbon metabolism while also changing the type and spatial distribution of ecosystem services [6]. The reduction in ecological space may lead to an imbalance in carbon metabolism, resulting in increased carbon emissions and the degradation of carbon storage functions, thus lowering the value of ecosystem services. Therefore, carbon metabolism and ecosystem service value interact and influence each other, with healthy ecosystems regulating carbon metabolism through their carbon sink functions [7]. To achieve the "dual-carbon" goal, in-depth studies of the spatial and temporal relationship between these factors provide a theoretical foundation for the development of more scientifically informed environmental policies and carbon reduction strategies [8].

Carbon is the material basis for various human socio-economic activities and a product of emissions. It is a crucial indicator of the coupled development and utilization of regional water, soil, and energy resources. Carbon metabolism refers to the flow of carbon between the biosphere, atmosphere, and urban systems, encompassing both carbon emission and absorption processes [9,10]. Current academic research on carbon metabolism primarily focuses on land use types with differing carbon sink functions at the urban level. Accounting methods vary based on their specific applications [11]; among these, carbon emission factor methods, model estimation methods, and remote sensing and mapping estimation methods have become the predominant approaches in the relevant research [11,12]. This study primarily focuses on the characteristics of the spatial distribution of carbon metabolism [13], the processes of carbon flow [14–16], the underlying driving mechanisms [17], and scenario modeling [18].

Building on the UN Millennium Ecosystem Assessment's research on ecosystem valuation and ecological well-being, Constanza and other scholars have delved deeper into the value of environmental services [19]. Researchers predominantly employ the equivalent factor method, the benefit transfer method, and the ecosystem service value (ESV) index method to calculate ESV in the context of land use change. Research has concentrated on the spatiotemporal patterns of ecosystem service values (ESVs) [20], simulation predictions [21], trade-offs and synergistic relationships among ESVs [3,22], and the spatiotemporal correlations of ESVs [23]. In recent years, scholars have increasingly focused on the correlation between ESV and carbon emissions [24,25]. Frequent anthropogenic activities significantly affect the sustainable provision of ecosystem services. Research findings have revealed a significant spatial correlation between carbon emissions and ecosystem service value (ESV). However, most studies have focused on provincial and urban scales, with limited attention to local spatial scales within the national territory. In the context of China's unique national conditions, it is particularly important to investigate the spatial relationship between carbon metabolism and the ESV at the county scale.

The report of the 18th National Congress of the Communist Party of China (CPC) introduced the concept of shaping a beautiful ecological space, an intensive and efficient production space, and a livable and moderate PLE space as part of the construction of a beautiful China. The term 'PLE space' refers to the following three interconnected space types: production, living, and ecological spaces [26]. The division of PLE space is not only a representation of the current situation but also emphasizes the evolution of spatial patterns over time. By analyzing the temporal and spatial dynamics of PLE spaces, it becomes possible to reveal the conversion relationships between different space types, the shrinkage or expansion of ecological spaces, and the dynamic changes in production and living spaces. This approach is crucial for understanding the impact of land use changes on carbon emissions and carbon sequestration [15]. Furthermore, the PLE space classification system considers not only economic development and production needs but also emphasizes the balance between social well-being and ecological protection [27]. PLE space can systematically encompass all land functions and coordinate the diverse needs of society, the economy, and the environment, thus promoting integrated management. This comprehensive perspective holds significant practical value for decision making in carbon emission management, ecological restoration, and resource allocation.

Even though relevant research has produced rich results and the theoretical foundations have been increasingly improved, existing studies have mainly focused on the driving mechanisms, spatial and temporal characteristics, and changes in patterns of carbon emissions in large-scale regions. There is a lack of in-depth exploration of the flow of carbon in small-scale areas and its relationship with land use under the 'Ecology–Production–Life' spatial system. In particular, there is a lack of integrated research on the process of carbon metabolism and the ecosystem value of land use in PLE spaces. Neglecting the PLE space classification system in studies of carbon metabolism and ecosystem service value introduces several limitations. The PLE space framework offers an integrated, cross-sectoral, and cross-disciplinary perspective that enables a comprehensive assessment of the synergistic effects between carbon metabolism and ecosystem services. It avoids the constraints imposed by relying on a single land use classification. Furthermore, the PLE space framework enhances the analysis of carbon emission reduction and ecological protection strategies across different regions while revealing the complex relationship between human activities and ecosystem functions, particularly during land use changes and urbanization. Therefore, studies that do not incorporate the PLE space framework are unable to fully capture the dynamic interactions between carbon metabolism and ecosystem services. Furthermore, the single carbon emission factor method commonly employed by research institutes has inherent limitations, failing to fully capture land use changes and carbon flows within ecosystems. This approach primarily relies on fixed emission factors and overlooks multiple sources of carbon emissions, including agriculture, energy consumption, livestock respiration, and straw burning. To address these limitations, research should integrate multiple accounting pathways to enhance the accuracy of carbon emission assessments.

As fundamental units of urban and rural development, counties are characterized by extensive agricultural activities and significant ecosystem service values [28]. Their land use is primarily composed of farmland, forests, and wetlands, which play a vital role in carbon sequestration and emissions reduction. Effective carbon emission management at the county level should integrate energy optimization with ecosystem services to enhance the carbon absorption capacity of agricultural and natural ecosystems [29]. By delineating ecological functional zones and implementing differentiated strategies, a balance between agricultural production and ecological conservation can be achieved, fostering sustainable low-carbon development. In recent years, Bin County has introduced various ecological protection policies to advance green development and environmental sustainability. As part of Harbin Municipal Government's 14th Five-Year Plan for Ecological and Environmental Protection, Bin County has actively participated in initiatives targeting key areas such as soil pollution prevention and control, groundwater protection, and rural ecological and environmental management.

Therefore, building on existing research, this study establishes a spatial classification system of 'Ecology–Production–Life' (PLE) in Bin County and comprehensively evaluates carbon metabolism and the ecosystem service value (ESV) in both horizontal and vertical dimensions (Figure 1). The vertical assessment focuses on accounting for carbon metabolism and ESV quantities in unit areas simultaneously, revealing both factors' spatial and temporal evolution characteristics. Conversely, horizontal measurements more effectively track the flow of carbon and ESV between different ecosystems throughout the study period. The study proposes, for the first time, a dual-dimensional evaluation framework, which provides a framework for the in-depth analyses of carbon metabolism and ESV in both vertical and horizontal directions, as opposed to the traditional method that focuses only on changes in carbon metabolism or ESV in a single dimension. This method is used to study the flow of regional substances in different ecosystems in different directions, especially in the agricultural counties in the black soil area. It can solve the dual problems of its economic development and ecological protection to address the limitations of previous research methods.

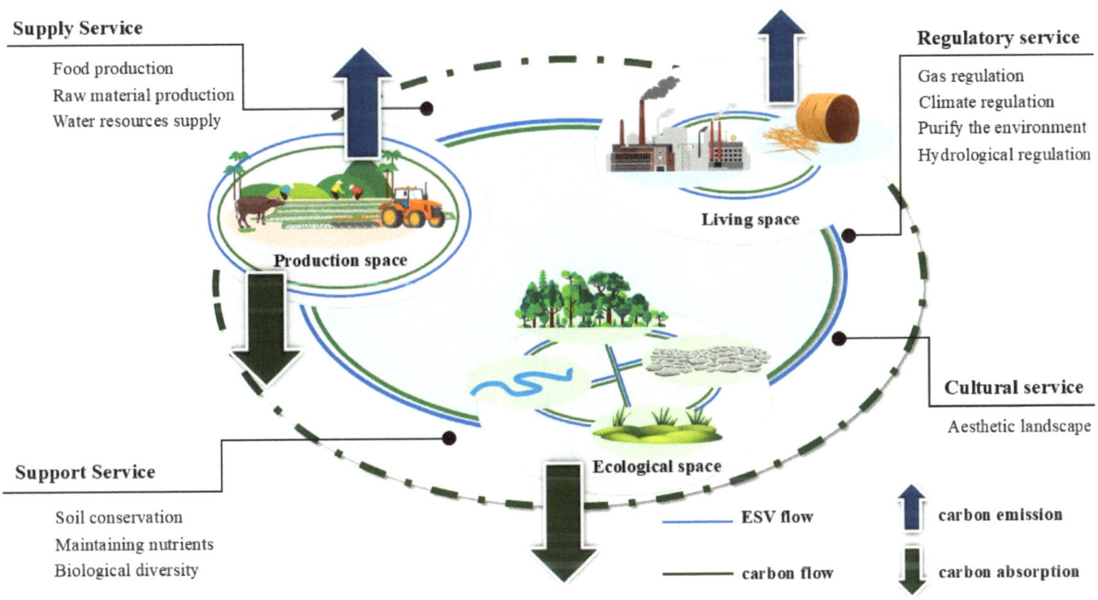

Figure 1. Spatial carbon metabolism and ESV horizontal and vertical transfer processes.

This study constructs a dual-dimensional evaluation for carbon metabolism and ecosystem service value (ESV) based on the "Ecology–Production–Life" (PLE) classification framework, analyzing their spatial and temporal dynamics in both horizontal and vertical dimensions. The flow of carbon and ESV across different ecosystems is quantified to reveal their spatial and temporal evolution, while spatial autocorrelation analysis is employed to explore the interaction mechanisms between carbon metabolism and ESV. The comprehensive assessment model integrates vertical and horizontal dimensions to provide a holistic understanding of carbon metabolism and ESV. Additionally, future strategies for sustainable development in Bin County are proposed for different sub-regions, offering a scientific basis and policy recommendations for promoting ecological sustainability in agricultural counties within the black soil region, with Bin County as a representative case.

2. Materials and Methods

2.1. Study Area

Bin County, situated in the southern part of Heilongjiang Province, covers 3843 km^2. Its geographical coordinates range from 126°55′41″ to 128°19′17″ (east longitude) and from 45°30′37″ to 46°01′20″ (north latitude). Renowned as the eastern suburban garden of Harbin City, Bin County is rich in resources. However, the rapid economic development in the region has exacerbated the conflict between agricultural economic growth and land ecological protection. Inappropriate developmental practices have significantly increased carbon emissions and environmental degradation. Agricultural production is a significant industry in Bin County, and as societal demands have grown, crop yields have risen correspondingly. This increase has been accompanied by a dramatic increase in the use of pesticides and fertilizers and heightened intensity in developing and cultivating agricultural land. Unfortunately, the absence of rational planning and regulatory policies has resulted in an imbalance in the carbon dynamics of Bin County. Therefore, exploring the spatial and temporal evolution of carbon metabolism and ecosystem service values across the PLE space in Bin County and their intercorrelation is crucial in better understanding the ecological impacts in the region (Figure 2).

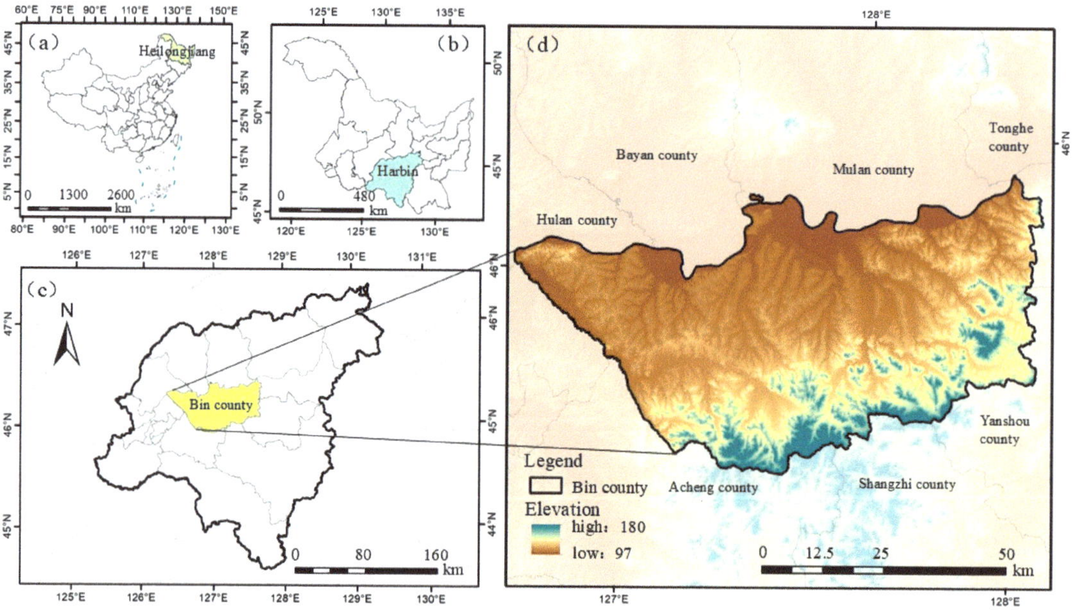

Figure 2. (**a**) Location map of Heilongjiang province. (**b**) Location map of Harbin city. (**c**) Location map of Bin county. (**d**) Map of the research area.

This study's selection of Bin County, a typical agricultural county, is significant and representative. (1) As a prominent agricultural county in the black soil region, Bin County has undergone extensive reclamation and utilization, resulting in frequent ecological and environmental issues. Consequently, there is an urgent need for a scientifically informed approach to sustainable development. (2) Using the Harbin Municipal Government's '14th Five-Year Plan' for ecological environmental protection as a policy guideline, it was pointed out that Bin County should carry out ecological protection work in areas such as rural ecological environment management. (3) The northeast black soil region is one of the only remaining 'three major black soil regions' in the world, which is known as the 'Giant Panda of Arable Land' because of its fertile soil and high food production capacity. As a typical county in the black soil area, Bin County has apparent advantages in cropland resources. However, Bin County is facing the dilemma of green development, and its research results can provide reference and guidance for other similar regions in the world, as well as new research perspectives and suggestions for managers formulating policies to protect the black soil area.

2.2. Data Sources

The data for this study include both spatial and statistical data. Detailed information on the data used in this study is provided in Table 1.

Table 1. Spatial data and statistics required for the research.

Data	Specific Indicators	Brief Introduction	Source
Spatial data	Land use data	Land use data for Bin County in 2000, 2010, and 2020, including the following seven land types: cropland, forest land, grassland, watershed, other land, and urban and rural land	Data Centre for Resource and Environmental Sciences, Chinese Academy of Sciences (DRC-CAAS) (https://www.resdc.cn/, accessed on 23 December 2024)
Statistical data	Agricultural data	Includes areas under crop cultivation, production of major crops, total mechanical power, irrigated areas, central livestock rearing, etc.	"China Energy Yearbook", "China Urban Statistical Yearbook", "China County Statistical Yearbook", and "Bin County Statistics Yearbook" from 2000 to 2020 (https://www.stats.gov.cn/sj/ndsj/, accessed on 23 December 2024)
	Energy data	Includes different industrial energy consumption	
	Demographic data	Includes rural resident population and urban resident population	

Table source: created by the author.

2.3. Methods

2.3.1. Production–Ecology–Life Classification System and Carbon Metabolism–ESV Dual-Dimensional Evaluation Framework

The case selected for analysis in this research is Bin County, located in Harbin City, where the distinction among PLE space is based on the core functional elements of land types [30]. Since different land types embody varying degrees of ecological, production, and living functions, the differentiation of the PLE space should rely on the primary functions of the land. For instance, cropland production is considered its core function as a cornerstone of national food security [31]. While cropland also possesses ecological functions, such as providing regulatory benefits through extensive agricultural planting, this study classifies cropland as cropland production space, emphasizing its primary role. Conversely, the ecological function of grassland is typically regarded as its core function, with the production function manifesting in animal husbandry activities. The ecological space of grassland is delineated according to its core function category and includes water bodies and woodland. Additionally, construction land is categorized into urban and rural living spaces to meet the study's requirements, with other construction land being grouped with urban land due to its minimal area. Therefore, concerning the existing classification standards, a "Production–Life–Ecology" classification system for Bin County is constructed based on the scientific requirements [32] (Table 2).

Table 2. Classification system of PLE space in Bin County.

PLE Space Type Based on PLE Dominant Function			**Land Use Classification**	
First Category	Secondary Category	Code	First Category	Secondary Category
Production space	Cropland production space	C	Cropland	Paddy field, dry land
Ecological space	Forest ecological space	F	Woodland	Wooded land, shrubland, open woodland
	Grassland ecological space	G	Grassland	High- and medium-cover grassland
	Water ecological space	W	Water	Rivers, canals, lakes, reservoirs, ponds, mudflats
	Other ecological Space	Un	Unused land	Marshland, bare land
Living space	Urban living space	Ur	Urban construction land	Townland, other building land
	Rural living space	R	Rural settlement land	Rural settlements

The research methodology employed in this study is illustrated in Figure 3. A spatial classification system of 'ecology–agriculture–urban' in Bin County is the foundation for this research, which utilizes land use vector data from Bin County between 2010 and 2020. A land transfer matrix is calculated based on land use/cover change using ArcGIS spatial analysis tools, facilitating the spatial visualization of land transfer scenarios. Carbon metabolism and ecosystem service value (ESV) are assessed comprehensively through a dual-dimensional evaluation. The horizontal dimension relies on the land use transfer matrix, enabling the effective tracking of carbon and ESV flows across different ecosystems via a carbon flow model and an ecosystem value gain/loss model. The vertical dimension evaluates the amount of carbon metabolism and ESV per unit area. Carbon metabolism is determined by calculating carbon absorption and emissions, while ESV is derived using the value per unit area equivalent factor method. The evolution of carbon metabolism density and ESV is then visualized through spatial analysis tools. Finally, spatial interactions between carbon metabolism and ESV are analyzed using bivariate autocorrelation to examine the spatial distribution of various clustering patterns, thereby identifying areas of environmental pressure and ecological potential within the study area.

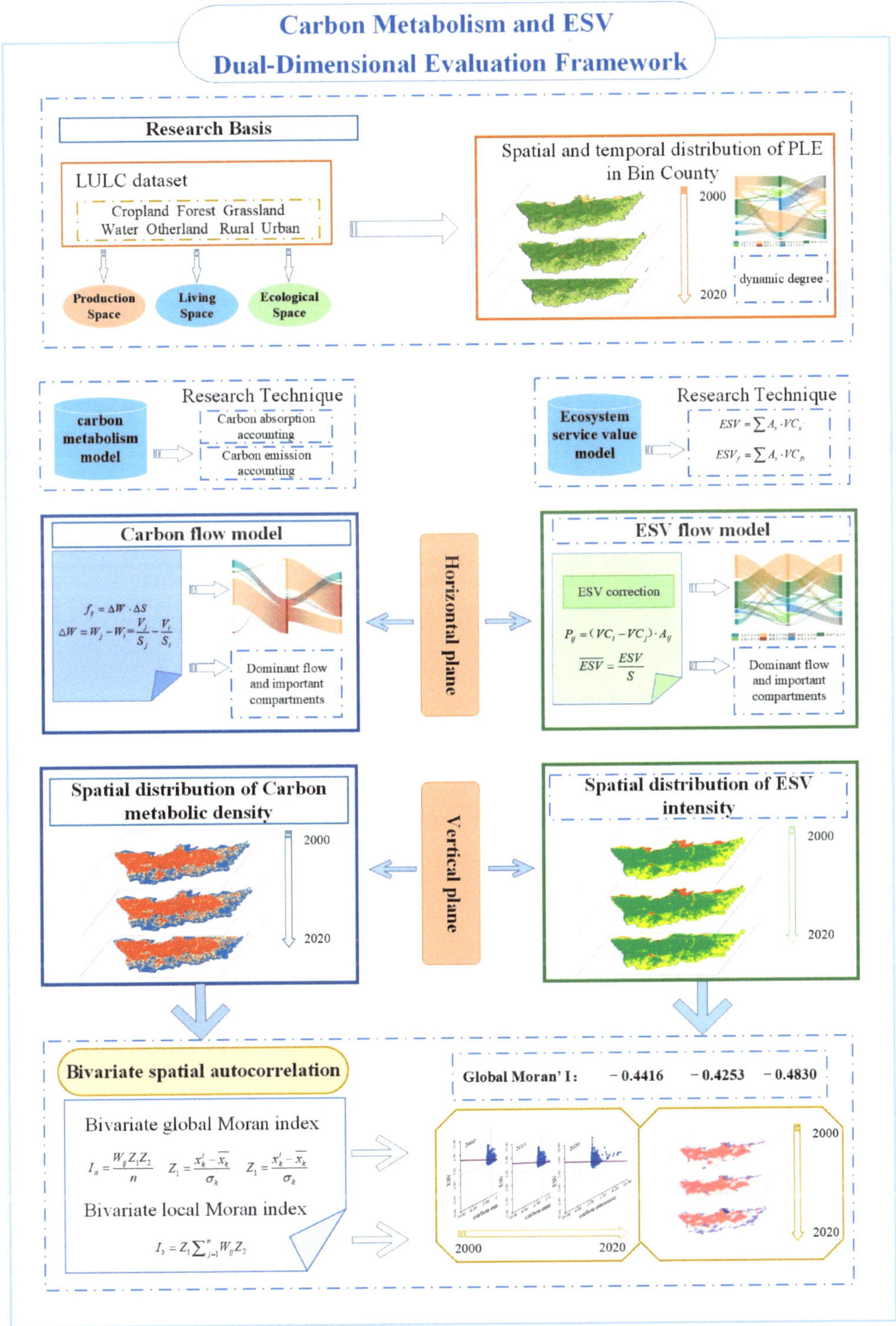

Figure 3. Research methodology flow chart.

2.3.2. Accounting for Carbon Metabolism in PLE Space

Regional carbon metabolism encompasses both carbon uptake and emission processes. This study primarily addresses the carbon fluxes associated with the following seven spatial land types: cropland, forest land, grassland, water bodies, urban land, rural land, and other land. The specific research methods are detailed in Table 3 [33,34].

Table 3. Carbon metabolism calculation methodology.

PLE Space Site Types	Carbon Emissions Calculator	Carbon Absorption Calculator	Carbon Emission/Carbon Absorption Factor
Cropland production space	Agricultural activities, livestock respiration	Estimation of carbon absorption in crop growth	—
Forest ecological space	—	Land use carbon sink factor approach	−0.578
Grassland ecological space	—	Land use carbon sink factor approach	−0.021
Water ecological space	—	Land use carbon sink factor approach	−0.252
Other ecological Space	—	Land use carbon sink factor approach	−0.005
Rural living space	Rural residents breathing, straw burning	Indirect estimation of carbon emissions	—
Urban living space	Respiration, energy consumption of urban residents	Indirect estimation of carbon emissions	—

(1) Estimation of carbon absorption

According to the Bin County PLE space classification system, cropland production space can contribute to carbon absorption and the various types of ecological space that function as carbon sinks. In this context, the ecological space employs the land use carbon sink coefficient method to calculate carbon absorption, as expressed by the following formula:

$$E = \sum \lambda_i = \sum S_i \cdot \delta_i \tag{1}$$

where E represents the total carbon emission (t); λi denotes the carbon emission (or carbon absorption) generated by the i th land use (t); Si refers to the land area (hm^2) associated with the i th land use, which includes various types of ecological space; and δi is the carbon emission (or absorption) coefficient of the i th land use (t/hm^2).

Given that the primary land type of agricultural production space is cropland, the indirect estimation method for carbon absorption through crop growth is employed [23]. In the calculation process, the main crops selected for this study, reflecting the actual conditions in Bin County, are rice, maize, and beans. The calculation formula is as follows:

$$C_d = C_f \cdot D_w = C_f \cdot \frac{Y_w}{H_i} \tag{2}$$

where C_d represents the carbon absorption of cropland; C_f denotes the carbon absorption rate of different crops; Y_w indicates the economic yield of various crops; D_w refers to the biological yield of other crops; and H_i is the economic coefficient of the i th crop [35] (Table 4).

Table 4. Economic coefficient and carbon absorption rate of main crops.

(Farm) Crops	H	Cf
Rice	0.45	0.4144
Maize	0.40	0.4709
Beans	0.34	0.4500

(2) Estimation of carbon emissions

Carbon emissions from agricultural production space, urban living space, and rural living space are accounted for separately. Considering the actual conditions in Bin County and the availability of data, the spatial carbon emissions from agricultural production are primarily calculated based on the agricultural inputs used in production activities. These inputs include fertilizer application, agricultural machinery, the use of agricultural films, and agricultural irrigation as follows:

$$E_a = C \cdot k_c + M \cdot k_M + W \cdot k_w + F \cdot k_f + G \cdot k_g \quad (3)$$

$$M = S \cdot D \quad (4)$$

where E_a represents the carbon emission from cropland; C denotes the total amount of fertilizer applied; M is the cultivated area of agricultural land; W represents the total power of agricultural machinery; F indicates the amount of agricultural film used; G is the irrigated area of agricultural land; S refers to the total area of agricultural land; and D is the replanting index. The different emission factors corresponding to various agricultural activities are provided in Table 5 [36].

Table 5. Carbon emission factors for each type of agricultural activity.

Activity	Fertilizer Application	Agricultural Machinery	Agricultural Films	Agricultural Cultivation	Agricultural Irrigation
Ratio	0.8956 t/t	0.00018 t C/kW	5.18 kg/kg	0.01647 t C/hm^2	0.26648 t C/hm^2

Carbon emissions from urban living spaces are primarily calculated based on urban residents' respiratory carbon emissions and energy-related carbon emissions. The respiratory carbon emissions of urban residents are determined by multiplying the urban population by the human respiratory carbon emission coefficients [37]. Energy carbon emissions are calculated using the following formula [38,39]:

$$E_t = \sum E_{ti} = \sum E_{ni} \cdot \theta_i \cdot f_i \quad (5)$$

$$CE_t = P_c / P_s \cdot E_t \quad (6)$$

where E_t represents the carbon emissions from construction land; E_{ti} denotes the carbon emissions from various fossil energy sources; E_{ni} is the consumption of different fossil energy sources; θ_i is the conversion coefficient of various fossil energy sources to standard coal; and f_i is the carbon emission coefficient of various fossil energy sources, with the conversion coefficient for standard coal provided in Table 6. Additionally, CE_t refers to the total carbon emissions of the county; P_c represents the total population density within the county; and P_s is the total population within the provincial area.

Carbon sources from rural living spaces are primarily calculated based on the carbon emissions from straw combustion and the respiration of rural residents. The calculation formula is as follows:

$$C_s = \sum_{i=1}^{n} P_i \cdot S_c \cdot \theta_i \cdot a \cdot b \tag{7}$$

where n represents the crop type (species); P_i is the yield of the ith crop (kg); S_c is the grain-to-straw ratio of the i th crop; θ_i is the carbon emission coefficient for straw combustion of the i th crop; a is the ratio of open burning of straw, with a value of 0.165; and b is the efficiency of straw combustion, with a value of 0.8 [40].

Table 6. Standard coal conversion coefficients and carbon emission coefficients of various energy sources.

Energy Type	Standard Coal Conversion Factor (kgce/kg, kgce/10^6 KJ, kgce/KW·h)	Carbon Emission Factor (tC/t)
Raw coal	0.7143	0.7559
Coke	0.9714	0.855
Crude oil	1.4286	0.5857
Petrol	1.4714	0.5538
Paraffin	1.4714	0.5714
Diesel oil	1.5714	0.5912
Fuel oil	1.4286	0.6185
Liquefied petroleum gas	1.7143	0.5042
Natural gas	1.2143	0.4483
Electricity	0.1229	2.5255

(3) Carbon metabolism density and carbon flow modeling

Based on the accounting of carbon emissions and absorption processes across various types of space, carbon fluxes on the vertical scale are established through the overall spatial cross-section at a single point in time. Consequently, the carbon fluxes of the PLE space are calculated by aggregating the total carbon emissions and carbon absorption within these spaces. In accordance with previous research methods [41,42], net carbon emission per unit area is defined as net carbon metabolism density. To quantify the carbon metabolism capacity of different spaces and to measure the resulting carbon fluxes and their directional changes within the PLE space, this study employs the difference in carbon metabolism density alongside the area of PLE space transfer. The calculation formula is as follows:

$$C_t = C_e + C_a \tag{8}$$

$$C = \sum_{i=1}^{n} \frac{S_i P_i}{S} \tag{9}$$

where C_t represents the total carbon flux in the PLE space; C_e denotes the carbon emission in the PLE space; C_a is the carbon absorption in the PLE space; S_i refers to the area of each type of space i within the grid; and P_i is the net carbon emission in each space. Additionally, C is the density of carbon metabolism; S represents the total area of the grid; S_i denotes the area corresponding to each type of PLE space i within the grid; and P_i indicates the carbon flux from each space.

$$f_{ij} = \Delta W \cdot \Delta S \tag{10}$$

$$\Delta W = W_j - W_i = \frac{V_j}{S_j} - \frac{V_i}{S_i} \tag{11}$$

here, f_{ij} represents the carbon flow from site j to site i (kg C); ΔW denotes the difference in carbon metabolism density (kg C/hm^2); ΔS is the area transferred from site j to site i (hm^2); W_i and Wj are the densities of net carbon flow in the horizontal direction at sites i and j (kg C/hm^2); V_i and V_j are the net carbon flows in the vertical direction at sites i and j (kg C); and S_i and S_j are the areas (hm^2) of sites i and j. If $f_{ij} > 0$, it indicates a positive carbon transfer that contributes to the balance of carbon metabolism; if $f_{ij} < 0$, it signifies a negative carbon transfer that exacerbates the imbalance in carbon metabolism.

2.3.3. Estimation of the Value of Ecosystem Services in PLE Space

(1) Correction for ESV per unit area

Revision of the economic value of food yields. The average grain production in Bin County from 2000 to 2020 was 3898.67 kg/hm^2, and the unit price of grain was based on the average price in Bin County for 2020, which was 2.52 CNY/kg. This reference was employed to determine the equivalence of other ecosystem services by utilizing the economic value of one equivalent as one-seventh of the market value of the average grain yield in the current year [43]. The calculation formula is as follows:

$$D = \frac{1}{7} \cdot \sum_{i=1}^{n} \left(\frac{P_i \cdot G_i}{S_i} \right) \tag{12}$$

where D denotes the standard equivalent ecosystem service value of PLE space in Bin County (CNY/ha); P_i represents the unit price of the i th crop in the current year (CNY); G_i indicates the annual production of the i th crop; and S_i signifies the planted area of the i th crop.

Land use type adjustment. In Bin County, the cropland production space is dryland and paddy farming, and the average value of these two equivalence factors is taken as the base equivalence. The forest ecological space in Bin County mainly consists of coniferous forests, mixed coniferous and broad forests, broad-leaved forests, shrub forests, etc., and the average of the four equivalence factors of these four woodland types is taken as the base equivalence. The ecological space of grassland in Bin County mainly refers to scrubland, grassland, and meadow, and the average value of the three equivalence factors is taken as the basis of its equivalence. The water ecological space in Bin County mainly refers to wetlands, and the wetland equivalent factor was used. The primary land type of living space in Bin County is construction land, which has a negative impact on the ecosystem due to the massive use of water resources and the production of pollutants that are harmful to humans. Based on previous research, the hydrological regulation of urban and rural living spaces is assigned a value of −7.55. The ecological space of unused land in Bin County mainly consists of desert and bare land, and the average value of these two equivalent factors was used [44].

Correction for regional coefficients of variation. Due to geographical and environmental condition variations across different regions, biomass levels also differ; higher biomass generally indicates stronger ecosystem service functions. In light of the specific circumstances of the study area, this research employs the net primary production potential instead of biomass to adjust the equivalent factor. The particular calculation formula [45] is as follows:

$$Q_t = \frac{NPP}{NPP_{mean}} \tag{13}$$

$$NPP = 3000\left[1 - e^{-0.0009695(V-20)}\right] \tag{14}$$

$$V = \frac{1.05R}{\sqrt{1 + (1 + 1.05/L)^2}} \quad (15)$$

$$L = 3000 + 25t + 0.05t^3 \quad (16)$$

where Q_t represents the regional coefficient of variation in year t; NPP denotes the net primary production potential of natural vegetation (t/(hm^2·a)); NPP_{mean} indicates the average net primary productivity across all types of vegetation; V is the actual annual evapotranspiration (mm); L is the average yearly evapotranspiration (mm); t refers to the average annual temperature (°C); and R is the average yearly precipitation (mm).

The average regulation coefficient of PLE space in Bin County from 2000 to 2020 is calculated to be 0.88.

(2) Determination of the value of ecosystem services per unit area

The coefficient of ecosystem service value per unit area was obtained after correcting for the economic value of grain yield, land use type, and regional variation coefficient, as presented in Table 7.

Table 7. ESV coefficients per unit area of the production–living–ecological space (CNY/hm^2).

Type of PLE Space	C	F	G	W	Ur	R	Un
ESV coefficients	5543.91	27,645.86	16,931.14	124,604.61	−10,596.59	−10,596.59	912.29

(3) Models for measuring the value of ecosystem services

Based on the actual conditions in Bin County and incorporating the standard equivalence and value coefficients derived from the calculations outlined in the above formula, the ecosystem service value of PLE space in Bin County can be determined using the following formulas [46]:

$$ESV = \sum A_x \cdot VC_x \quad (17)$$

$$ESV_f = \sum A_x \cdot VC_{fx} \quad (18)$$

where ESV represents the value of ecosystem services; A_x denotes the area of the land category; VC_x indicates the value coefficient; ESV$_f$ refers to the value of the f th ecosystem service function; and VC_{fx} is the value coefficient of the f th ecosystem service function.

The relationship between the value of ecosystem services and their intensity is as follows:

$$\overline{ESV} = \frac{ESV}{S} \quad (19)$$

where ESV denotes the ecosystem service value, $\overline{ESV}$ is the ecosystem service value intensity, and S is the grid area.

(4) Ecosystem service value flow gain/loss modeling

To reflect the spatial transfer flow of ecosystem service value and the direction of quantitative change while revealing the internal gains and losses of ecosystem service value [47], this research applies a value gain and loss model based on the land use transfer matrix. This approach quantifies ecosystem service value gains and losses within the study area. The calculation formula is as follows:

$$P_{ij} = \left(VC_i - VC_j\right) \cdot A_{ij} \quad (20)$$

where P_{ij} represents the ecosystem service value gain or loss resulting from the conversion of land use type i to type j (CNY); VC_i and VC_j are the ecosystem service value coefficients for land types i and j, respectively; and A_{ij} denotes the area of land use type i converted to type j (hm^2).

2.3.4. Bivariate Spatial Autocorrelation Analysis

Bivariate spatial autocorrelation refers to the clustering of the high and low values of two variables in space, which indicates whether the two variables are correlated spatially and assesses the degree of spatial autocorrelation [48]. The bivariate global Moran index quantifies the level of autocorrelation between the two variables within the specified space. The bivariate global Moran index is calculated as follows:

$$I_a = \frac{W_{ij} Z_1 Z_2}{n} \tag{21}$$

$$Z_1 = \frac{x_k^i - \overline{x_k}}{\sigma_k} \tag{22}$$

$$Z_2 = \frac{x_r^i - \overline{x_r}}{\sigma_r} \tag{23}$$

where I_a represents the bivariate global Moran's index; Z_1 and Z_2 are the standardized values of element i and element j for the k and r attributes, respectively; σ_k and σ_r denote the variances of the k and r attributes, respectively; W_{ij} is the weight matrix of the elements; and n is the number of study subjects.

The bivariate local Moran's index assesses the degree of association between two variables within a regional space. It is calculated as follows:

$$I_b = Z_1 \sum_{j=1}^{n} W_{ij} Z_2 \tag{24}$$

where I_b represents the bivariate local Moran's index, W_{ij} is the weight matrix of the elements, and n is the number of study objects.

3. Results Analysis

3.1. Analysis of the Evolution of Spatial and Temporal Patterns in PLE Space

In this study, land use information for Bin County in Harbin City was processed using ArcGIS, utilizing national land use data from 2000, 2010, and 2020. The data were classified into seven types of production–living–ecological (PLE) space (Figure 4). The results indicate that between 2000 and 2020, land use types in Bin County were predominantly characterized by cropland production space and forest ecological space, with the overall area remaining relatively stable and exhibiting minor local changes. Specifically, the area of agricultural production space experienced a slight decline from 2000 to 2010, followed by a continued decrease from 2010 to 2020. Forest ecological space, primarily concentrated in the southeast, remained stable overall. The area of grassland ecological space saw a significant increase from 2000 to 2010 but decreased in the northwest from 2010 to 2020. Additionally, the water ecological space remained stable during 2000–2010 but experienced a significant decline from 2010 to 2020, negatively impacting the overall ecosystem. Urban living space is concentrated in township centers, indicating population concentration, while rural living space, though more significant, is irregularly distributed, showing little change over the study period.

By combining the PLE space transfer heat map (Figure 5) with the Sankey diagram (Figure 6), it is evident that between 2000 and 2010, cropland production space in Bin

County primarily flowed into forest ecological space and rural residents' living space, covering areas of 99.53 km^2 and 46.77 km^2, respectively. Concurrently, forest ecological space converted to cropland production space, with an area of 125.87 km^2, indicating a gradual increase in the demand for agricultural land. Additionally, the change in grassland primarily originated from cropland production space, with an area of 37.16 km^2. In contrast, the area transferred out was relatively small, reflecting the local management unit's focus on ecological protection. Rural living space transformed 51.19 km^2 into cropland production space, whereas cropland also transferred 46.77 km^2 into rural living space. The transfer in and out of areas was roughly equivalent, indicating that the development of agricultural production and residents' living conditions has remained stable.

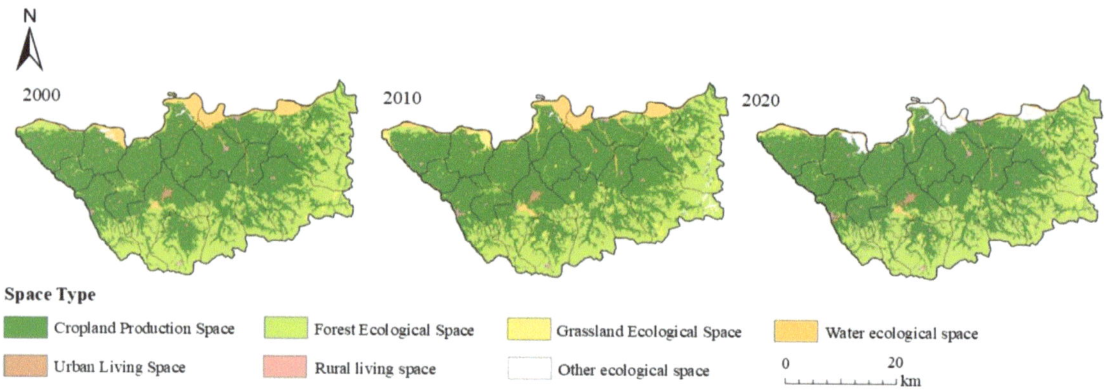

Figure 4. Spatial and temporal distribution of PLE in Bin County from 2000 to 2020.

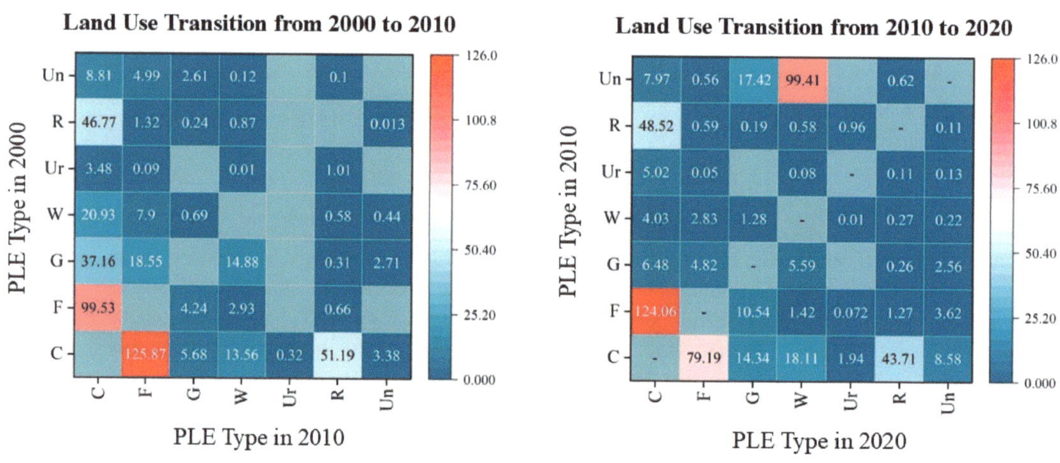

Figure 5. Heat map of land use transfer in Bin County from 2000 to 2020.

From 2010 to 2020, cropland production space in Bin County primarily flowed into forest ecological space and rural residents' living space, covering 124.06 km^2 and 48.52 km^2, respectively. Meanwhile, forest ecological space shifted to cropland production space by 79.19 km^2. Compared to the previous period, the area flowing into forest ecological space increased by 1.24 times, while the area of forest ecological space converting to cropland production space decreased by 1.58 times. During this period, grassland ecological space mainly transitioned to other ecological spaces and cropland production space, with areas of 17.42 km^2 and 14.34 km^2, respectively. Compared to the previous stage, the outflow area of the grassland ecological space showed a significant increase, highlighting that the importance of grassland ecological protection has yet to receive sufficient attention. Additionally, the inflow area from rural living space to cropland production space decreased from 51.19 km^2 to 43.71 km^2, and land transfers to various types of ecological space also declined. The conversion to forest ecological space increased from 0.66 km^2 to 1.27 km^2. This suggests that government and management have begun to focus on ecological protection despite the impact of urban development on production space.

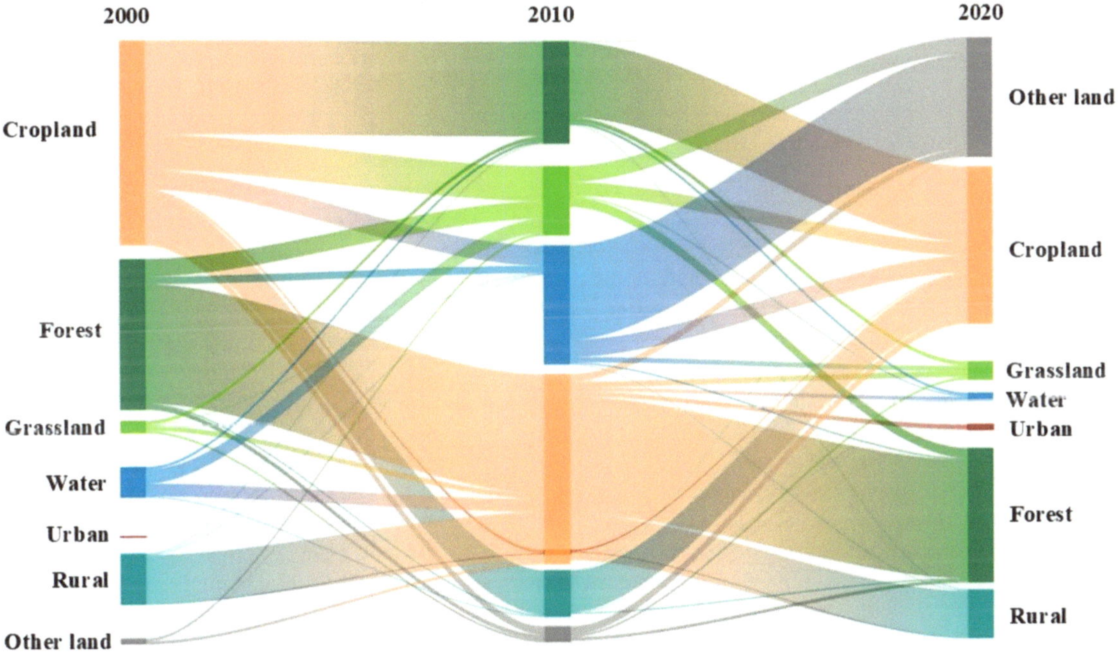

Figure 6. Sankey diagram of land use transfer in PLE space in Bin County from 2000 to 2020.

From Table 8, an analysis of the attitude toward individual PLE spaces in Bin County reveals that the attitude towards other ecological spaces from 2000 to 2020 is 9.34%. This is followed by grassland ecological space and urban living space, with attitudes of 7.11% and 4.62%, respectively. The most significant adverse change in land use attitude during this period occurred in water ecological space, which experienced a decrease of 15.15%. This was followed by cropland production space, which experienced a slight negative change of −0.21%.

Table 8. Dynamicity of the production–living–ecological space.

Attitude	Type of PLE Space	2000–2010	2010–2020	2000–2020
Homogeneous attitude	C	−0.07%	−0.13%	−0.21%
	F	−0.42%	0.45%	0.01%
	G	40.90%	−3.21%	7.11%
	W	−0.09%	−5.99%	−15.15%
	Ur	5.48%	2.01%	4.62%
	R	−0.41%	0.43%	0.01%
	Un	11.87%	59.52%	9.34%
Integration of dynamic attitudes		0.19%	0.44%	0.43%

Significant differences in the mobility of the three types of PLE space were observed during the study period. Grassland ecological space exhibited the highest mobility, at 40.90%, followed by other ecological spaces and urban living spaces, with mobilities of 11.87% and 2.48%, respectively, between 2000 and 2010. Meanwhile, different types of space displayed negative changes. In the subsequent period from 2010 to 2020, the motivation for other ecological spaces rose to the forefront, with urban living space following at 2.01%. Conversely, water body ecological space experienced the most significant negative change, while grassland ecological space and cropland production space also demonstrated downward trends. The shrinkage of water body ecological space was particularly drastic during this period.

In the context of Bin County, this change is closely related to implementing the policy of returning cropland to forest and grassland. Between 2000 and 2010, the motivation of grassland ecological space was as high as 40.90 percent, indicating that the grassland ecological area achieved significant growth during this period. From 2010 to 2020, the combined motivation of other ecological spaces also increased, indicating that environmental projects were still being promoted. Nevertheless, the dynamics of water ecological space show a negative change, suggesting that although environmental management has achieved initial results, we still need to pay attention to the existing problems to achieve the integrated and coordinated development of urban and rural areas and natural ecology.

3.2. Analysis of the Spatial and Temporal Evolution of Carbon Metabolism in PLE Space

(1) Carbon metabolism calculation results

By using land use data from Bin County, the carbon emissions and carbon absorption of PLE space over the past 20 years were estimated, with the results presented in Tables 4 and 5. The data reveal that carbon emissions from urban residents' living spaces from 2000 to 2020 consistently represented a dominant share of total carbon emissions, with an increasing proportion each year. Specifically, the emissions were 204.2×10^4 t C, 654.1×10^4 t C, and 607.1×10^4 t C, contributing 69.83%, 72.99%, and 90.44% of the total carbon emissions, respectively. As shown in Table 9, the study area exhibited a trend of total carbon emissions that increased and declined from 2000 to 2020. The peak was reached in 2010, with total emissions in Bin County at 896.16×10^4 t C, an increase of 3.07 times compared to 292.41×10^4 t C in 2000. After this peak, total emissions began to decline, with emissions in 2020 recorded at 671.25×10^4 t C, reflecting a decrease of 224.90×10^4 t C compared to 2010. Regarding carbon sinks, the total carbon sinks in Bin County in 2000, 2010, and 2020 were 42.75×10^4 t C, 107.94×10^4 t C, and 124.80×10^4 t C, respectively. While the total carbon sink in Bin County shows an increasing trend, it primarily comprises contributions from cropland, forest, grassland, water, and other ecological spaces. Notably, the carbon sink from cropland land production space accounted for 82.35%, 93.28%, and

94.18% of the total carbon sink in those years. The carbon sinks from other ecological spaces, such as grasslands and water bodies, decreased after peaking in 2010.

Table 9. Carbon emissions and carbon absorption of production–living–ecological space from 2000 to 2020.

Type of PLE Space	2000		2010		2020	
	Carbon Emissions /t C	Carbon Absorption /t C	Carbon Emissions /t C	Carbon Absorption /t C	Carbon Emissions /t C	Carbon Absorption /t C
C	838,303.89	352,024.58	2,370,254.82	1,006,794.54	593,803.90	1,175,286.13
W	-	70,452.82	-	67,481.93	-	70,542.97
G	-	30.89	-	157.19	-	106.70
W	-	4953.79	-	4909.89	-	1969.40
Ur	2,042,356.62	-	6,540,966.70	-	6,072,356.61	-
R	43,417.52	-	50,337.53	-	46,356.65	-
Un	-	4.26	-	9.31	-	64.70

(2) Characterization of the spatial distribution of carbon metabolism density

The carbon flux for each land use within the PLE space in Bin County was determined by calculating the carbon absorption and emissions associated with each land use type. By utilizing land use data, ArcGIS 10.7 software was employed to classify the carbon metabolism density values for Bin County from 2000 to 2020 with natural breaks. This classification resulted in the following five levels: lower, low–medium, medium, high, and extreme. The spatial distribution of carbon metabolism density in the PLE space for the years 2000, 2010, and 2020 was, thus, mapped (Figure 7). Notably, areas with extreme carbon metabolism density values are primarily concentrated in the central region of Bin County. These regions mainly rely on agricultural production and human activities as the primary sources of carbon emissions. The spatial area of cropland is concentrated and widely distributed, making it the primary contributor to carbon emissions and resulting in the highest carbon metabolism density values. In contrast, areas with lower carbon metabolism density values are predominantly located in the peripheral regions of the county, particularly within ecologically sustainable forest spaces. However, these areas have begun to decrease in size year after year, following a brief period of modest growth. Between 2000 and 2020, only minimal changes were observed in low-to-medium, medium, and extreme carbon metabolism density regions, with the overall number of regions remaining stable. The distribution of extreme and medium regions primarily surrounds high carbon metabolism density zones, exhibiting a diffuse trend and indicating the potential for transformation into low carbon metabolism density classes. Therefore, carbon reduction and greening strategies could be implemented in these areas to support future low-carbon development goals.

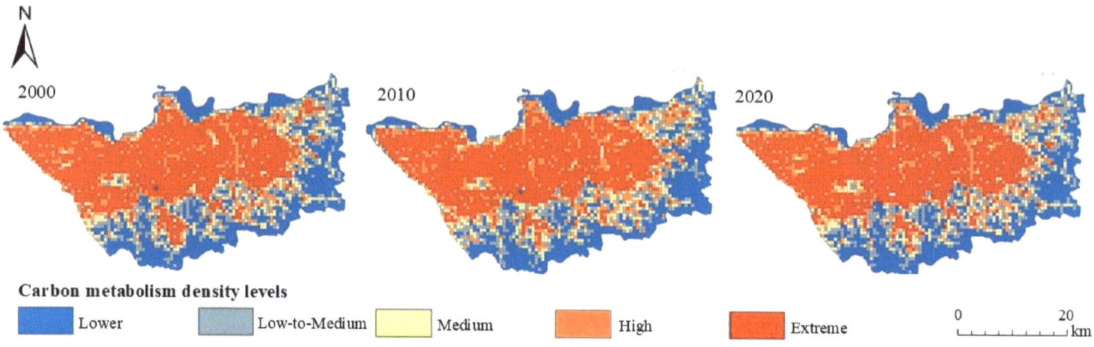

Figure 7. Spatial distribution map of carbon metabolism density in Bin County from 2000 to 2020.

(3) Carbon flow process analysis

According to the carbon metabolism density and land use transfer matrix, the carbon flow in the horizontal direction of various types of PLE space land in Bin County from 2000 to 2020 was obtained, as shown in Figure 8.

Based on the Sankey map of horizontal carbon flow in PLE space in Bin County (Figure 9), three primary carbon flow pathways were identified from 2000 to 2010, all indicating negative carbon flow. These include shifts from cropland production to urban living space and rural living space to urban living space, with transfer amounts of 188.77×10^4 t C and 54.97×10^4 t C, in that order. In this period, cropland production and forest land ecological space were the main carbon flow output compartments, while rural and urban living spaces were the primary input compartments. Positive carbon flow between 2000 and 2010 was primarily observed from urban living space to cropland production space and from cropland production space to forest ecological space, with transfer amounts of 8.4×10^4 t C and 1.5×10^4 t C, correspondingly. From 2010 to 2020, two main carbon flow pathways were identified as follows: cropland production space with negative carbon flow shifted to urban living space, and cropland production space with positive carbon flow shifted to forest land ecological space, with transfer amounts of 210.43×10^4 t C and 6.7×10^4 t C, in sequence. Compared to the previous period, interactions among carbon metabolism compartments were stronger between 2010 and 2020, marked by an increase in overall carbon transfer and a diversification of carbon flow pathways. The positive carbon metabolism compartments also contributed more significantly to positive carbon flows.

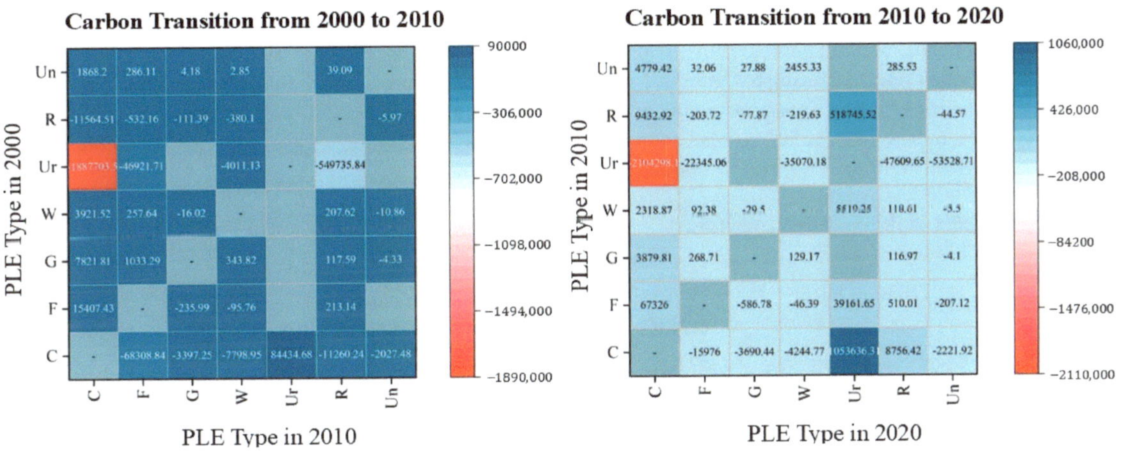

Figure 8. Heat map of carbon flow in Bin County from 2000 to 2020.

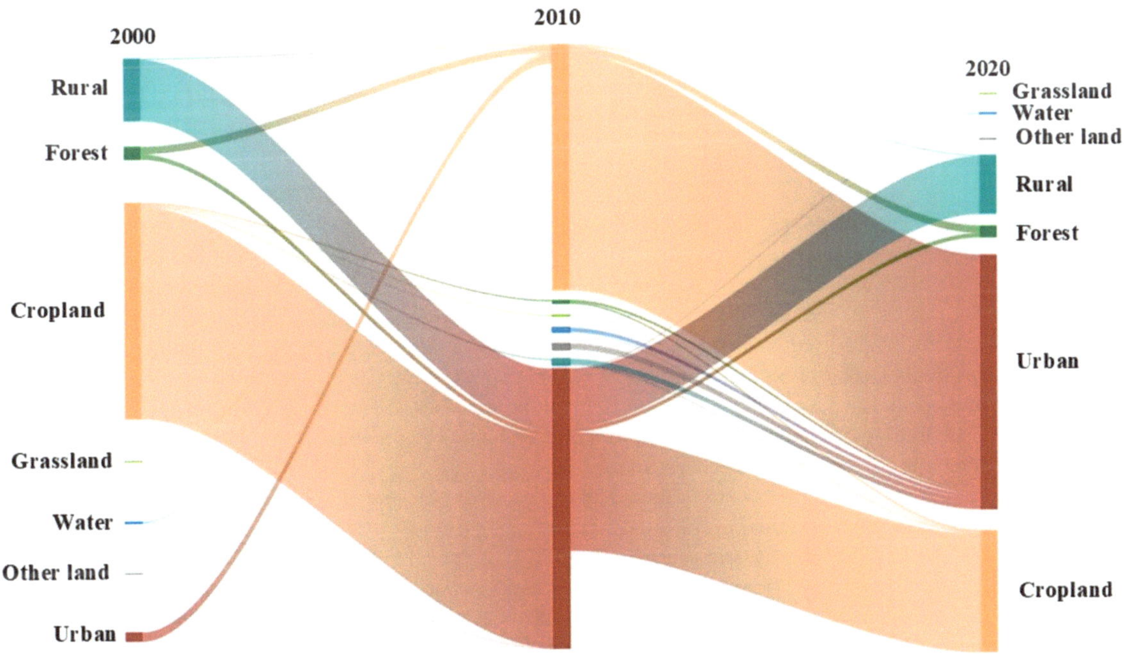

Figure 9. Sankey diagram of carbon flows in PLE space in Bin County from 2000 to 2020.

As shown in Table 10, the net carbon transfer in Bin County between 2000 and 2020 was consistently negative. This indicates that land use changes have adversely affected the region's carbon balance, resulting in an imbalanced carbon metabolism. The land conversion was more significant when shifting from negative to positive carbon metabolism compartments. From 2000 to 2010, positive carbon transfer was the lowest, amounting to only 11.60×10^4 t C. The primary source of this positive carbon transfer was the shift from urban living space to cropland production space, which accounted for 8.44×10^4 t C, marking it as the critical compartment and pathway for positive carbon transfer. The amount of negative carbon transfer was more than 22 times greater than that of positive carbon transfer. The shift in cropland production space to urban living space resulted in a negative carbon transfer of 188.77 × 10^4 t C, contributing 72.77%, with the net carbon transfer reaching a maximum of 247.82 × 10^4 t C. The land use changes driving negative carbon transfer during both periods primarily stemmed from this shift. This indicates that the high carbon emissions from urban living spaces are a significant factor obstructing the balance of regional carbon metabolism, peaking at 210.43 × 10^4 t C from 2010 to 2020, which accounted for 91.87% of the harmful carbon transfer.

Table 10. The dominant carbon transfer in the evolution of spatial patterns.

Exchange Value	2000–2010		2010–2020	
	Transitions (104 tC)	**Direction**	**Transitions (104 tC)**	**Direction**
Beneficial carbon flow	11.60	+	171.76	+
Harmful carbon flow	−259.41	−	−229.04	−
Net carbon flow	−247.82	−	−57.28	−
H1	8.44	Ur → C	105.36	Ur → C
Contribution ratio	72.81%	/	61.34%	/
H2	−188.77	C → Ur	−210.43	C → Ur
Contribution ratio	72.77%	/	91.87%	/

Note: "+" represents positive carbon flow direction; "−" represents negative flow direction.

3.3. *Analysis of the Spatial and Temporal Evolution of Ecosystem Service Values (ESV) in PLE Space*

(1) ESV Calculation Results

As shown in Table 11, the total ecosystem service value of the PLE space in Bin County decreased from 2000 to 2020, with a reduction of CNY 1.432 billion. The ecological service values for 2000, 2010, and 2020 were CNY 6.984 billion, CNY 5.227 billion, and CNY 5.419 billion, respectively, indicating a decline in overall ecosystem service value. The values for the ecological water space and cropland production space experienced the most significant reductions, amounting to CNY 1.475 billion and CNY 26 million, respectively. This phenomenon primarily results from the diminished values of production and ecological spaces. As the urban population increases, the demand for urban land rises, leading to a reduction in the areas of cropland and water bodies, which subsequently affects the overall value of ecosystem services.

The primary contributor to the ecological value of PLE space in Bin County is forest ecological space, with its contribution rate increasing from 59.56% to 60.77%. Despite the decline in overall ecosystem service value (ESV), its positive impact on the ecosystem remains significant. The contribution rate of cropland production space rose to 23.23% between 2000 and 2010. However, during 2010–2020, although the contribution rate continued to increase, the total ESV decreased, indicating the weakening of its ecological service capacity. The contribution of grassland ecological space increased throughout the study period, though its overall impact was minimal. In contrast, water ecological space experienced a significant decline in contribution rate from 2000 to 2010, accompanied by a decrease in ESV. This suggests that degradation has impaired its ecological function. Urban and rural living spaces both showed negative contribution rates, highlighting their limited role in providing ecosystem services. Meanwhile, the contribution rate of other ecological space, though relatively small, exhibited a slow growth trend, indicating potential for further ecological value. Overall, the value of ecosystem services in the PLE space of Bin County has decreased over these 20 years, indicating increasingly severe ecological issues that require urgent attention.

Table 11. Summary table of the value of ecosystem service in PLE during 2010–2020.

Ecosystem	2000		2010		2020		ESV Changes (CNY Billion)	
	ESV (CNY Billion)	ESV Contribution Rate (%)	ESV (CNY Billion)	ESV Contribution Rate (%)	ESV (CNY Billion)	ESV Contribution Rate (%)	2000–2010	2010–2020
C	12.68	18.16	12.59	23.23	12.42	22.37	−0.09	−0.17
F	33.70	48.25	32.28	59.56	33.74	60.77	−1.42	1.46
G	0.25	0.36	1.27	2.34	0.86	1.55	1.02	−0.41
W	24.49	35.07	9.33	17.21	9.74	17.54	−15.16	0.41
Ur	−0.08	−0.12	−2.06	−0.24	−0.15	−0.27	−0.05	−0.02
R	−1.21	−1.73	−1.16	−2.14	−1.21	−2.81	0.05	−0.05
Un	0.01	0.01	0.02	0.04	0.12	0.22	0.01	0.1
Total amount	69.84	100.00	52.27	100.00	55.52	100.00	−15.64	1.32

(2) Characterization of the spatial distribution of ecosystem service values (ESVs)

To analyze the spatial distribution and changes in ecosystem service values (ESVs) in Bin County from 2000 to 2020, this study employed ArcGIS software to classify the ESVs into the following five levels based on natural breaks: low, low to medium, medium, high, and extreme. The spatial distribution of PLE space ESVs in Bin County for 2000, 2010, and 2020 was then determined (Figure 10).

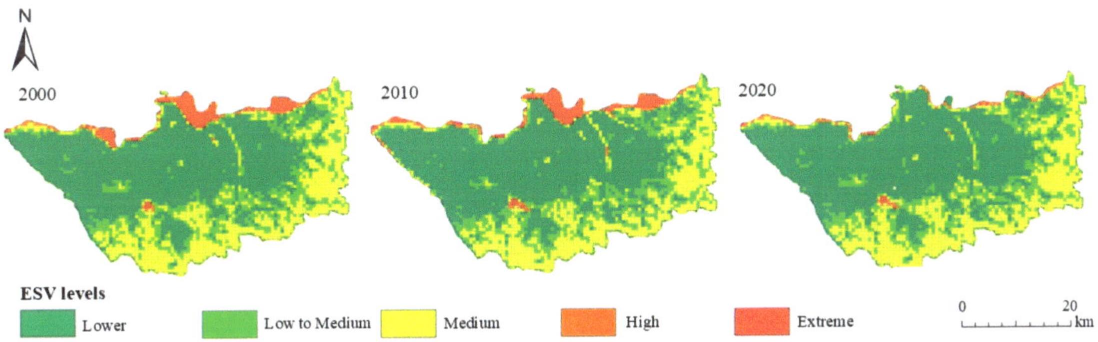

Figure 10. Spatial distribution map of ecosystem service value intensity in Bin County from 2000 to 2020.

The overall layout of ESV distribution in Bin County remained relatively consistent between 2000 and 2010. The central and western regions predominantly exhibited low-to-medium ESVs, mainly due to the extensive distribution of cropland production space in these areas, which has a relatively low ecological service value. By 2020, the most noticeable increase in lower ESVs occurred in Minhe Township, where watershed regions were transformed into unused land between 2010 and 2020, contributing to a decline in overall ecosystem service value. In summary, the spatial distribution pattern of ecosystem service values (ESVs) in the PLE space of Bin County from 2000 to 2020 was characterized by higher values in the northwest and central regions, while the southeast exhibited lower values. This pattern remained consistent over the two decades. Furthermore, the area classified as having extreme ESVs in the PLE space significantly diminished, reducing the overall ecosystem service value and highlighting the urgent need to address ecological issues.

(3) Analysis of the flow of ecosystem service value (ESV) gains and losses

This study employs the ecosystem service value flow gain–loss model to analyze the impact of changes in the PLE space pattern on ecosystem service values. This method quantitatively assesses the flow of ecosystem service values among different land uses due to transfers between them. Utilizing land use transfer data for PLE space in Bin County and the ecosystem service values associated with each type of PLE space, the flow of ecosystem service value gains and losses during each period was calculated.

In this model, the horizontal axis represents the changes in value for the transferred areas of each PLE space land use type. In contrast, the vertical axis reflects the changes in value for the regions replenished for each PLE space land use type. This approach enables a comprehensive understanding of how land use dynamics influence the overall ecosystem service values in Bin County (Figure 11).

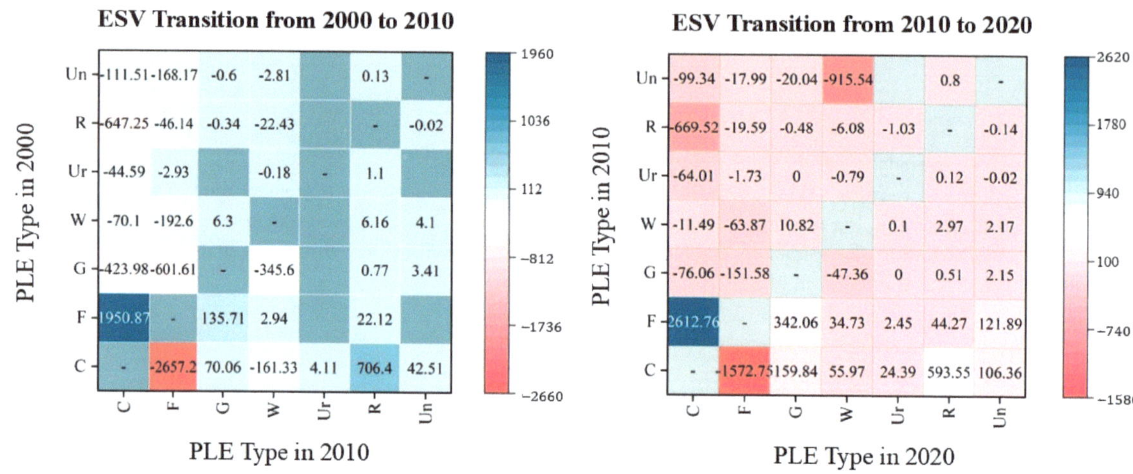

Figure 11. Heat map of ESV gains and losses in Bin County from 2000 to 2020.

Combining the Sankey diagram of ecological value loss and gain in PLE space in Bin County illustrates each period's dominant paths and key compartments of ecological value changes (Figure 12). From 2000 to 2020, four main paths of ecological value loss and gain were identified. In both periods, the flow of ecological value from forest ecological space to cropland production space served as the primary loss path, with losses amounting to CNY 265.720 billion in the first period and CNY 157.275 billion in the second. The forest ecological space emerged as a crucial compartment for ecological value gains, with positive flows of ecological value from other spaces to the forest ecological space. The value gains during the two periods were CNY 211.163 billion and CNY 315.816 billion, respectively. Notably, from 2010 to 2020, the ecological value flow path from water body ecological space to unused land ecological space became the second major pathway, resulting in a loss of CNY 91.554 billion, significantly higher than other flow paths. This increase can be attributed to the higher ecological value potential of water body ecological space compared to unused land ecological space, with the loss of land use area contributing to the overall decline in ecological value.

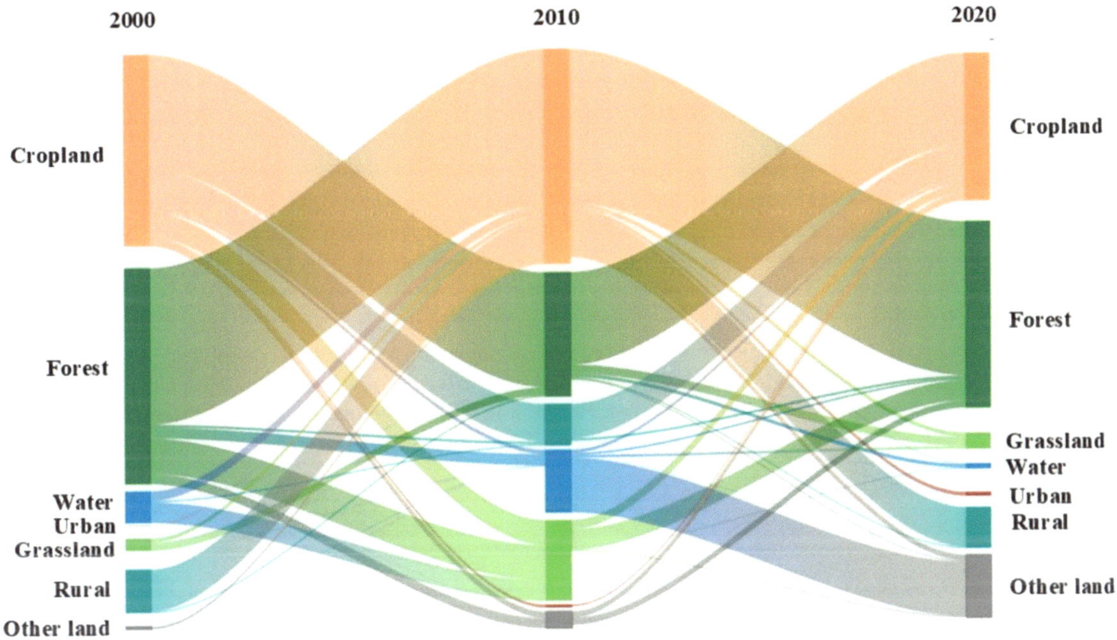

Figure 12. Sankey diagram of ecological value gains and losses in PLE space in Bin County.

As shown in Table 12, from 2000 to 2010, the ecosystem service value (ESV) gained was CNY 295.667 billion, while the losses amounted to CNY 549.939 billion, resulting in a total net loss of CNY 254.272 billion. The conversion of cropland production space to forest ecological space was the primary source of ESV gain during this period, accounting for 68.98% of the total gain. Conversely, the loss of various types of ecological space, particularly forest ecological space, was the main driver of ESV loss, contributing to approximately 48.32% of the total loss.

Table 12. The dominant ESV transfer is in the evolution of spatial patterns.

Exchange Value	2000–2010		2010–2020	
	Transitions (Billion CNY)	Direction	Transitions (Billion CNY)	Direction
ESV gain	2956.67	+	4117.90	+
ESV loss	−5499.39	−	−3739.40	−
total profit	−2542.72	−	378.50	+
H1	1950.87	C → F	2612.76	C → F
Contribution ratio	65.98%	/	63.45%	/
H2	−2657.20	F → C	−1572.75	F → C
Contribution ratio	48.32%	/	42.06%	/

Note: "+" represents positive carbon flow direction; "−" represents negative flow direction.

From 2010 to 2020, the ESV gain increased to CNY 411.790 billion, while losses decreased to CNY 373.940 billion. This period exhibited a smoother overall value flow compared to the previous decade, resulting in a net gain of CNY 378.50 billion. The ecological value gain and loss patterns remained consistent with the earlier period, primarily involving the interconversion between forest ecological space and cropland production space. Notably, 63.45% of the value gain originated from converting cropland production space to forest ecological space. In comparison, 42.06% of the value loss stemmed from converting forest ecological space to cropland production space, leading to a significant loss of CNY 157.275 billion.

In summary, Bin County, as a typical agricultural region in the northeast black soil area, experiences significant negative impacts on ecological and environmental benefits due to agricultural production practices. While recent ecological protection policies implemented by the government have positively influenced the horizontal flow of ecological value, the total ecosystem service value (ESV) continues to decline. This decline is primarily attributed to the gains in PLE space ESV over the past 20 years, mainly from expanding forest ecological space. However, human activities such as land reclamation and urban expansion have encroached upon other ecological areas, leading to substantial ESV losses.

Although policies such as returning farmland to forests have provided some compensation for the ecological disturbances and damages caused by human activities, the overall benefits have not been sufficient to offset the losses. Consequently, the ESV in the region continues to decline. Moving forward, it is crucial to enhance the protection of forest ecosystems and increase the forested land area, fostering a trajectory of high-quality regional ecosystem development.

3.4. Analysis of the Correlation Between Carbon Metabolism and Ecosystem Service Value in PLE Space

The study utilized GeoDa 1.18 software to analyze the ecosystem service value (ESV) and carbon metabolic density data for Bin County at three key time points, 2000, 2010, and 2020, using a 1 km × 1 km grid cell framework. This analysis treated ESV as the independent variable, while carbon metabolic density values were the dependent variable.

According to Table 13 and Figure 13, the bivariate Moran's index for Bin County in 2000, 2010, and 2020 was negative, with the specific values of −0.4416, −0.4253, and −0.4830, respectively. The p-values corresponding to these negative Moran's indices were all less than 0.01, and the absolute z-values exceeded 65. These statistical results indicate a significant negative correlation between ESV and carbon metabolic density in spatial distribution. Specifically, high ESVs were associated with low carbon metabolic intensity, while low ESVs correlated with high carbon metabolic density. This finding suggests that the distribution of carbon metabolic density and ecosystem service value among neighboring units in the PLE space of Bin County is not random or discrete but exhibits significant spatial heterogeneity.

Table 13. Global Moran's index of bivariate spatial autocorrelation in the production–living–ecological space.

Year	Global Moran' I	p-Value	Z-Score
2000	−0.4416	<0.001	−79.9794
2010	−0.4253	<0.001	−65.4576
2020	−0.4830	<0.001	−74.8719

Based on the bivariate LISA plot (Figure 14), the high–low clusters in 2000 and 2010 were predominantly located in the central and northwestern regions of Bin County. This area primarily consists of plains, where the high–low clusters are distributed in a fragmented manner, showing a high degree of spatial coincidence with cropland production. In 2020, the distribution of high and low clusters remained similar to that of the previous two periods; however, the low and high clusters began to expand toward the south and east of Bin County. In these areas, mountains and hills limit agricultural use, resulting in a predominance of forest land associated with lower carbon emission intensity and higher ecosystem service value. Notably, the low–low clustering in 2020 exhibited a significantly larger distribution area compared to the earlier periods. This increase can be attributed to changes in other ecological spaces, primarily unused land, which generates lower carbon emissions and possesses a relatively low ecosystem service value.

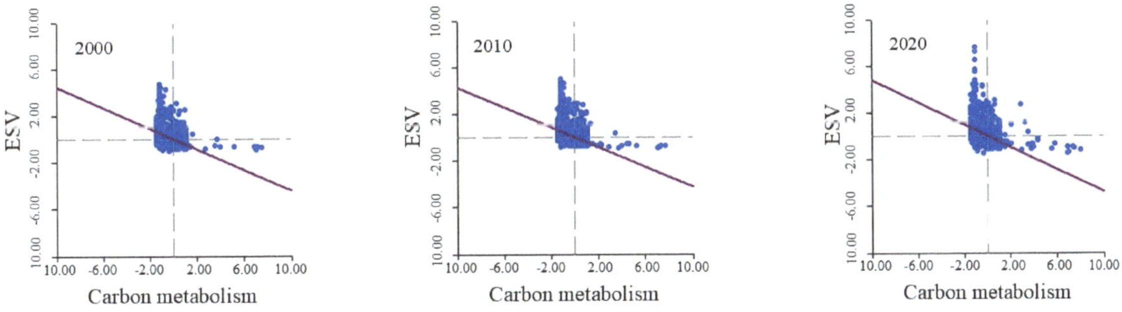

Figure 13. Bivariate Moran's index for the production–living–ecological space from 2000 to 2020.

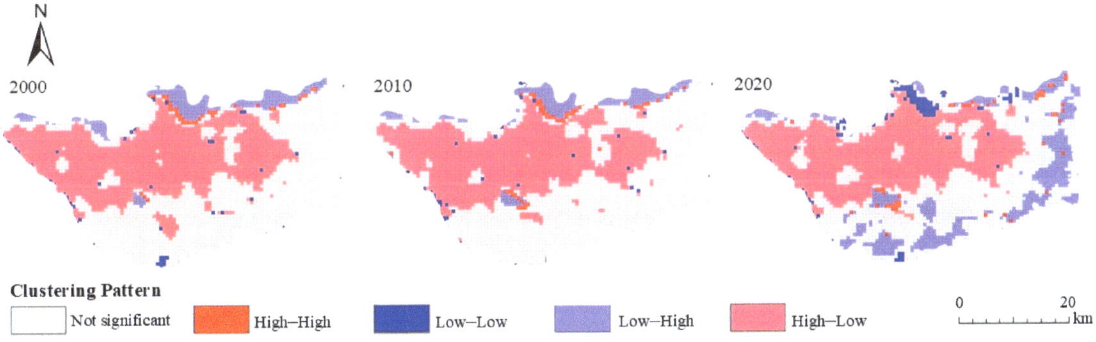

Figure 14. LISA cluster graph of carbon metabolism and ESV in the PLE during 2000–2020.

In summary, a clear negative spatial correlation between carbon emission intensity and ecosystem service value exists within the PLE space of Bin County. Specifically, when a region's ecosystem service value significantly increases, when a neighboring areas' carbon emissions tend to decrease, and vice versa. This phenomenon is rooted in the study's PLE space categorization methodology, which identifies water and forest ecological spaces with higher ecosystem service values. The key features of these spatial types—such as water bodies and woodlands—serve as crucial conduits for carbon dioxide assimilation into living organisms. In promoting the "dual carbon target", the involvement of these essential features, woodlands, and water bodies is critical for achieving carbon neutrality. Therefore, enhancing ecosystem service values and reducing carbon emissions should be considered as complementary and simultaneous to ecological protection strategies vital for the region's sustainable development.

4. Discussion

4.1. Changes in Carbon Metabolism and Regional Ecosystem Service Value Distribution Under Long Time Series and Analysis of Causes

Changes in the spatial patterns of different land types reflect shifts in their primary functions. At the same time, socio-economic activities also influence terrestrial ecosystems' carbon cycling processes, thereby affecting ecosystem services' supply capacity. In the northeastern black soil region, the intensification of agricultural activities has directly or indirectly impacted carbon metabolism within the ecosystem. From 2000 to 2020, total carbon emissions in the study area increased significantly, resulting in a growing imbalance in carbon metabolism. The analysis of carbon flow paths reveals that this phenomenon can primarily be attributed to spatial structural changes driven by rapid urbanization, increased agricultural activities, and energy consumption. Consistent with findings from previous studies, many urbanized counties have experienced a sharp rise in carbon emissions due to the expansion of construction land. However, unlike other studies focused on urban agglomerations or development areas, research in agricultural counties within the northeast black soil region indicates that the extensive use of fertilizers, pesticides, and other agricultural inputs in cropland production has led to severe pollution and a marked increase in carbon emissions. This trend poses significant challenges to the sustainable development of agriculture and can have detrimental effects on ecological and environmental benefits. Addressing these issues will be crucial for promoting sustainable agricultural practices and mitigating the ecological impacts of intensified agricultural production.

Total ecosystem service values (ESVs) decreased from 2000 to 2020. Higher ESVs were concentrated in the southern regions of the study area, primarily due to the positive impact of extensive forest cover on ESV intensity. However, a significant loss of ecological value in grassland ecological space occurred between 2010 and 2020. This decline can be attributed to both natural and anthropogenic factors. Natural factors include the windy conditions typical of spring in the northeast black soil region, which can lead to wind erosion and subsequent grassland degradation. Anthropogenic factors encompass excessive cultivation and fertilization practices that may diminish soil fertility in the black soil area, adversely affecting grassland vegetation growth. While recent ecological protection policies implemented by the government have positively impacted the ecosystem value of the northeast black soil region, the overall trend remains downward. In 2023, the Harbin municipal government issued the "14th Five-Year Plan" for ecological environmental protection, introducing a range of policies promoting green development and environmental sustainability. The key measures include enhancing the prevention and control of agricultural surface pollution, encouraging cleaner agricultural practices, and improving the management of livestock and poultry waste. Additionally, the plan emphasizes the construction of rural waste treatment facilities and optimizing domestic sewage treatment systems to address rural ecological challenges. These initiatives provide essential policy guidance for implementing future ecological protection and restoration projects.

In this study, Bin County, located in the northeast black soil region, was employed as a case study to evaluate carbon metabolism and ecosystem service values using the "PLE space" classification system. The research also investigated the global applicability and advantages of governance strategies for agricultural countries. As one of the three major black soil regions globally, the northeast black soil region offers a representative context due to its distinctive agroecological characteristics and land use conflicts. This study demonstrates how Bin County, as a predominantly agricultural region, optimizes the functional configuration of production, living, and ecological spaces through refined zoning strategies, effectively balancing the demands of food production and ecological protection. The strengths of this research approach include the following: (1) Applicability: The zoning governance model for agroecosystems is adaptable to regions with diverse climatic conditions, land use patterns, and agricultural activities. (2) Scalability: The assessment framework integrates carbon metabolism and ecosystem service evaluations, enabling the tracking of carbon flow pathways and value dynamics, making it suitable for ecological and economic analyses across various scales. The findings from the Bin County case study provide actionable strategies for promoting low-carbon development in similar agroecosystems. Furthermore, they offer direct guidance for global cropland management and contribute to achieving carbon neutrality goals.

4.2. Recommendations for Zonal Carbon Reduction and Sink Enhancement Based on Clustering Model Conditions

Based on the results of the correlation study between carbon metabolism and ecosystem service value of PLE space, it can be found that the potential of ecological value enhancement and environmental pressure is different for different space types. Based on the results of the previous study, the area is divided according to the differences in clustering patterns. Meanwhile, combined with the study area's natural, economic, and social factors, targeted zoning management measures are formulated to rationally utilize different ecosystems' ecological efficiency and resource allocation to provide theoretical support for the healthy development of Bin County's socio-economy (Figure 15).

(1) Production–living integrated management zone (high–low clustering): This area is characterized by high carbon emission intensity alongside low ecosystem service value. To address this, the comprehensive management of agricultural cropland is essential to ensure the coordinated development of agriculture and ecological protection. This should involve optimizing land use through scientific planning and implementing efficient agricultural production methods. For instance, promoting green agricultural technologies and precision agriculture can enhance the output efficiency of cropland while minimizing negative environmental impacts. Additionally, encouraging farmers to adopt sustainable farming practices is crucial, including reducing the use of chemical fertilizers and pesticides. Promoting organic and ecological agriculture can further enhance agriculture's environmental service functions. Strengthening soil protection and improvement measures for cropland production is also vital to prevent soil degradation and erosion, ensuring the long-term sustainability of agricultural output. Furthermore, inefficient or degraded cropland should be transitioned to ecological protection areas or subjected to ecological restoration as necessary, thereby improving the comprehensive benefits derived from land resources. By managing agricultural cropland comprehensively and optimizing the balance between ecological and construction land, the sustainable use of land resources in the region can be achieved, promoting the synergistic development of agriculture, ecology, and the economy.

(2) Production–ecology synergistic development zone (high–high clustering area): This area exhibits high carbon emission intensity alongside a significant value of ecosystem services. The key measures to be implemented include optimizing water resource management, mainly focusing on the irrigation management of paddy fields. The adoption of water-saving irrigation technologies, such as drip and sprinkler irrigation, are essential in reducing water waste, enhancing irrigation efficiency, and lessening the demand for water resources in the context of agricultural mechanization. Improvements in agricultural mechanization should involve promoting energy-efficient machinery and equipment to decrease carbon emissions. Additionally, introducing biomass energy sources, such as biomass pellets and biodiesel, is recommended to replace traditional fuels and reduce carbon emissions. Strengthening the restoration and protection of ecosystems surrounding water bodies is also crucial. Maintaining the integrity and functionality of these ecosystems can enhance their capacity to absorb carbon, contributing to ecological health and climate mitigation efforts.

(3) Ecological conservation zone (low–high clustering area): This area has a high ecosystem service value and minimal carbon emissions, indicating that current ecological achievements should be consolidated. The key measures include developing eco-industries by introducing eco-agriculture, eco-tourism, and eco-breeding. These initiatives will leverage the region's rich environmental resources to enhance economic benefits while reducing carbon emissions. Additionally, optimizing the energy

structure and land use is essential. Promoting the use of clean energy and reducing reliance on fossil fuels can help lower carbon emissions and protect the ecological environment. Strengthening environmental management and monitoring is also critical; establishing a robust environmental management framework can facilitate the effective monitoring and assessment of the ecological environment. This approach should enable the timely identification and resolution of environmental issues, ensuring the stability and security of ecological land.

(4) Ecological potential zone (low–low clustering area): This area has low carbon emissions and similarly low ecosystem service value. The main measures to be implemented include ecological restoration and reconstruction efforts, such as afforestation, grassland restoration, and wetland rehabilitation, to improve vegetation cover and enhance carbon storage. Additionally, promoting eco-tourism can capitalize on unused land's natural landscape and ecological resources, thereby increasing its economic value. This may involve developing eco-tourism attractions and establishing eco-leisure resorts. Constructing ecological parks and green spaces on unutilized land can enhance green space coverage and improve the overall ecological environment. Finally, promoting the use of renewable energy is essential. This can be achieved by establishing renewable energy facilities, such as wind farms and photovoltaic power stations, on unused land to reduce reliance on fossil fuels.

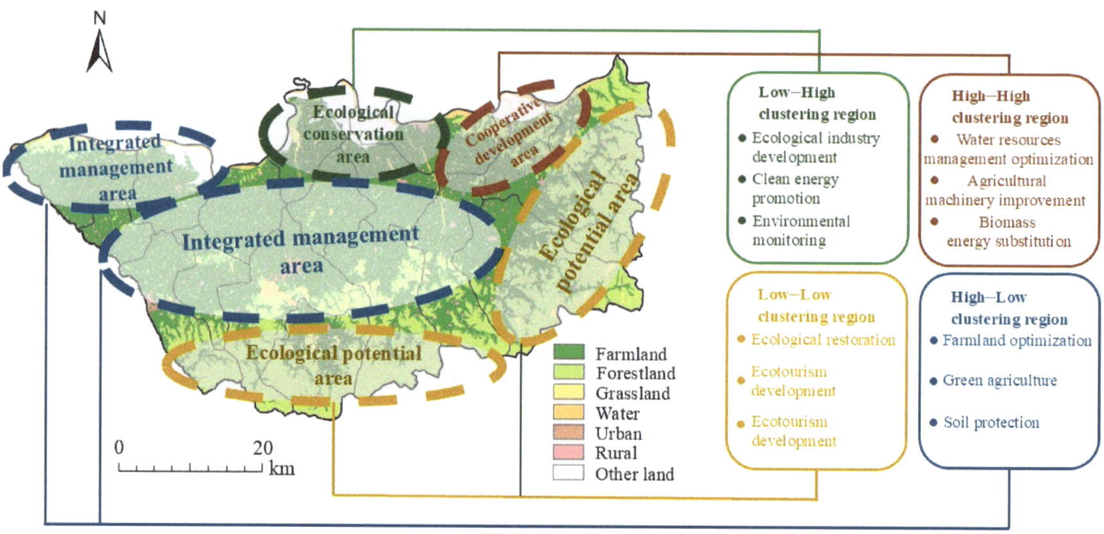

Figure 15. Development of zoning management measures in the study area based on distinct clustering patterns.

4.3. Limitations and Future Enhancements

Existing research has separately examined carbon metabolism and ecosystem service value (ESV) within horizontal and vertical dimensions at the county scale, confirming the spatial correlation between regional carbon metabolism and ESV. This research employs various statistical methods for data analysis; however, certain limitations persist.

Carbon emission coefficients were selected based on previous studies, which may exhibit regional variations in calculating carbon metabolism data. Future research should adjust the carbon emission coefficient calculation model to reflect these differences. Additionally, while the study utilized the carbon emission coefficient method, more refined accounting models must be developed to address the diversity of carbon emission sources, including agricultural production activities, livestock respiration, and residential respiration. The equivalent factor method was employed to measure the ecosystem service value. Although this approach is suitable for large-scale ESV accounting, it remains somewhat subjective. We hope that future studies will conduct more in-depth research on ESVs at the county level. Researchers should seek more comprehensive and precise data sources to enhance future investigations' scientific rigor and accuracy.

5. Conclusions

The analysis of changes in the PLE space pattern indicates that the overall area of PLE space in Bin County, located in the northeast black soil region, remained relatively stable during the study period, with only minor local changes. Transformations primarily shifted from production and living spaces to production spaces. This shift can be attributed to economic development, which has caused an increase in the demand for agricultural land in Bin County. Meanwhile, a decline in ecological spaces, such as grasslands and water bodies, reflects the insufficient emphasis on ecological importance.

The analysis of the spatial and temporal evolution of carbon metabolism reveals that, despite a decreasing trend in carbon emissions, the net carbon transfer between compartments was negative. Additionally, the density of carbon metabolism exhibited a pattern of high values in the central areas and lower values in the surrounding regions. These findings suggest that agricultural production and human activities have disrupted carbon metabolism in Bin County, necessitating future regulation with a focus on low-carbon development strategies.

The spatial and temporal evolution of ecosystem service value (ESV) demonstrated a distribution pattern characterized by higher values in the central and western regions and lower values in the southeastern part during the study period. While the flow of ecological value had a positive influence, the overall decline in ESV was associated with policies such as farmland-to-forest conversions. However, anthropogenic activities, including over-cultivation and grazing, have caused significant disturbances and the degradation of ecosystems.

The clustering characteristics of carbon metabolism and ESV in Bin County revealed significant negative spatial correlations. The imbalance in carbon metabolism caused by changes in the PLE space pattern was directly reflected in the decline in ecosystem service value. Thus, enhancing ecosystem service value and regulating regional carbon metabolism should be pursued in tandem as complementary ecological protection strategies.

Author Contributions: Conceptualization, Y.S. and Y.C.; methodology, Y.S.; software, Y.S.; validation, Y.S. and Y.C.; formal analysis, Y.S.; investigation, Y.S.; resources, Y.S.; data curation, L.W.; writing—original draft preparation, Y.S.; writing—review and editing, Y.S., Y.C. and Y.Z.; visualization, Y.S., Y.C. and Y.Z.; supervision, Y.C.; project administration, L.W.; funding acquisition, L.W. All authors have read and agreed to the published version of the manuscript.

Funding: The Open Topics of the Think Tank on Ecological Civilization Construction and Green Development in Higher Education Institutions of Heilongjiang Province (ZKKF2022181); Philosophy and Social Science Research Project of Heilongjiang Province in 2021 (21GLB061).

Data Availability Statement: https://www.resdc.cn/Datalist1.aspx?FieldTyepID=1,3, accessed on 10 September 2024; http://www.chinabx.gov.cn/hebbx/tjxxgk/tongji.shtml, accessed on 3 October 2024.

Acknowledgments: We are grateful to the editors and reviewers for their valuable comments and suggestions.

Conflicts of Interest: The authors declare no conflicts of interest.

References

1. Gao, P.; Yue, S.; Chen, H. Carbon emission efficiency of China's industry sectors: From the perspective of embodied carbor emissions. *J. Clean. Prod.* **2021**, *283*, 124655. [CrossRef]
2. Xie, G.D.; University of Chinese Academy of Sciences. On the Carbon Sink Capacity of China's Ecosystems and Its Enhancemen Pathways. *Environ. Prot.* **2023**, *51*, 12–16. [CrossRef]
3. Cai, W. Identifying Ecosystem Services Bundles for Ecosystem Services Trade-Off/Synergy Governance in an Urbanizing Region *Land* **2022**, *11*, 1593. [CrossRef]
4. Baral, H.; Guariguata, M.R.; Keenan, R.J. A proposed framework for assessing ecosystem goods and services from planted forests *Ecosyst. Serv.* **2016**, *22*, 260–268. [CrossRef]
5. Bremer, L.L.; Farley, K.A.; Chadwick, O.A.; Harden, C.P. Changes in carbon storage with land management promoted by paymen for ecosystem services. *Environ. Conserv.* **2016**, *43*, 397–406. [CrossRef]
6. Fu, B.J.; Zhang, L.W.; Xu, Z.H.; Zhao, Y.; Wei, Y.P.; Skinner, D. Ecosystem services in changing land use. *J. Soils Sediments* **2015**, *15* 833–843. [CrossRef]
7. Wu, L.; Chen, Y.; Feylizadeh, M.R.; Liu, W. Estimation of China's macro-carbon rebound effect: Method of integrating Data Envelopment Analysis production model and sequential Malmquist-Luenberger index. *J. Clean. Prod.* **2018**, *198*, 1431–1442 [CrossRef]
8. Liu, X.; Li, S. A comparison analysis of the decoupling carbon emissions from economic growth in three industries of Heilongjiang province in China. *Environ. Sci. Pollut. Res.* **2021**, *28*, 65200–65215. [CrossRef] [PubMed]
9. Tang, X.; Woodcock, C.E.; Olofsson, P.; Hutyra, L.R. Spatiotemporal assessment of land use/land cover change and associated carbon emissions and uptake in the Mekong River Basin. *Remote Sens. Environ.* **2021**, *256*, 112336. [CrossRef]
10. Zhu, E.; Deng, J.; Zhou, M.; Gan, M.; Jiang, R.; Wang, K.; Shahtahmassebi, A. Carbon emissions induced by land-use and land-cover change from 1970 to 2010 in Zhejiang, China. *Sci. Total Environ.* **2019**, *646*, 930–939. [CrossRef]
11. Zhang, Y.; Xia, L.; Fath, B.D.; Yang, Z.; Yin, X.; Su, M.; Liu, G.; Li, Y. Development of a spatially explicit network model of urban metabolism and analysis of the distribution of ecological relationships: Case study of Beijing, China. *J. Clean. Prod.* **2016**, *112* 4304–4317. [CrossRef]
12. Tian, S.; Wang, S.; Bai, X.; Luo, G.; Li, Q.; Yang, Y.; Hu, Z.; Li, C.; Deng, Y. Global patterns and changes of carbon emissions from land use during 1992–2015. *Environ. Sci. Ecotechnol.* **2021**, *7*, 100108. [CrossRef] [PubMed]
13. Xia, C.; Li, Y.; Xu, T.; Chen, Q.; Ye, Y.; Shi, Z.; Liu, J.; Ding, Q.; Li, X. Analyzing spatial patterns of urban carbon metabolism and its response to change of urban size: A case of the Yangtze River Delta, China. *Ecol. Indic.* **2019**, *104*, 615–625. [CrossRef]
14. Chen, S.Q.; Chen, B. Determining carbon metabolism in urban areas though network environ theory. *Procedia Environ. Sci.* **2012** *13*, 2246–2255. [CrossRef]
15. Lin, G.; Jiang, D.; Fu, J.; Cao, C.; Li, X.; Li, Z.; Huang, S. Carbon Flow Analysis of the Evolution of "Production-Living-Ecological Space" Pattern: A Case Study of Tangshan City. *Sci. Technol. Rev.* **2020**, *38*, 107–114.
16. Xia, L.; Fath, B.D.; Scharler, U.M.; Zhang, Y. Spatial variation in the ecological relationships among the components of Beijing's carbon metabolic system. *Sci. Total Environ.* **2016**, *544*, 103–113. [CrossRef] [PubMed]
17. Shan, W.; Jin, X.; Yang, X.; Gu, Z.; Han, B.; Li, H.; Zhou, Y. A framework for assessing carbon effect of land consolidation with life cycle assessment: A case study in China. *J. Environ. Manag.* **2020**, *266*, 110557. [CrossRef] [PubMed]
18. Fryer, J.; Williams, I.D. Regional carbon stock assessment and the potential effects of land cover change. *Sci. Total Environ.* **2021** *775*, 145815. [CrossRef] [PubMed]
19. Costanza, R.; d'Arge, R.; de Groot, R.; Farber, S.; Grasso, M.; Hannon, B.; Limburg, K.; Naeem, S.; O'Neill, R.V.; Paruelo, J.; et al The value of the world's ecosystem services and natural capital. *Nature* **1997**, *387*, 253–260. [CrossRef]
20. Qi, F.; Liu, J.; Gao, H.; Fu, T.; Wang, F. Characteristics and spatial–temporal patterns of supply and demand of ecosystem services in the Taihang Mountains. *Ecol. Indic.* **2023**, *147*, 109932. [CrossRef]
21. Zhang, J.; Lu, X.; Qin, Y.; Zhang, Y.; Yang, D. Can Urbanization-Driven Land-Use and Land-Cover Change Reduce Ecosystem Services? A Case of Coupling Coordination Relationship for Contiguous Poverty Areas in China. *Land* **2024**, *13*, 82. [CrossRef]
22. Xu, Q.; Yang, Y.; Yang, R.; Zha, L.-S.; Lin, Z.-Q.; Shang, S.-H. Spatial Trade-Offs and Synergies between Ecosystem Services in Guangdong Province, China. *Land* **2024**, *13*, 32. [CrossRef]
23. Yang, M.; Gao, X.; Siddique, K.H.M.; Wu, P.; Zhao, X. Spatiotemporal exploration of ecosystem service, urbanization, and their interactive coercing relationship in the Yellow River Basin over the past 40 years. *Sci. Total Environ.* **2023**, *858*, 159757. [CrossRef] [PubMed]
24. Chen, Y.; Lu, H.; Li, J.; Xia, J. Effects of land use cover change on carbon emissions and ecosystem services in Chengyu urban agglomeration, China. *Stoch. Environ. Res. Risk Assess.* **2020**, *34*, 1197–1215. [CrossRef]
25. Zhang, R.; Yu, K.; Luo, P. Spatio-Temporal Relationship between Land Use Carbon Emissions and Ecosystem Service Value in Guanzhong, China. *Land* **2024**, *13*, 118. [CrossRef]

26. Kong, D.; Chen, H.; Wu, K. Characteristics of Evolution, Ecological and Environmental Effects, and Influencing Factors of China's "Production-Living-Ecological Space". *J. Nat. Resour.* **2021**, *36*, 1116–1135.
27. Liu, Q.; Yang, D.; Cao, L. Evolution and Prediction of the Coupling Coordination Degree of Production–Living–Ecological Space Based on Land Use Dynamics in the Daqing River Basin, China. *Sustainability* **2022**, *14*, 10864. [CrossRef]
28. Yang, H.; Yan, S.; An, N.; Yao, Q. Using Ecological Footprint Analysis to Evaluate Sustainable Development in Lushan County, China. *Land* **2024**, *13*, 1081. [CrossRef]
29. Wang, X.; Liu, X.; Wu, Y.; Chen, R.; Wang, S. Dynamic Assessment and Change Analysis of Ecosystem Service Value Based on Physical Assessment Method in Cili County, China. *Forests* **2023**, *14*, 869. [CrossRef]
30. Yang, Y.; Bao, W.; Li, Y.; Wang, Y.; Chen, Z. Land Use Transition and Its Eco-Environmental Effects in the Beijing–Tianjin–Hebei Urban Agglomeration: A Production–Living–Ecological Perspective. *Land* **2020**, *9*, 285. [CrossRef]
31. Long, H.; Liu, Y.; Wu, X.; Dong, G. Spatio-temporal dynamic patterns of farmland and rural settlements in Su–Xi–Chang region: Implications for building a new countryside in coastal China. *Land Use Policy* **2009**, *26*, 322–333. [CrossRef]
32. Zhou, D.; Xu, J.; Lin, Z. Conflict or coordination? Assessing land use multi-functionalization using production-living-ecology analysis. *Sci. Total Environ.* **2017**, *577*, 136–147. [CrossRef] [PubMed]
33. West, T.O.; Marland, G. A synthesis of carbon sequestration, carbon emissions, and net carbon flux in agriculture: Comparing tillage practices in the United States. *Agric. Ecosyst. Environ.* **2002**, *91*, 217–232. [CrossRef]
34. Xia, C.; Li, Y.; Xu, T.; Ye, Y.; Shi, Z.; Peng, Y.; Liu, J. Quantifying the spatial patterns of urban carbon metabolism: A case study of Hangzhou, China. *Ecol. Indic.* **2018**, *95*, 474–484. [CrossRef]
35. Huang, Q.; Zhang, Y. Decoupling and Decomposition Analysis of Agricultural Carbon Emissions: Evidence from Heilongjiang Province, China. *Int. J. Environ. Res. Public Health* **2022**, *19*, 198. [CrossRef]
36. Wu, X.; Ge, Y.; Wang, Y.; Liu, D.; Gu, B.; Ren, Y.; Yang, G.; Peng, C.; Cheng, J.; Chang, J. Agricultural carbon flux changes driven by intensive plastic greenhouse cultivation in five climatic regions of China. *J. Clean. Prod.* **2015**, *95*, 265–272. [CrossRef]
37. Wang, Y.; Jeong, S.; Hang, Y.; Wang, Q. Determinants of net energy-related CO2 emissions in China: A source-to-sink decomposition analysis. *Environ. Impact Assess. Rev.* **2023**, *98*, 106979. [CrossRef]
38. Ye, H.; Wang, K.; Zhao, X.; Chen, F.; Li, X.; Pan, L. Relationship between construction characteristics and carbon emissions from urban household operational energy usage. *Energy Build.* **2011**, *43*, 147–152. [CrossRef]
39. Kennedy, C.; Steinberger, J.; Gasson, B.; Hansen, Y.; Hillman, T.; Havránek, M.; Pataki, D.; Phdungsilp, A.; Ramaswami, A.; Mendez, G.V. Methodology for inventorying greenhouse gas emissions from global cities. *Energy Policy* **2010**, *38*, 4828–4837. [CrossRef]
40. Li, S.; Wang, Z. The Effects of Agricultural Technology Progress on Agricultural Carbon Emission and Carbon Sink in China. *Agriculture* **2023**, *13*, 793. [CrossRef]
41. Chen, S.; Xu, B.; Chen, B. Unfolding the interplay between carbon flows and socioeconomic development in a city: What can network analysis offer? *Appl. Energy* **2018**, *211*, 403–412. [CrossRef]
42. Xia, L.; Zhang, Y.; Yu, X.; Li, Y.; Zhang, X. Hierarchical structure analysis of urban carbon metabolism: A case study of Beijing, China. *Ecol. Indic.* **2019**, *107*, 105602. [CrossRef]
43. Hasan, S.S.; Zhen, L.; Miah, M.G.; Ahamed, T.; Samie, A. Impact of land use change on ecosystem services: A review. *Environ. Dev.* **2020**, *34*, 100527. [CrossRef]
44. Xie, G.; Zhang, C.; Zhen, L.; Zhang, L. Dynamic changes in the value of China's ecosystem services. *Ecosyst. Serv.* **2017**, *26*, 146–154. [CrossRef]
45. Xie, G.; Zhang, C.; Zhang, L.; Chen, W.; Li, S. Improvement of the Evaluation Method for Ecosystem Service Value Based on Per Unit Area. *J. Nat. Resour.* **2015**, *30*, 1243–1254.
46. Wang, F.; Fu, W.; Chen, J. Spatial–Temporal Evolution of Ecosystem Service Value in Yunnan Based on Land Use. *Land* **2022**, *11*, 2217. [CrossRef]
47. Wu, D.; Liu, J.; He, T.; Wang, S.; Wang, R. Profit and loss analysis on ecosystem services value based on land use change in Yellow River Delta. *Trans. Chin. Soc. Agric. Eng.* **2009**, *25*, 256–261.
48. Swank, J.M.; Mullen, P.R. Evaluating Evidence for Conceptually Related Constructs Using Bivariate Correlations. *Meas. Eval. Couns. Dev.* **2017**, *50*, 270–274. [CrossRef]

MDPI AG
Grosspeteranlage 5
4052 Basel
Switzerland
Tel.: +41 61 683 77 34

Land Editorial Office
E-mail: land@mdpi.com
www.mdpi.com/journal/land

www.ingramcontent.com/pod-product-compliance
Lightning Source LLC
Chambersburg PA
CBHW041416210926
51850CB00013B/3288
* 9 7 8 3 7 2 5 8 4 1 6 1 5 *